ArcGIS Engine
地理信息系统开发从入门到精通

邱洪钢 张青莲 熊友谊 编著

第二版

人民邮电出版社
北京

图书在版编目（CIP）数据

ArcGIS Engine地理信息系统开发从入门到精通 / 邱洪钢，张青莲，熊友谊编著. -- 2版. -- 北京：人民邮电出版社，2013.4
ISBN 978-7-115-30462-9

Ⅰ．①A… Ⅱ．①邱… ②张… ③熊… Ⅲ．①地理信息系统－应用软件－软件开发 Ⅳ．①P208

中国版本图书馆CIP数据核字(2012)第307742号

内 容 提 要

本书讲解基于ArcGIS Engine 10开发平台，介绍了相关的开发技术和工程应用，并用C#语言编程实现了工程实例。

本书分4篇共16章，第一篇基础篇（第1～9章）集中介绍了 ArcGIS Engine 基础知识，包括开发基础组件对象模型、ArcGIS Engine 介绍、基于.NET 的 ArcGIS Engine 的开发、ArcGIS Engine 中的控件、框架控件介绍、控件使用实例等，为以后应用 ArcGIS Engine 的各种接口，快速地实现系统的开发打下坚实的基础；第二篇应用提高篇（第 10 章）介绍了空间分析,通过学习这些高级应用可以使读者得心应手地完成各种 GIS 系统的开发；第三篇综合实例篇（第 11 章～第 12 章）用两个综合例子将前面讲解的知识点串起来，让读者将学习的知识点融合起来，以便可以胜任项目开发的角色；第四篇常见疑难解答与经验技巧集萃（第13～16章），本篇将一些开发过程中常见的异常、数据库连接与释放、数据加载以及一些经验技巧作了介绍，本篇的例子主要是对开发过程中经常碰到的问题和实战技巧进行了汇总解答，以便帮助读者提高工作效率。

本书从开发者的角度，全面讨论了 ArcGIS Engine 开发的知识，让读者了解和掌握 ArcGIS Engine 开发的实战技术，无论是想对 ArcGIS Engine 入门还是对 ArcGIS Engine 感兴趣的 GIS 人员，都能从本书中得到提高。

◆ 编　著　邱洪钢　张青莲　熊友谊
　责任编辑　张　涛

◆ 人民邮电出版社出版发行　北京市丰台区成寿寺路11号
　邮编　100164　电子邮件　315@ptpress.com.cn
　网址　http://www.ptpress.com.cn
　北京九州迅驰传媒文化有限公司印刷

◆ 开本：787×1092　1/16
　印张：17.5　　　　　　　　　2013年4月第2版
　字数：558千字　　　　　　　2025年1月北京第20次印刷
ISBN 978-7-115-30462-9

定价：55.00元（附光盘）
读者服务热线：(010)81055410　印装质量热线：(010)81055316
反盗版热线：(010)81055315

第二版说明

本书第一版出版后，收到了大量读者反馈，本人的几位校友也买了这本书，并就项目中碰到的问题、Bug 向我咨询，也有读者咨询一些功能的实现。反馈的意见主要是以下 4 类：

（1）书上代码太多，示例讲解不深；

（2）书上小例子很简洁，很适合入门学习，一个小例子讲解一个小功能；

（3）书上的很多介绍，也参考了 ArcGIS 的相关文件；

（4）光盘中的代码单独运行可以执行，但是功能函数复制到项目中经常出错，或者运行光盘代码可以，但是同样编写在自己的项目中就出现一大堆的 Bug。

在这里首先感谢读者对本书的关注，本书的起因是本人在以前开发项目过程中，一边看英文资料，一边实现功能，同时也做了大量的笔记，当时学习 Engine 开发就是看英文帮助，在指导研究生写 WebGis 方面的毕业论文时，发现学习和实践者对一些基础的理论、平台的接口和类很难弄清楚，因此，萌发了写一本书的想法，本人在白天上班，晚上熬夜，经过一年的努力完成了本书的编写。

Engine 平台帮助文件、资料目前基本还是英文为主，因此，接口、类的介绍都是从英文帮助中翻译过来，日常工作中，都是看英文资料理解了然后编写代码，所示翻译的效果局限于本人的英语水平，敬请读者谅解。

Engine 平台的帮助中提供了大量的小例子代码，本人在学习的过程中经常去看这些小例子程序片段，对于接口、类的使用非常清晰，因此，本书的完整小例子也是功能尽量少，这样初学者学习起来容易上手，但同时也带来了一些冗余重复代码（每个小示例的代码中都有加载 MXD 函数）占用了一些篇幅。

对于示例都是以讲解、示范接口与类的使用，因此没有很系统的一个大项目来展示 Engine 的开发，对于这点我认为，每个项目都是由这些接口和类组成的，只有掌握了这些接口和类，就能自由发挥，开发出一个大项目来，至于项目的需求、设计、架构等，读者可以查阅这方面专门的书籍，更有针对性。我相信大家掌握了一个项目开发的各项技能，不管是多大的项目都是这些技能的翻版、组装。

有读者提出，运行书中代码可以执行，但在自己的项目中同样编写代码却出现大量 Bug。这种情况很正常，GIS 系统的基础就是电子地图数据，本书的数据是 ESRI 随平台发布的世界数据，数据是正常的，读者自己项目中用的数据可能会存在一些问题，有问题的数据，Engine 接口、类使用时必然出现 Bug，我一位校友就碰到该问题，我帮他调试成功后，查明原因就是数据问题。

此次第二版中，将这些冗余的重复代码去掉，只保留了接口、类演示的核心功能函数程序。因为开发方面的学习，都是需要经过大量的编码、经历过很多 Bug 后才能真正掌握，希望读者自己动手编写。

因此建议本书的读者：

初学者，根据书上的代码自己重新输入一遍；

有经验的开发者，把本书当做工具书（当然本书的编写没有涵盖 Engine 的全部功能），对某些接口、类按需参照翻阅。

前　　言

本书是一本指导读者快速入门 ArcGIS Engine 并提高的书籍，着重介绍了 ArcGIS Engine 的基本结构、开发技术和使用中的一些技巧，通过本书的学习，读者可以对 ArcGIS Engine 的许多具体功能有个较清晰的了解并加以应用。

本书内容

本书的例子采用 C#编写，大部分使用 MapControl 控件来展示地图，本书没有将所用到的各个接口图绘制出来，这些接口图可以在 ArcGIS Engine 的帮助文件中获得。为了帮助读者学习到重点知识以便应用到实战中，结合自己用 ArcGIS Engine 开发实战经验，选择一些重要的类和接口进行详细介绍，当读者熟习本书的内容后，就会很清楚地找到所需要的接口和类，便于提高学习效率。

本书是在第 1 版的基础上修订的。分 4 篇共 16 章，主要内容为：第 1 章开发基础，主要讲解了组件对象模型、ArcGIS Engine 介绍、基于.NET 的 ArcGIS Engine 的开发等；第 2 章讲解了 ArcGIS Engine 中的控件、框架控件、控件使用实例等；第 3 章几何对象和空间坐标系，主要讲解了 Geometry 对象、点对象、空间坐标系及变换等；第 4 章地图组成及图层控制，主要讲解了地图对象、图层对象、屏幕显示对象、图层控制等；第 5 章地图制图，包括地图标注、专题图制作、地图打印输出等；第 6 章空间数据管理，主要包括 SDE 及空间数据、空间数据库及组织、空间数据模型、Geodatabase 的使用与开发等；第 7 章数据编辑包括捕捉功能设计与实现、要素编辑、高级编辑等；第 8 章栅格数据，主要包括访问和创建栅格数据、栅格数据配准、栅格数据处理、栅格图层渲染等内容；第 9 章三维可视化，主要讲解了数据的三维显示、三维分析等；第 10 章空间分析，主要包括空间查询、空间几何图形的集合运算、空间拓扑运算、网络及网络分析等；第 11 章符号库管理系统的开发，包括系统设计、符号管理工具实现等；第 12 章讲解空间数据管理系统；第 13 章～第 16 章将一些开发过程中常见的异常、数据库连接与释放、数据加载以及一些经验技巧作了归纳介绍，本篇的例子主要是对开发过程中经常碰到的问题和实战技巧进行了汇总，以便帮助读者提高工作效率。

本书特色

- 资深技术开发工程师亲自执笔。笔者深入理解了 ArcGIS Engine 内涵、精髓，结合自己丰富培训经验，并结合大量的一线工程实践，潜心编写而成。
- 软件版本采用当前最为流行的 ArcGIS Engine 版本。在知识点讲解过程中穿插了新功能的讲述与应用。
- 知识全面、系统，科学安排内容层次架构，由浅入深，循序渐进，适合读者的学习规律。
- 理论与实践应用紧密结合。基础理论知识穿插在知识点的讲述中，言简意赅、目标明确，目的使读者知其然，亦知其所以然，达到学以致用的目的。
- 知识点+针对每个知识点的小实例+综合实例的讲述方式，可以使读者快速地学习掌握 ArcGIS Engine 软件操作及应用该知识点解决实践中的问题。综合实例部分，深入细致剖析工程应用的流程、细节、难点、技巧，可以起到融会贯通的作用。

- 常见问题解答与技巧集萃。针对读者学习过程中容易遇到的问题，笔者实践过程中总结了实战技巧，本书在最后安排了"常见问题解答与技巧集萃"部分，将零星点滴的经验、技巧、难点一一分析，最大程度地贴近和满足读者的需要。

读者对象

本书从开发者的角度，全面讨论了 ArcGIS Engine 开发的知识，让读者了解和掌握 ArcGIS Engine 开发的实战技术，无论是想对 ArcGIS Engine 入门还是对 ArcGIS Engine 感兴趣的 GIS 人员，都能从本书中得到提高。

本书的例子程序使用 Visual Studio.NET 的 C#开发工具，书中的 ArcGIS Engine 软件和地图数据均来自 ESRI 公司的产品，在此表示衷心的感谢。本书的实例程序中介绍了许多实战技巧，且全部来自学习和工作实践中获得的经验，由于水平有限，书中难免会存在谬误和不足之处，欢迎读者指正。

本书有邱洪钢、张青莲、熊友谊、张文金、何云松、李霓、曹冬梅主编，参与编写的还有郝旭宁、李建鹏、赵伟茗、刘钦、于志伟、张永岗、周世宾、姚志伟、曹文平、张应迁、张洪才、汪海波、李成等。

谨以此书献给我的父母及家人，他们的支持是我人生最大的财富，也是我编写本书的最大动力。

编者

目 录

第一篇 基 础 篇

第1章 开发基础 ………………… 2
1.1 组件对象模型 ………………… 2
1.2 ArcGIS Engine 介绍 ………… 2
   1.2.1 ArcGIS Engine 的体系结构 ……………… 2
   1.2.2 ArcGIS Engine 的类库 …… 2
   1.2.3 ArcGIS Engine10 新特性介绍 ……………… 4
1.3 .NET 平台概述 ……………… 4
   1.3.1 .NET 平台介绍 …………… 4
   1.3.2 .NET FrameWork 4.0 新特性 ………………… 5
1.4 基于.NET 的 ArcGIS Engine 的开发 ………………………… 5
1.5 本章小结 …………………… 9

第2章 ArcGIS Engine 中的控件 … 10
2.1 制图控件介绍 ……………… 10
   2.1.1 地图控件 ………………… 10
   2.1.2 页面布局控件 …………… 12
2.2 3D 控件介绍 ……………… 13
   2.2.1 场景控件——SceneControl …………… 13
   2.2.2 Globe 控件 ……………… 15
   2.2.3 SceneControl 和 GlobeControl 的异同 …… 16
2.3 框架控件介绍 ……………… 19
   2.3.1 图层树控件——TOCControl ………………… 19
   2.3.2 工具栏控件——ToolbarControl ……………… 20
2.4 控件使用实例 ……………… 21
   2.4.1 GIS 系统常用功能集合 … 21
   2.4.2 布局控件与地图控件关联 …………………… 28

   2.4.3 布局控件中属性设置与绘制元素 …………… 29
2.5 本章小结 …………………… 31

第3章 几何对象和空间坐标系 … 32
3.1 Geometry 对象 ……………… 32
3.2 Envelope 对象 ……………… 41
3.3 Curve 对象 ………………… 41
   3.3.1 Segment 对象 …………… 42
   3.3.2 Path 对象 ………………… 46
   3.3.3 Ring 对象 ………………… 46
   3.3.4 PolyCurve 对象 ………… 46
3.4 点对象 ……………………… 47
3.5 线对象 ……………………… 57
3.6 面对象 ……………………… 57
3.7 空间坐标系及变换 ………… 58
3.8 本章小结 …………………… 59

第4章 地图组成及图层控制 …… 60
4.1 地图对象 …………………… 60
   4.1.1 IMap 接口 ………………… 60
   4.1.2 IGraphicsContainer 接口 … 62
   4.1.3 IActiveView 接口 ……… 63
   4.1.4 IActiveViewEvents 接口 … 64
   4.1.5 IMapBookmark 接口 …… 64
   4.1.6 ITableCollection 接口 …… 64
4.2 图层对象 …………………… 64
   4.2.1 ILayer 接口 ……………… 65
   4.2.2 要素图层 ………………… 65
   4.2.3 CAD 文件 ………………… 68
   4.2.4 TIN 图层 ………………… 69
   4.2.5 GraphicsLayer …………… 69
4.3 屏幕显示对象 ……………… 69
4.4 页面布局对象 ……………… 70
4.5 地图排版 …………………… 71

4.5.1 Page 对象 ·················· 71
4.5.2 SnapGrid 对象 ············ 71
4.5.3 SnapGuides 对象 ·········· 71
4.5.4 RulerSettings 对象 ········ 72
4.6 Element 对象 ······················ 72
4.6.1 图形元素 ··················· 72
4.6.2 框架元素 ··················· 75
4.7 MapGrid 对象模型 ················ 75
4.7.1 MapGrid 对象 ············ 76
4.7.2 MapGridBorder 对象 ····· 76
4.8 MapSurround 对象 ················ 77
4.8.1 图例对象 ··················· 77
4.8.2 指北针对象 ················ 78
4.8.3 比例尺对象 ················ 78
4.8.4 比例文本对象 ·············· 80
4.9 Style 对象 ·························· 80
4.10 添加、删除图层数据 ············ 81
4.10.1 矢量数据的添加 ········· 81
4.10.2 栅格数据的添加 ········· 82
4.10.3 删除图层数据 ············ 82
4.11 图层控制 ························· 82
4.11.1 图层间关系的调整 ······ 83
4.11.2 图层显示状态的控制 ··· 83
4.12 本章小结 ························· 83

第 5 章 地图制图 ······················ 84
5.1 地图标注 ··························· 84
5.2 符号及符号库 ······················ 85
5.2.1 颜色对象 ··················· 85
5.2.2 Symbol 对象 ·············· 90
5.3 专题图制作 ························ 105
5.3.1 SimpleRenderer 专题图 ··· 105
5.3.2 ClassBreakRenderer
专题图 ······················ 105
5.3.3 UniqueValueRenderer
专题图 ······················ 105
5.3.4 ProportionalSymbolRenderer
专题图 ······················ 105
5.3.5 ChartRenderer 专题图 ····· 105
5.3.6 DotDensityRenderer
专题图 ······················ 106
5.4 地图打印输出 ····················· 113
5.4.1 Printer 对象 ·············· 113
5.4.2 Paper 对象 ·············· 114

5.4.3 在控件中打印输出 ········ 114
5.4.4 地图的转换输出 ··········· 114
5.4.5 ExportFileDialog 对象 ···· 115
5.5 本章小结 ························· 115

第 6 章 空间数据管理 ················ 116
6.1 SDE 及空间数据 ·················· 116
6.1.1 SDE 介绍 ················· 116
6.1.2 空间数据 ················· 117
6.2 空间数据库及组织 ··············· 118
6.2.1 混合型空间数据库 ······· 119
6.2.2 集成型空间数据库 ······· 120
6.3 空间数据模型 ····················· 122
6.3.1 矢量模型（vector model）··· 122
6.3.2 栅格模型（raster model）··· 123
6.3.3 数字高程模型（DEM，Digital Elevation Model）········ 123
6.3.4 面向对象的数据模型
（Object-Oriented
Data Model）············· 124
6.3.5 混合数据模型（Hybrid Model）···················· 124
6.4 Geodatabase 体系结构 ·········· 125
6.4.1 Geodatabase 介绍 ········ 125
6.4.2 Geodatabase 的体系结构··· 125
6.5 Geodatabase 对象模型 ·········· 126
6.5.1 Geodatabase 中的主要类 ··· 126
6.5.2 Geodatabase 中的
其他常用类 ··············· 127
6.6 Geodatabase 的使用与开发 ····· 127
6.6.1 空间数据库连接 ·········· 127
6.6.2 创建新的数据集 ·········· 127
6.6.3 空间数据的入库 ·········· 128
6.7 本章小结 ························· 132

第 7 章 数据编辑 ······················ 133
7.1 简介 ······························· 133
7.2 捕捉功能设计与实现 ············ 133
7.3 要素编辑 ························· 136
7.3.1 开始编辑 ················· 136
7.3.2 结束编辑 ················· 137
7.3.3 图形编辑 ················· 138
7.4 高级编辑 ························· 142
7.5 本章小结 ························· 142

第 8 章 栅格数据 144

- 8.1 简介 144
- 8.2 访问和创建栅格数据 144
- 8.3 栅格数据配准 145
- 8.4 栅格数据处理 145
  - 8.4.1 栅格数据转换 145
  - 8.4.2 栅格数据变换 147
  - 8.4.3 栅格数据叠置分析 147
  - 8.4.4 栅格数据与矢量数据叠加分析 147
- 8.5 栅格图层渲染 148
  - 8.5.1 RasterRGBRenderer（栅格 RGB 符号化）........ 148
  - 8.5.2 RasterUniqueValueRenderer（唯一值符号化）........ 148
  - 8.5.3 RasterClassfyColorRamp-Renderer（分类符号化）... 149
  - 8.5.4 RasterStretchColorRamp-Renderer 150
  - 8.5.5 RasterDiscreteColorRenderer（点密度符号化）........ 151
- 8.6 本章小结 151

第 9 章 三维可视化 152

- 9.1 简介 152
- 9.2 数据的三维显示 152
  - 9.2.1 DEM 数据的加载 152
  - 9.2.2 叠加纹理数据 152
  - 9.2.3 分层设色 153
- 9.3 三维分析 157
  - 9.3.1 三维场景属性查询 157
  - 9.3.2 坡度分析 157
  - 9.3.3 通视分析 157
  - 9.3.4 剖面图绘制 157
- 9.4 本章小结 160

第二篇 应用提高篇

第 10 章 空间分析 162

- 10.1 简介 162
- 10.2 空间查询 162
  - 10.2.1 基于属性查询 162
  - 10.2.2 基于空间位置查询 163
  - 10.2.3 要素选择集 164
- 10.3 空间几何图形的集合运算 165
- 10.4 空间拓扑运算 166
- 10.5 空间关系运算 176
  - 10.5.1 IRelationalOperator 接口 176
  - 10.5.2 IproximityOperator 接口 185
- 10.6 网络及网络分析 188
  - 10.6.1 主要对象类 189
  - 10.6.2 类之间的相互关系 190
- 10.7 本章小结 191

第三篇 综合实例篇

第 11 章 符号库管理系统的开发 194

- 11.1 简介 194
- 11.2 系统设计 194
  - 11.2.1 主程序界面设计 195
  - 11.2.2 点状符号 195
  - 11.2.3 线状符号 196
  - 11.2.4 面状符号 197
- 11.3 符号管理工具实现 198
- 11.4 本章小结 226

第 12 章 空间数据管理系统 227

- 12.1 简介 227
- 12.2 空间数据管理框架设计 227
- 12.3 空间数据管理实现 227
- 12.4 本章小结 252

第四篇 常见疑难解答与经验技巧集萃

第 13 章 空间数据库连接与释放 254

- 13.1 Shapefile 文件 254
- 13.2 Coverage 数据格式 254
- 13.3 Geodatabase 数据格式 255

13.4 ArcSDE（Enterprise Geodatabase）
数据库连接 ················ 255
13.5 TIN 不规则三角网 ············ 255
13.6 栅格数据 ··················· 256
13.7 CAD 数据 ·················· 256
13.8 一般关系表 ················· 256
13.9 ArcSDE 客户端负载连接方式 ··· 257
13.10 ArcSDE 连接 Oracle 数据库 ··· 257
13.11 ArcSDE 连接释放 ············ 258
13.12 自动关闭空闲 SDE 连接 ······ 258

第 14 章 空间数据库加载 ············ 259
14.1 通过设置属性加载个人
数据库 ···················· 259
14.2 通过名称加载个人数据库 ····· 260
14.3 SDE 数据库 ················ 260
14.4 分图层加载 CAD 图层 ······· 261
14.5 整幅 CAD 图的加载 ········· 262

第 15 章 程序出错和异常 ············ 263
15.1 释放资源异常问题 ··········· 263
15.2 表结构操作错误 ············· 263

15.3 要素编辑的错误 ············· 263
15.4 Network I/O Error 异常 ······· 264
15.5 数据插入错误 ··············· 264
15.6 索引被占用异常 ············· 264
15.7 SDE 导入空间数据错误 ······· 264
15.8 HRESULT:0x80040228 异常 ···· 265
15.9 HRESULT:0x80040213 异常 ···· 265
15.10 HRESULT:0x80040205 ········ 265
15.11 HRESULT:0x80010105
(RPC_E_SERVERFAULT) ······ 265

第 16 章 其他经验技巧 ·············· 266
16.1 ArcEngine 中的先闪烁
后刷新现象 ················ 266
16.2 ArcEngine 中几种数据的删除
方法和性能比较 ············ 266
16.3 数据游标 ·················· 268
16.4 投影变换 ·················· 268
16.5 ITopologicalOperator ········· 268
16.6 缓冲区查询 ················ 269
16.7 插入记录效率 ·············· 269

第一篇

基 础 篇

第 1 章　开发基础
第 2 章　ArcGIS Engine 中的控件
第 3 章　几何对象和空间坐标系
第 4 章　地图组成及图层控制
第 5 章　地图制图
第 6 章　空间数据管理
第 7 章　数据编辑
第 8 章　栅格数据
第 9 章　三维可视化

第 1 章　开发基础

1.1 组件对象模型

COM 即组件对象模型，是关于如何建立组件，以及如何通过组件建立应用程序的一个规范，说明了如何动态交替更新组件。组件对象模型（COM）是微软公司为计算机工业的软件生产更加符合人类的行为方式开发的一款新的软件开发技术。在 COM 构架下，人们可以开发出各种各样的、功能专一的组件，然后将它们按照需要组合起来，构成复杂的应用系统，因此可以将系统中的组件用新的替换掉，以便随时进行系统的升级和定制，也可以在多个应用程序中重复利用一个组件。

COM 是开发软件组件的一种方法。组件实际上是一些小的二进制可执行程序，它们可以给应用程序、操作系统以及其他的组件提供服务。组件可以在运行时刻，在不被重新链接或编译应用程序的情况下被卸下或替换。Microsoft 的许多技术，如 Activex、Directx、Ole 等都是基于 COM 而建立起来的，并且 Microsoft 的开发人员也在大量使用 COM 组件来定制他们的应用程序及操作系统。

ESRI 选择 COM 作为 ArcGIS 组件技术的原因是因为，COM 是一项成熟的技术，能提供良好的性能，目前有很多开发工具支持，而且有很多组件可用于扩展 Engine 的功能。因此，基于 Engine 开发应理解 COM 技术，需要理解的层次取决于开发的深度。

1.2 ArcGIS Engine 介绍

1.2.1 ArcGIS Engine 的体系结构

Arc Engine 是一个简单的、独立于应用程序的 Arc Objects 编程环境，开发人员用于建立自定义应用程序的嵌入式 GIS 组件的一个完整类库。Arc Engine 由一个软件开发包和一个可以重新分发的为 ARCGIS 应用程序提供平台的运行时（runtime）组成。

Arc Engine 功能层次由以下 5 个部分组成。
- 基本服务：由 GIS 核心 Arc Objects 构成，如要素几何体和显示。
- 数据存取：Arc Engine 可以对许多栅格和矢量格式进行存取，包括强大而灵活的地理数据库。
- 地图表达：包括用于创建和显示带有符号体系和标注功能的地图的 Arc Objects，以及包括创建自定义应用程序的专题图功能的 Arc Objects。
- 开发组件：用于快速应用程序开发的高级用户接口控件和高效开发的一个综合帮助系统。
- 运行时选项：Arc Engine 运行时可以与标准功能或其他高级功能一起部署。

1.2.2 ArcGIS Engine 的类库

System 类库是 Engine 中最底层的类库。包含给构成 ArcGIS 的其他类库提供服务的组件。库中包含了大量可供开发者调用的接口。AoInitializer 对象也包含在 System 类库中，提供给开发者初始化和注销 Arc Engine。应用程序不能扩展此类，可通过类库中包含的接口来扩展 ARCGIS 系统。

SystemUI 类库：主要定义了 ArcGIS 系统中所使用的用户界面组件类型。这些用户界面组件可以在 ArcGIS Engine 中进行扩展。开发者可利用接口来扩展 UI 组件。

Geometry 类库：包含了核心几何对象，如点、线、多边形及其几何类型和定义。除了这些实体外，就是作为多边形、多义线的组成部分的几何图形，它们是组成几何图形的子要素，如 Segment、Path、Ring 等。Polyline、Polygon 是由一系列相连接的片段组成的，片段的类型如 CircularArc、Line、BezireCurve 等，每个片段是由两个不同的点组成的：起点和终点，以及一个定义两点之间弯曲度的元素类型组成。所有的几何图形对象都支持 Buffer、Clip 等几何操作，几何子要素不可以扩展。

Display 类库：包含了支持向输出装置绘制符号体系的组件，除了负责实际输出图像的主要显示对象外，还包含了表示符号和颜色的对象，它们用来控制在显示上绘制实体的属性。还包含了在与显示交互时提供给用户的可视化反馈的对象。

Server 类库：包含了用于获取到 ArcGIS Server 的连接的对象，使用 GISServerConnection 对象来访问 ArcGIS Server。通过此对象来获取 ServerObjectsManager 对象，开发人员可以操作 ServerContext 对象，用于处理运行于服务器上的 ArcObjects。还可以通过 GISClient 类库与 ArcGIS Server 进行交互。

Output 类库：包含了生成输出所必须的对象，通常是从地图或页面布局输出到打印机、绘图仪，或导出到文件中。

Geodatabase 类库：包含了所有与数据访问相关的定义的类型，为地理数据提供了编程 API，是建立在标准工业关系型和对象关系数据库技术之上的地理数据库。Geodatabase 类库提供了比 ArcObjects 架构中更高级的数据源提供者实现的接口，可以通过扩展地理数据库以支持特定类型的数据对象。

GISClient 类库：包含了操作远程 GIS 服务的对象，这些 Web 服务可以由 ArcIMS 或 ArcGIS Server 提供。GISClient 提供了以无态方式直接或通过 Web 服务目录操作 ArcGIS Server 对象的通用编程模型。在 ArcGIS Server 上运行的 ArcObjects 组件不能通过 GISClient 接口来访问。要直接访问在服务器上运行的 ArcObjects，应使用 Server 类库中的功能。

DataSourcesFile 类库：包含了适用于地理数据库应用程序接口（APIs）所支持的矢量数据格式的工作空间工厂和工作空间。开发者不能扩展 DataSourcesFile 类库。

DataSourcesGDB 类库：包含了适用于存储在 RDBMS 中的地理数据库所支持的矢量和栅格数据格式的工作空间工厂和工作空间。开发者不能扩展此类库。

DataSourcesOleDB 类库：包含了用于 Microsoft OLE DB 数据源的 GeoDatabase API 实现。此类库只能用在 Microsoft Windows 操作系统上。这些数据源包括支持数据提供者和文本工作空间的所有 OLE DB。此类库不能扩展。

DataSourcesRaster 类库：包含了用于栅格数据源的 GeoDatabaseAPI 实现，这些数据源包括 ArcSDE 支持的关系型数据库管理系统，以及其支持的 RDO 栅格文件格式。当需要支持新的栅格格式时，可以通过扩展 RDO 实现，而不能直接扩展 DataSourcesRaster。

GeoDatabaseDistributed 类库：包含了支持分布式地理数据库的访问。不能扩展此类库。

Carto 类库：该类库支持地图的创建和显示，PageLayout 对象是地图及其底图元素的容器。地图元素包括指北针、图例、比例尺等。Map 对象包括地图上所有图层都有的属性，如空间坐标系、地图比例尺以及地图图层的操作。此对象可以加载各种类型的图层，不同类型的图层由相应的对象处理，如 FeatureLayer 对象处理矢量数据，RasterLayer 对象处理栅格数据，TinLayer 对象处理 TIN 数据等。通常图层都有一个相关的 Renderer 对象，来控制数据在地图中的显示方式。Renderer 通常用 Display 类库中的符号来进行绘制，Renderer 只是将特定符号与待绘制实体属性相匹配。

Location 类库：包含了与位置数据操作相关的对象，位置数据可以是路径事件或者地理编码的位置，开发者可以创建自己的地理编码对象。线性参考功能提供对象用于向线性要素添加事件，并可用各种绘制方法来绘制这些事件。

NetworkAnalysis 类库：该类库支持应用网络的创建和分析，提供了用于在地理数据库中加载网络数据的对象，并提供对象用于分析加载到地理数据库中的网络。

Controls 类库：包含了用于应用软件开发的控制器，包括通过控制器来使用命令和工具。ArcGIS Controls 通过封装 ArcObjects，并提供粗粒度的 API 简化了开发的过程，开发者也可以同时对 ArcObjects 进行细粒度的访问。

GeoAnalyst 类库：包含了核心空间分析的操作，这些操作可以通过空间分析和 3D 分析扩展模块来使用。

3Danalyst 类库：包含了用于进行数据 3D 分析以及支持 3D 数据显示的对象。Scene 对象是 3Danalyst 类库中的主要对象之一，是数据的容器。Camera 和 Target 对象规定在考虑要素位置与观察者关系时场景如何浏览等。

GlobeCore 类库：包含了用于进行球体数据分析以及支持球体数据显示的对象。该类库中有一个开发控件及与其一起使用的命令和工具。该开发控件可以与 Controls 类库中的对象协同使用。为了使用这个类库中的对象，需要 3D Analyst 扩展模块的许可，或者 ArcGIS Engine 运行时 3D 分析选项的许可。

SpatialAnalyst 类库：包含了用于进行栅格与矢量空间分析的对象。使用该类库中的对象，需要 ArcGIS 空间分析扩展模块的许可，或者 ArcGIS Engine 运行时空间分析选项的许可。

1.2.3 ArcGIS Engine10 新特性介绍

在 ArcGIS 10 中，将 ArcGIS Desktop、ArcGIS Engine、Net sdks 以及 ArcGIS Server 上的 AO 整合成一个 SDK。ArcGIS 10 版本开始，除了支持原有的操作系统外，还增加了 Windows 7 的支持。应用许可方面，之前的版本 ArcGIS Engine 每台机器支持单个用户，ArcGIS 10 版本开始使用浮动许可机制，允许应用程序在一个拥有一定许可的 License 服务器上取许可。

此外，在编辑方面也做了比较大的改变，新版本的捕捉环境允许自定义工作中使用，并不仅限于编辑器中的工具，捕捉环境可以通过 ArcMap 中获取或者通过 ArcGIS Engine 的 esricontrols 中获取。TOC 窗体和视图在 ArcGIS 10 版本中重新架构，以及高级 MapTips 的表达，更加详细信息，读者可以在 esrichina 中文官网查看。

1.3 .NET 平台概述

1.3.1 .NET 平台介绍

Web 服务是由简单网页构成的静态服务网站，发展到可以交互执行一些复杂步骤的动态服务网站，这些服务可能需要一个 Web 服务调用其他的 Web 服务，并且像一个传统软件程序那样执行命令。这就需要和其他服务整合，需要多个服务能够一起无缝地协同工作，需要能够建立与设备无关的应用程序，需要能够容易地协调网络上的各个服务的操作步骤，容易地建立新的用户化的服务。

微软推出的.NET 技术正是为了满足这一需求而开发的。.NET 将 Internet 本身作为构建新一代操作系统的基础，并对 Internet 和操作系统的设计思想进行了延伸，使得开发人员能够创建出与设备无关的应用程序，很容易地实现 Internet 连接。

.NET 系统由以下 5 部分组成。

（1）.NET 开发平台。

.NET 开发平台包括.NET Framework 和 Visual Studio.net。Visual Studio.net 是一套完整的开发

工具，用于生成 ASP Web 应用程序、Web Services、桌面应用程序和移动应用程序等。多种开发语言使用同一集成开发环境，该环境允许它们共享工具并有助于创建混合语言解决方案。.NET Framework 是生成、部署和运行 Web 服务及应用程序的平台。其具有两个主要组件：通用语言运行时（Common Language Runtime）和.NET Framework 类库。通用语言运行时是.NET Framework 的基础，提供了内存管理、线程管理和远程处理等核心服务，并严格检查类型安全。.NET Framework 类库是一个综合的面向对象的类型集合，可以使用它开发多种应用程序。

（2）.NET 服务器。

.NET 服务器提供了广泛聚合和集成 Web 服务的服务器，是搭建.NET 平台的后端基础。

（3）.NET 基础服务。

.NET 基础服务提供密码认证、日历、文件存储、用户信息等功能。

（4）.NET 终端设备。

.NET 终端设备提供 Internet 连接，并实现 Web 服务的终端设备的前端基础。个人计算机、个人数字助理（PDA）设备以及各种嵌入式设备，将在这个领域发挥作用。

（5）.NET 用户服务。

.NET 用户服务即是满足人们各种需求的服务，是.NET 的最终目标，也是.NET 的价值体现。

1.3.2 .NET FrameWork 4.0 新特性

.NET4 框架引入一个改进的安全模式，.NET4 框架不会自动使用当前版本公共语言运行库来运行使用以前版本的.NET 框架构建应用程序，因此，在.NET4 框架下运行以前的应用程序，你必须使用你的 Visual Studio 项目的属性中制定的目标.NET 框架来重新编译你的应用程序，或者在应用程序的配置文件中使用元素来指定支持的运行时刻库。这里只介绍和代码相关的新特性，其他的内容，读者可以自行查阅官网的介绍。

1.4 基于.NET 的 ArcGIS Engine 的开发

本节通过一个例子程序介绍基于.NET 的 ArcGIS Engine 开发过程，从而为后续章节学习打下基础。本例子的样例数据采用 ArcGIS 安装目录下的"World.mxd"数据文件，本书的安装目录"D:\Program Files\ArcGIS\DeveloperKit10.0\Samples\data\World"。为方便本书样例使用，将 Word 文件夹复制到 E 盘根目录下。

在应用程序编写代码之前，应先把应用程序用到的 ArcGIS 控件和其他的 ArcGIS Engine 库引用装载到开发环境之中。

> **重点提示**
>
> （1）安装好 Engine 后，在 VS2010 的工具箱中，自动会增加"ArcGIS Windows Forms"选项卡标签，无需要按下面的步骤重新创建"ESRI"选项卡，本实例重新创建的目的是，演示当选择项丢失时，可以通过该方式重新加载到"ArcGIS Windows Forms"选项卡标签中。
>
> （2）本实例添加的"ESRI"选项卡中，每个控件名字都比"ArcGIS Windows Forms"选项卡中多了"AX"前缀，但控件功能及使用是一样的。
>
> （3）本实例中"添加引用"内容，也是为了演示目的，让读者知道从哪里添加引用，从工具箱中拖曳控件到 Form 窗体上后，解决方案的引用中，自动添加相关的内容，本书第一版中有读者反馈代码运行出错等问题，其中就是因为引用丢失，所示代码编译出错。

（1）启动 Visual Studio.NET，从"新建项目"对话框中创建一个新的 Visual C# "Windows 窗体应用程序"项目，如图 1-1 所示。

（2）将项目命名为"Sample"，并选择"位置"保存该项目。

（3）单击"视图"菜单，选择"工具箱"子菜单项，如图 1-2 所示。

▲图 1-1 新建项目

▲图 1-2 选择工具箱

（4）在"工具箱"空白处单击鼠标右键，在弹出的快捷菜单中选择"添加选项卡"，然后在新增选项卡上输入"ESRI"作为选项卡标签，如图 1-3 所示。

（5）在"工具箱"的"ESRI"标签上单击鼠标右键，然后从弹出的快捷菜单中选择"选择项"，如图 1-4 所示。

▲图 1-3 添加选项卡

▲图 1-4 选择项

（6）在弹出的"选择工具箱项"对话框中选择".NET Framework 组件"选项卡，选中

"AxMapControl"、"AxPageLayoutControl"、"AxTOCControl"和"AxToolbarControl"等复选框,单击"确定"按钮,将所选择的控件添加到工具栏上,如图1-5所示。

(7)在"项目"菜单中选择"添加引用"项,在弹出的"添加引用"对话框中,如图1-6所示;双击"ESRI.ArcGIS.Carto"、"ESRI.ArcGIS.Display"、"ESRI.ArcGIS.Geometry"、"ESRI.ArcGIS.System"、"ESRI.ArcGIS.SystemUI"和"ESRI.ArcGIS.Utility"等选项,单击"确定"按钮,如图1-7所示。

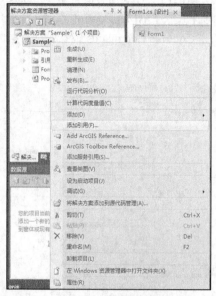
▲图1-5 选择工具箱项

▲图1-6 添加引用菜单

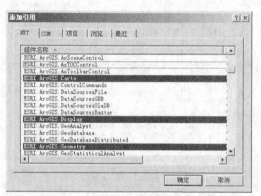
▲图1-7 添加引用

(8)打开.NET窗体,双击"工具箱"中"ESRI"标签栏中的AxMapControl控件,将MapControl加入到窗体上,再将AxToolbarControl也添加到窗体上。

(9)在窗体上双击显示窗体代码窗口(或按"F7"键),加载图层到MapControl中,在Form_Load事件中添加地图文档加载代码,代码如下:

```
private void Form1_Load(object sender, EventArgs e)
{
        InitializeComponent();
        string path = @"D:\World\";
        string fileName = @"World.mxd";
```

```
//加载地图文件
axMapControl1.LoadMxFile(path+fileName);
//将地图全屏最大化
axMapControl1.Extent = axMapControl1.FullExtent;
}
```

（10）设置 ToolbarControl 与 MapControl 控件关联。

设置工具控件与地图控件关联，使用工具栏来操作地图。在 .NET 窗体上选中 ToolbarControl 控件，单击 "ActiveX –Properties…"，弹出 "属性" 对话框，或者鼠标右键单击 ToolbarControl 控件，选择 "属性" 菜单项，在 "Buddy" 下拉列表中选择关联 "axMapControl1"，如图 1-8 所示。

选择 "Items" 选项卡，单击 "Add" 按钮，在弹出的 "Controls Commands" 对话框中添加 "Pan"、"Zoom In" 和 "Zoom Out"，如图 1-9 所示。

▲图 1-8 设置工具控件与地图控件关联

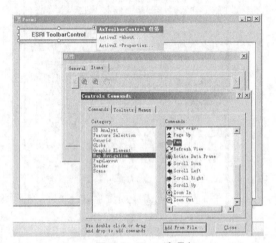

▲图 1-9 Items 选项卡

（11）在 Visual Studio.NET 中按 "F5" 键运行程序，结果如图 1-10 所示。

▲图 1-10 运行程序界面

> **重点提示**
>
> ArcGIS Engine 10 版本开始有一些明显的产品架构上的变更，新架构最明显的优势就是独立存在的 ArcGIS Desktop 和 Engine runtime，可以将这两个产品安装在不同的目录下，可以分别对这两个产品打补丁等。基于这一架构的变化，要求 Engine 应用程序以及自定义组件等绑定到计算机上指定的产品。例如，绑定到 Desktop 和 Engine 两种产品上，则如下所示。

（12）添加绑定产品代码，本例绑定 "EngineOrDesktop"，如图 1-11 所示。

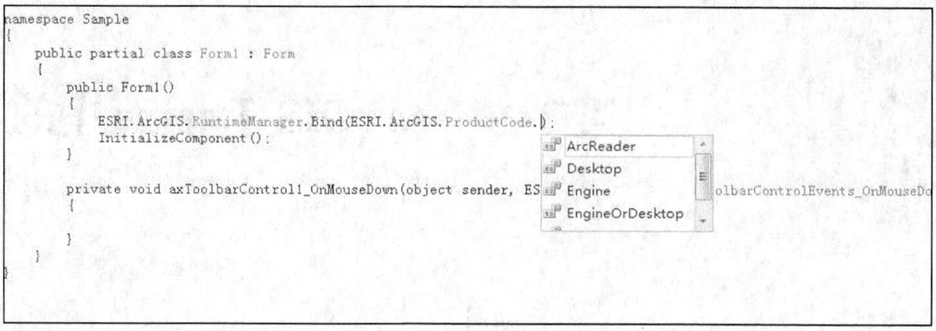

```
namespace Sample
{
    public partial class Form1 : Form
    {
        public Form1()
        {
            ESRI.ArcGIS.RuntimeManager.Bind(ESRI.ArcGIS.ProductCode.);
            InitializeComponent();                                  ArcReader
        }                                                           Desktop
        private void axToolbarControl1_OnMouseDown(object sender, ES Engine      olbarControlEvents_OnMouseDo
        {                                                           EngineOrDesktop
        }
    }
}
```

▲图 1-11　运行程序界面

（13）在 Visual Studio.NET 中按"F5"键运行程序可以显示。

1.5　本章小结

本章对全书的知识点作了一个概要的介绍，让读者有个基本印象。介绍了组件对象模型、ArcGIS Engine 的体系结构、ArcGIS Engine 的类库，以及 ArcGIS Engine、.Net 平台新特性。最后做了一个简单的小程序，演示了基于.NET 的 ArcGIS Engine 的开发过程，并给出了新版本的一些差异，以使读者能对 ArcGIS Engine 有一个感性认识。

第 2 章　ArcGIS Engine 中的控件

为了快速构建一个 GIS 应用程序，ArcGIS Engine 给开发者提供了一些可视化控件，如制图控件、3D 控件、框架控件等。ArcGIS 控件可以通过两种方式建立应用程序，其一，ArcGIS 控件可以嵌入到现有的应用程序中以增强制图功能；其二，ArcGIS 控件可用于创建新的独立应用程序。

制图控件，如 MapControl、PageLayoutControl，其中 MapControl 控件主要用于地理数据的显示和分析，PageLayoutControl 用于生成一幅成品地图。MapControl 封装了 Map 对象，而 PageLayoutControl 则封装了 PageLayout 对象。这两个控件都实现了 IMxContents 接口，因此不仅可以读取 ArcMap 创建的地图文档，而且可以将自身的地图内容写到一个新的地图文档中。

三维控件，如 GlobeControl、SceneControl 都具有导航功能，允许终端用户操作三维视图，而不必使用控件命令或自定义命令。通过设置 Navigate 属性，用户就可以操作三维视图，如前后左右移动、放大缩小等。

框架控件，如 TOCControl、ToolbarControl，需要与其他的控件协作使用。例如在 TOCControl 控件属性页中设置与 MapControl 关联，单在 MapControl 中删除一个图层时，该图层也从 TOCControl 中删除。

本章详细介绍这 6 个控件，并给出 C#的实际使用。

2.1　制图控件介绍

2.1.1　地图控件

MapControl 控件封装了 Map 对象，并提供了其他的属性、方法和事件，用于管理控件的外观、显示属性和地图属性，管理、添加数据图层，装载地图文档，显示、绘制跟踪图层。MapControl 上存在着诸如 TrackRectangle、TrackPolygon、TrackLine 和 TrackCircle 等帮助方法，用于追踪或"橡皮圈箍（rubber banding）"显示上的几何图形（Shape）。VisibleRegion 属性可用于更改 MapControl 显示区内的几何图形。MapControl 控件实现的主要接口有 IMapControlDefault、IMapControl2、IMapControl3、IMapControlEvents2 等，如图 2-1 所示。

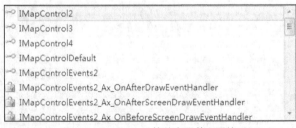

▲图 2-1　MapControl 控件实现的主要接口

2.1.1.1 IMapControlDefault 接口

IMapControlDefault 接口是地图控件默认接口。当将 MapControl 控件拖曳到容器上时，会自动创建一个 axMapControl1 的对象，该对象全部继承父类接口的属性和方法。下面的代码演示了如何使用这个接口。

```
IMapControlDefault mapControlDefault;
MapControl mapControl;
MapControl = axMapControl1.Object as IMapControlDefault;
```

2.1.1.2 IMapControl2 接口

IMapControl2 接口提供了一系列的属性和方法，如设置控件外观，设置 Map 对象或控件的显示属性，添加和管理数据图层、地图文档，在控件上绘制图形和返回几何对象等，如图 2-2 所示。

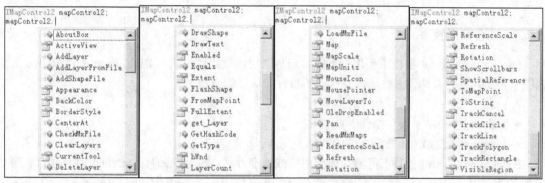

▲图 2-2　IMapControl2 接口提供的属性和方法

2.1.1.3 IMapControl3 接口

该接口继承自 IMapControl2，并增加了 8 个属性和一个方法。
CustomProperty：设置自定义控件属性。
DocumentFilename：返回 MapControl 装入的地图文档的文件名。
DocumentMap：返回 MapControl 最后装入的地图名称。
KyeIntercept：返回或设置 MapControl 截取键盘按键信息。
Object：返回 MapControl 控件。

```
IMapControl2 mapControl2;
mapControl2=axMapControl1.Object as IMapControl2 ;
mapControl2 =axMapControl1.GetOcx() as IMapControl2 ;
IMapControl3 mapControl3;
mapControl3 = axMapControl1.Object as IMapControl3;
mapControl3 = axMapControl1.GetOcx() as IMapControl3;
```

ShowMapTips：确定是否显示地图的 Map Tips。
TipDelay：设置 Map Tips 的延迟时间。
TipStyle：设置 Map Tips 的显示样式。
SuppressResizeDrawing()：当控件尺寸发生变化时阻止数据实时重绘。

2.1.1.4 IMapControlEvents2 接口

IMapcontrolEvents2 定义了 MapControl 能够处理的全部事件，如图 2-3 所示。图中，

OnBeforeScreenDraw 事件是屏幕绘制前触发的事件，OnViewRefreshed 是视频刷新触发事件。

▲图 2-3　MapControl 能够处理的全部事件

2.1.2　页面布局控件

PageLayoutControl 控件主要用于页面布局与制图。该控件封装了 PageLayout 类，提供了布局视图中控制元素的属性和方法，以及其他的事件、属性和方法。

- Printer 属性提供了处理地图打印的设置。
- Page 属性提供了处理控件的页面效果。
- Element 属性则用于管理控件中的地图元素。

PageLayoutControl 控件不能添加地图图层或地理数据，必须通过使用 MXD 文件来加载需要处理的数据。PageLayoutControl 控件主要实现 IPageLayoutControlDefault、IPageLayoutControl、IPageLayoutControl2、IPageLayoutControlEvents 等接口，如图 2-4 所示。

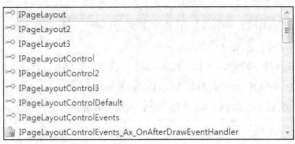

▲图 2-4　PageLayout Control 控件主要实现的接口

IPageLayoutControlDefault 接口

IPagelayoutControlDefault 是界面调用的默认接口。该接口是纯接口，因此可以在新版本中使用。接口的方法和属性都是相同的最高版本的主界面上的 PageLayoutControl，例如 IPageLayoutControlDefalut 相当于 IPageLayoutControl2，但在以后的新版本中可能成为 IPageLayoutControl3。通过使用 IPageLayoutControlDefault 接口，可以保证使用的 PageLayoutControl 是最新版本。

> **注意**　mapDocument 属于 carto 名称空间在属性，嵌入式互操作类型中改为 FALSE。

2.2 3D 控件介绍

SceneControl 和 GlobeControl 都是嵌入式的开发组件，可以通过开发环境增加到窗体中或者对话框中，它们都提供了设置属性，并且能直接单击控件来增加功能，当然这个属性页也提供了快捷的选择控件属性和方法，并允许开发者编写自己的代码来进行相关的操作。3D 控件主要用到两个类：**3DAnalyst** 和 **GlobeCore**。

1. 3DAnalyst 类

包含了 3D scenes 的对象，当然也包含了 ArcGlobe 应用程序运用的类库，主要的开发组件是在 GlobeCore 类库中的定义，定义文件为 esri3DAnalyst.olb，主要定义了 SceneControl、Scene、SceneGraph、3Dproperties、SceneExporter3D、SceneViewer、3Dsymbol 和 AnimationTrack 等。

2. GlobeCore 类

是 ArcGlobe 应用程序主要运用的类库，同时进行基于 global 数据的 ArcGIS 3D 分析。在 GlobeCore 类库中，定义了 2D 和 3D 数据在 globe 的表面显示及数据操作，定义文件为 esriGlobeCore.olb，主要定义了 GlobeControl、Globe、GlobeDisplay、GlobeCamera、GlobeViewer、GlobeLayerProperties、GlobeLayer 和 AnimationTrack 等。

3DAnalyst 和 GlobeCore 类库包括了 SceneControl 和 GlobeControl 开发组件，并且提供了基于 Scene 和 Globe 的命令并可协同工作。SceneControl 和 GlobeControl 也可以与 ToolbarControl 和 TOCControl 开发组件关联。它所定义的类库如图 2-5 所示。

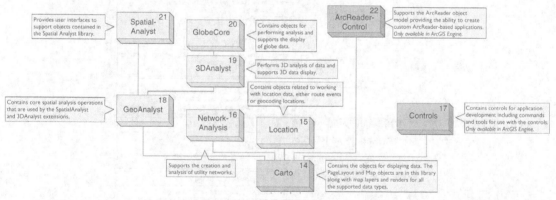

▲图 2-5　定义的类库

但是需要注意的是：SceneControl 和 GlobeControl 对图层的定义有所区别。Globe 的图层分为 esriGlobeLayerTypeDraped、esriGlobeLayerTypeFloating、esriGlobeDataElevation 等 3 种，更适合三维用户使用。而在 SCENE 中就没有这种概念，这也是很多 SCENE 用户转向 GLOBE 时比较迷茫的。Elevation 图层一经定义，所有的 Draped 图层将依附于 Elevation 图层表面上。如果要单独定义图层的 Elevation，就必须定义该图层为 Draped 图层，这一点就与 SCENE 一样了。

2.2.1 场景控件——SceneControl

SceneControl 是一个高性能的嵌入式的开发组件，提供给开发者建立和扩展 Scene 程序，当然也提供了基于 ArcScene 的功能给用户，以便进行绘图等操作。控件 SceneControl 相当于 ArcScene Desktop 应用程序中的 3D 视图，并且提供了显示和增加空间数据到 3D 的方法等。

　　SceneControl 是单一的开发进程，并且提供了粗粒度 ArcEngine 组件对象，也提供了强大纹理着色的功能。SceneControl 通过对象接口 ISceneViewer 来表现，ISceneViewer 接口提供了一个 Camera 对象，该对象由视角（Observer）和目标（Target）构成。SceneControl 控件提供一些属性和方法操作三维对象。例如，Camera、Scene、SceneGraph 和 SceneViewer 等属性。LoadSxFile 方法，用于导入 scene 文档。SceneControl 是进行三维开发最基本的控件，如图 2-6 所示。

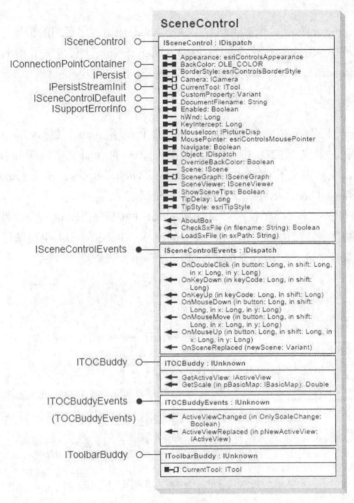

▲图 2-6　SceneControl 开发进程

　　3DAnalyst 类库包含了命令、工具和工具集，可与 SceneControl 控件共同工作来进行操作。例如，scene navigation 命令和工具用来移动对象到新的位置（camera），Scene 命令包含了 GUID 和命令描述，详细信息读者可以参考帮助中的 "Names and IDs of the ControlCommands" 部分。帮助信息中介绍了 Commands 和 tools 使用，例如：Pan, zoom, fly, set observer, select Toolbars, 我们可以看看下面所有 SceneControl 中已经提供了命令、工具，如图 2-7 所示。

第 2 章　ArcGIS Engine 中的控件

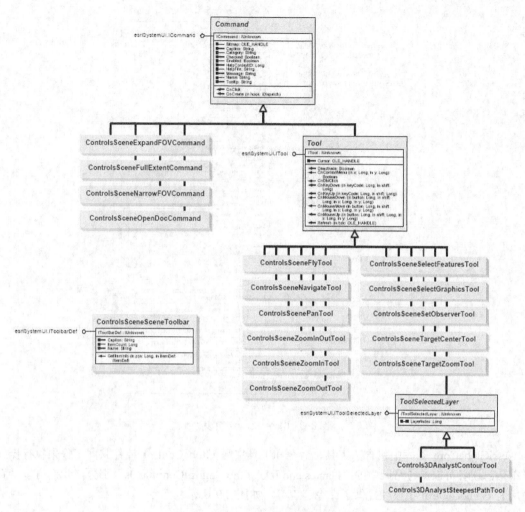

▲图 2-7　SceneControl 提供的命令、工具

2.2.2　Globe 控件

　　GlobeControl 是一个高性能的嵌入式的开发组件，提供给开发者建立和扩展 ArcGlobe 程序，当然也提供了基于 ArcGlobe 的功能给用户，以便进行绘图等操作。GlobeControl 显示 3D 视图，并能提供全球表现的位置，而且是基于 3D 数据。GlobeControl 控件对应于 ArcGlobe 桌面应用程序的三维视图。GlobeControl 封装了 GlobeViewer 对象，可以加载 ArcGlobe 应用程序创作的 Globe 文档。

　　GlobeControl 也是单一的开发进程，并且提供了粗粒度 ArcEngine 组件对象，当然也提供了强大纹理着色的 ArcEngine 组件。GlobeControl 通过对象接口来操作 IGlobe 视图，用户可以通过 IGlobeViewer 对象来操作 ArcGlobe 应用程序。IGlobeViewer 接象包含一个 GlobeDisplay，GlobeDisplay 又包含一个 Globe。GlobeControl 提供了经常使用的属性和方法，例如，GlobeControl 有 GlobeCamera、Globe、GlobeDisplay 和 GlobeViewer 等属性，当然 GlobeControl 也能执行一些方法或任务，例如，GlobeControl 有 Load3dFile 方法用于导入 globe 文档。GlobeControl 是进行三维开发最基本的控件，因为其提供了用户界面，所以更容易进行开发，当然使用对象模型也能很容易地理解及开发三维功能，如图 2-8 所示。

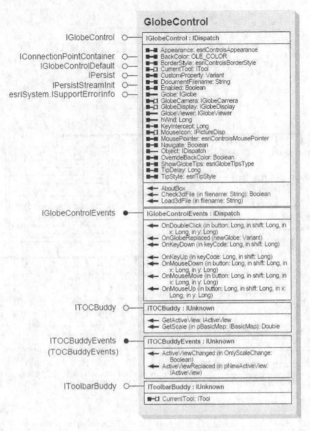

▲图 2-8　Globe Control 开发进程

esriGlobeCore 类库中包含了工具、命令和工具集与 GlobeControl 控件共同工作来执行操作，详细信息读者可以参考帮助中的"Names and IDs of the ControlCommands"部分，我们可以看看下面所有 GlobeControl 中已经提供了命令、工具，如图 2-9 所示。

2.2.3　SceneControl 和 GlobeControl 的异同

SceneControl 支持下面的主要特征：
3D 线符号有 Tubes、walls 和 textured lines；
TIN 数据显示和分析；
基于内存 Memory 的显示原理；
支持立体和平面视图；
表面分析工具，比如最短路径和等高线的生成 1；
Layer 支持，比如图层坐标转换；
输出到 3D 格式（vrml）；
动态阴影效果。

GlobeControl 支持下面的主要特征：
所有的数据源必须具有空间参考；
空间参考可以是地理坐标系统或工程坐标系统；
页面显示，提供多级显示机制，对于大数据量支持 caching 的方式；
超链接（hyperlinks）；

第 2 章 ArcGIS Engine 中的控件

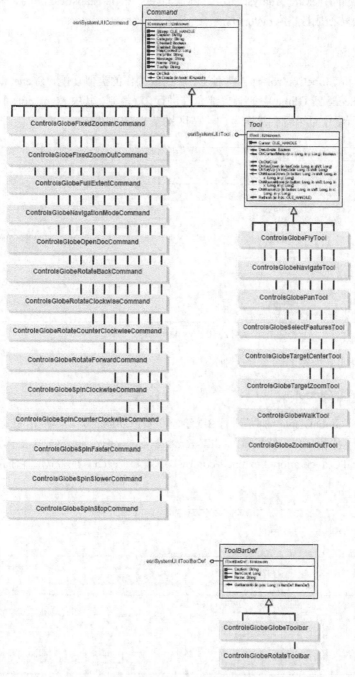

▲图 2-9 Globe Control 提供的命令、工具

导航和分析工具，比如量测、步行、放大、缩小、移动目标到中心、缩放目标；
地图符号支持光栅要素图层；
旋转工具栏；
各式各样的显示目标（正面朝上）、观测者位置和指北针、剪切控制面板。
可以创建一个模板图层，比如 MapServer layers、ArcIMS Image layers、feature annotation layers、WMS layers 和 MOLE(Military Overlay Editor) layers。对 Feature 透明度，在 GlobeControl 中有一些

数据不被支持，比如 Tracking Analyst 图层、测量 layers 和 Geostatistics 图层。另外，TINs 数据不直接支持，所以要转换成栅格（rasters）。

Scene 和 Globe

SceneControl 和 GlobeControl 有一些对于 3D 对象操作的类似功能，Scene 和 Globe 都支持 2D 和 3D 数据图层，Scene 或 Globe 中的对象都允许进行 3D 控制，其中 MapControl 中的 Map CoClass 相似。从图 2-10 所示中可以看到这两者的一些区别及相似的地方。

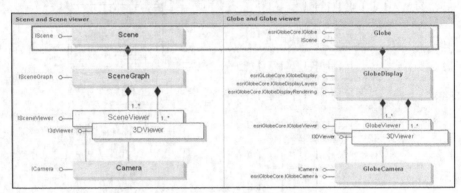

▲图 2-10　Scene 和 Globe 比较

下面进一步说明其内在的联系及区别。

SceneGraph 和 GlobeDisplay

SceneGraph 和 GlobeDisplay 都是进行 3D 绘制和着色功能的类，当然在 3D 制图程序方面还是有一些不同。对于视图窗口，可以进行 Add、remove、set active、refresh 等操作，也可以改变一些属性，比如 Vertical exaggeration、extent、contrast，以及对 caches 的控制。它们在模型中的位置，如图 2-11 所示。

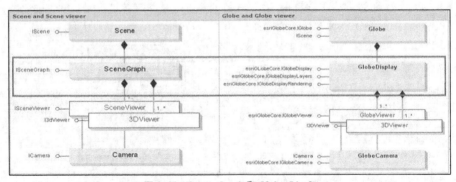

▲图 2-11　Scene Graph 和 Globe Display

SceneViewer 和 GlobeViewer

SceneViewer 和 GlobeViewer 是 SceneControl 和 GlobeControl 中的 3D 显示窗口，当然它们都各自支持自身的控件。这个当前的视图还可以通过 ActiveViewer 得到。I3DViewer 提供了一些共同的属性和方法来对 scene 和 globe 的视图进行操作，并且比如全屏等都可以在 Scene 和 Globe 中使用，它们都是非模态 3D 显示窗口，也能通过不同的视角来显示 3D 数据。它们在模型中的位置，如图 2-12 所示。

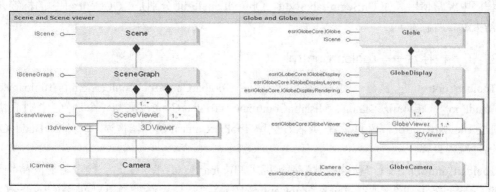

▲图 2-12　SceneViewer 和 Globe Viewer

Camera 和 GlobeCamera

Camera 和 GlobeCamera 分别控制着一个 scene viewer 和 Globe 中的不同的观测点。它们在模型中的位置，如图 2-13 所示。

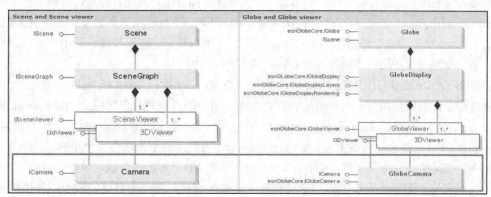

▲图 2-13　Camera 和 Globe Camera

2.3 框架控件介绍

2.3.1　图层树控件——TOCControl

TOCControl 是用来管理图层的可见性和标签的编辑。TOCControl 需要一个"伙伴控件"，或实现了 IActiveView 接口的对象协同工作。"伙伴控件"可以是 MapControl、PageLayoutControl、ReaderControl、SceneControl 或 GlobeControl。"伙伴控件"可以通过 TOCControl 属性页设置，或者在驻留 TOCControl 的容器被显示时用 SetBuddyControl 方法通过编程设置。TOCControl 的每个"伙伴控件"都实现了 ITOCBuddy 接口。TOCControl 用"伙伴控件"来显示其地图、图层和符号体系内容的一个交互树视图，并保持其内容与"伙伴控件"同步。TOCControl 通过 ITOCBuddy 接口来访问其"伙伴控件"。

TOCControl 的主要接口有两个，一个是 ITOCControl，另一个是 ITOCControlEvents。ITOCControl 接口是任何与 TOCControl 有关的任务的出发点，如设置控件的外观、设置"伙伴控件"、管理图层的可见性和标签的编辑等。ITOCControlEvents 是一个事件接口，它定义了 TOCControl

能够处理的全部事件，如 OnMouseDown、OnMouseMove、OnMouseUp 等，这些事件在构建独立应用程序中经常使用，如 OnBeginLabelEdit、OnEndLabelEdit 分别是 TOCControl 中的标签开始编辑、结束编辑时触发。

2.3.2　工具栏控件——ToolbarControl

ToolbarControl 包括 6 个对象及相关接口：ToolbarControl、ToolbarItem、ToolbarMenu、CommandPool、CustomizeDialog、MissingCommand。ToolbarControl 要与一个"伙伴控件"协同工作，通过 ToolbarControl 属性页设置，或者在驻留 ToolbarControl 的容器被显示时用 SetBuddyControl 方法通过编程设置。

ToolbarControl 的每个"伙伴控件"都实现了 IToolbarBuddy 接口，这个接口用于设置"伙伴控件"的 CurrentTool 属性。如通过设置 MapControl 作为其"伙伴控件"，当用户单击该 ToolbarControl 上的"拉框放大"工具时，该放大工具就会成为 MapControl 的 CurrentTool。放大工具的实现过程是：通过 ToolbarControl 获取其"伙伴控件"，然后在 MapControl 上提供显示终端用户拉动鼠标所画的框，并改变 MapControl 的显示范围。

ToolbarControl 一般要与一个"伙伴控件"协同工作，并有一个控件命令选择集，以便快速提供功能强大的 GIS 应用程序。ToolbarControl 不仅提供了部分用户界面，而且还提供了部分应用程序框架。ArcGIS Desktop 应用程序，如 ArcMap、ArcGlobe 和 ArcScene 等具有强大而灵活的框架，包括工具条、命令、菜单、泊靠窗口和状态条等用户界面组件，这些框架使终端用户可以通过改变位置、添加和删除这些用户界面组件来定制应用程序。

ArcGIS Engine 提供了几套使用 ArcGIS 控件的命令，以便执行某种特定动作，开发人员可通过创建执行特定任务的定制命令来扩展这套控件命令。所有的命令对象都实现了 ICommand 接口，ToolbarControl 在适当的时候要使用该接口来调用方法和访问属性。在命令对象被驻留到 ToolbarControl 后，就会立即调用 ICommand::OnCreate 方法，这个方法将一个句柄（Handle）或钩子（hook）传递给该命令操作的应用程序。命令的实现一般都要经过测试，以查看该钩子（hook）对象是否被支持，如果不支持则该钩子自动失效，如果支持，命令则存储该钩子以便以后使用。ToolbarControl 使用钩子（hook）来联系命令对象和"伙伴控件"，并提供了属性、方法和事件用于管理控件的外观，设置伙伴控件，添加、删除命令项，设置当前工具等。ToolbarControl 的主要接口有：IToolbarControl、IToolbarControlDefault、IToolbarControlEvents。

（1）IToolbarControl：该接口是任何与 ToolbarControl 有关的任务的出发点，如设置"伙伴控件"的外观，设置伙伴控件，添加或去除命令、工具、菜单等。

（2）IToolbarControlDefault：该接口是自动暴露的默认的 dispatch 接口，该接口的属性和方法与 ToolbarControl 的最高版本主接口的属性、方法相同。例如目前版本中的 IToolbarControlDefault 等同于 IToolbarControl，但在今后的新版本中，可能会变为 IToolbarControl2。在开发中使用 IToolbarControlDefault 接口，能够保证总是访问到最新版本的 ToolbarControl。

（3）IToolbarControlEvents：该接口是一个事件接口，定义了 ToolbarControl 能够处理的全部事件，如 OnDoubleClick、OnItemClick、OnKeyDown 等。

在 ToolbarControl 上可以驻留以下 3 类命令。

（1）实现了响应单击事件的 ICommand 接口的单击命令。用户单击事件，会导致对 ICommand::OnClick 方法的调用，并执行某种动作。通过改变 ICommand::Checked 属性的值，简单命令项的行为就像开关那样。单击命令是可以驻留在菜单中的惟一命令类型。

（2）实现了 ICommand 接口和 ITool 接口、需要终端用户与"伙伴控件"的显示进行交互的工具。ToolbarControl 维护 CurrentTool 属性。当终端用户单击 ToolbarControl 上的工具时，该工具就成为 CurrentTool。ToolbarControl 会设置"伙伴控件"的 CurrentTool 属性。当某个工具为 CurrentTool 时，该工具会从"伙伴控件"收到鼠标和键盘事件。

（3）实现了 ICommand 接口和 IToolControl 接口的工具控件。这通常是用户界面组件。ToolbarControl 驻留了来自 ItoolControl：hWnd 属性窗口句柄提供的一个小窗口，只能向 ToolbarControl 添加特定工具控件的一个例程。

可以使用 3 种方法向 ToolbarControl 添加命令，第 1 种是指定唯一识别命令的一个 UID，第 2 种是指定一个 progID，第 3 种是给 AddToolbarDef 方法提供某个现有命令对象的一个例程。下面给出样例代码。

```
//Add a toolbardef by passing a UID
UID uID = new UIDsClass();
uID.Value = "esriControls.ControlsMapNavigationToolbar";
axToolbarControl1.AddToolbarDef(uID,-1,false,0,esriCommandStyles.esriCommandStyleIcon-
Only);
//Add a toolbardef by passing a ProgID
string progID = "esriControls.ControlsMapNavigationToolbar";
axToolbarControl1.AddToolbarDef(progID,-1,false,0,esriCommandStyles.esriCommandStyleI-
conOnly);
//Add a toolbardef by passing an IToolbarDef
IToolBarDef toolBarDef = new ControlsMapNavigationToolbarClass();
axToolbarControl1.AddToolbarDef(toolBarDef,-1,false,0,esriCommasndStyles.esriCommands-
tyleIconOnly);
```

ToolbarControl 更新命令的使用。默认情况下，ToolbarControl 每半秒钟自动更新一次，以确保驻留在 ToolbarControl 上的每个工具条命令项的外观与底层命令的 Enabled、Bitmap、Caption 等属性同步。改变 UpdateInterval 属性可以更改更新频率。UpdateInterval 为 0 会停止任何自动发生的更新，可以通过编程调用 Update 方法以刷新每个工具条命令项的状态。

在应用程序中首次调用 Update 方法时，ToolbarControl 会检查每个工具条命令项的底层命令的 ICommand::OnCreate 方法是否已经被调用过，如果还没有调用过该方法，该 ToolbarCommand 将作为钩子被自动传递给 ICommand::OnCreate 方法。

2.4 控件使用实例

2.4.1 GIS 系统常用功能集合

（1）窗体创建介绍。

该实例是一个较全的 GIS 常用功能集合，本实例只是简单介绍了一些常用功能的开发，让大家对 GIS 的这些功能涉及的接口、类，有个大概的了解，后续的章节中对这些类和接口有更详细的介绍。

在 VS2010 中创建一个 Window 应用程序项目，命名为"ControlSample1"（本书所有的代码均为 C#语言编写）。在左边的工具箱中选择"所有 Windows 窗体"选项卡，如图 2-14 所示，在该选项卡中将"menuStrip"控件拖曳到"ControlSample1"窗体上，在"ArcGIS Windwos Forms"选项卡中拖曳两个"MapControl"、一个"ToolbarControl"控件到"ControlSample1"窗体上，在"所有 Windows 窗体"选项卡拖曳一个"Lable"、"TextBox"控件到"ControlSample1"窗体上，如图 2-15 所示。

▲图 2-14 工具箱

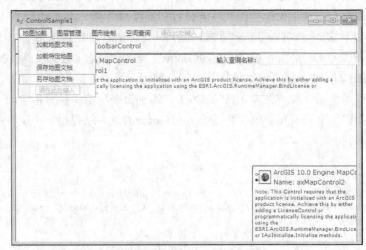
▲图 2-15 程序主界面

在"menuStrip"控件上,添加如下菜单,如图 2-16 所示。

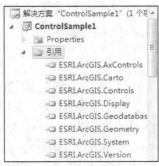
▲图 2-16 菜单项

(2)代码介绍。

参考第一章的 1.4 章节,添加对如下引用(后续章节实例不在重复介绍)如图 2-17 所示。

▲图 2-17 添加引用

在代码"Form1.cs"窗口中添加如下代码,引入名称空间:

```
using ESRI.ArcGIS.Controls;
using ESRI.ArcGIS.Carto;
using ESRI.ArcGIS.esriSystem;
```

```csharp
using ESRI.ArcGIS.Geometry;
using ESRI.ArcGIS.Display;
using ESRI.ArcGIS.Geodatabase;
```

在第一章的 1.4 章节介绍了绑定产品，本实例中需要绑定 license，具体代码如下：

```csharp
public ControlSample1()
{
    ESRI.ArcGIS.RuntimeManager.Bind(ESRI.ArcGIS.ProductCode.EngineOrDesktop);
    ESRI.ArcGIS.RuntimeManager.BindLicense(ESRI.ArcGIS.ProductCode.Engine);
    InitializeComponent();
}
```

图层加载功能将用到 MapControl 控件提供的 LoadMxFile 方法，该方法通过指定的*.Mxd 文档路径直接获取，该方法第一个参数是文件路径，第二个参数是 Mxd 文档中地图的名称或索引，第三个参数是密码，这里使用.NET 的 Type.Missing 字段通过反射进行调用获取参数的默认值。

加载 Mxd 文档时，可以通过 MapControl 控件的 CheckMxFile 方法进行验证 Mxd 文档路径是否有效，代码示例如下：

```csharp
//加载地图文档
private void loadMapDocument()
{
    System.Windows.Forms.OpenFileDialog openFileDialog;
    openFileDialog = new OpenFileDialog();
    openFileDialog.Title = "打开地图文档文件";
    openFileDialog.Filter = "map documents(*.mxd)|*.mxd";
    if (openFileDialog.ShowDialog() == DialogResult.OK)
    {
        string filePath = openFileDialog.FileName;
        if (axMapControl1.CheckMxFile(filePath))
        {
            axMapControl1.MousePointer = esriControlsMousePointer.esriPointerHourglass;
            axMapControl1.LoadMxFile(filePath, 0, Type.Missing);
            axMapControl1.MousePointer = esriControlsMousePointer.esriPointerDefault;
            loadEagleEyeDocument(filePath);
            axMapControl1.Extent = axMapControl1.FullExtent;
        }
        else
        {
            MessageBox.Show(filePath + "不是有效的地图文档");
        }
    }
}
```

加载 Mxd 文档可以通过 MapControl 控件的 LoadMxFile 方法，也可以通过 MapDocument 对象对地图文档中的地图逐个加载，MapDocument 提供了 Open 方法，用于打开一个地图文档文件，该方法第一个参数为地图文档文件的路径，第二个参数为密码，代码示例如下：

```csharp
//加载地图文档
private void loadMapDoc()
{
    mapDocument = new ESRI.ArcGIS.Carto.MapDocumentClass();
    try
    {
        System.Windows.Forms.OpenFileDialog openFileDialog;
        openFileDialog = new OpenFileDialog();
        openFileDialog.Title = "打开地图文档";
```

```csharp
    openFileDialog.Filter = "map documents(*.mxd)|*.mxd";
    if (openFileDialog.ShowDialog() == DialogResult.OK)
    {
        string filePath = openFileDialog.FileName;
        mapDocument.Open(filePath, "");
        for (int i = 0; i < mapDocument.MapCount; i++)
        {
            axMapControl1.Map = mapDocument.get_Map(i);
        }
        axMapControl1.Refresh();
    }
    else
    {
        mapDocument = null;
    }
}
catch (Exception e)
{
    MessageBox.Show(e.ToString());
}
}
```

一个 Mxd 文档中可以包含多个地图，loadMapDocument2 方法，演示如何读取 Mxd 文档中特定的地图，MapControl 控件提供了 ReadMxMaps 方法，用于获取 Mxd 文档中地图数组，找到特定地图，再进行加载。

```csharp
//加载地图文档中的特定地图
private void loadMapDocument2()
{
    System.Windows.Forms.OpenFileDialog openFileDialog;
    openFileDialog = new OpenFileDialog();
    openFileDialog.Title = "打开地图文档";
    openFileDialog.Filter = "map documents(*.mxd)|*.mxd";
    if (openFileDialog.ShowDialog() == DialogResult.OK)
    {
        string filePath = openFileDialog.FileName;
        if (axMapControl1.CheckMxFile(filePath))
        {
            IArray arrayMap = axMapControl1.ReadMxMaps(filePath, Type.Missing);
            int i;
            IMap map;
            for (i = 0; i < arrayMap.Count; i++)
            {
                map = arrayMap.get_Element(i) as IMap;
                if (map.Name == "Layers")
                {
                    axMapControl1.MousePointer = esriControlsMousePointer.esriPointerHourglass;
                    axMapControl1.LoadMxFile(filePath, 0, Type.Missing);
                    axMapControl1.MousePointer = esriControlsMousePointer.esriPointerDefault;
                    loadEagleEyeDocument(filePath);
                    axMapControl1.Extent = axMapControl1.FullExtent;
                    break;
                }
            }
        }
        else
        {
            MessageBox.Show(filePath + "不是有效的地图文档");
        }
    }
}
```

MaPDocument 对象还提供了地图文档对象的保存和另存,方法分别是 Save 和 SaveAs,在进行地图文档保存前,需要先对地图文档是否可读写进行判断,MapDocument 对象提供了 get_IsReadOnly 方法,完整示例如下:

```csharp
//保存地图文档
private void saveDocument()
{
    if (mapDocument == null)
    {
        MessageBox.Show("地图文档对象为空,请先加载地图文档");
    }
    else
    {
        if (mapDocument.get_IsReadOnly(mapDocument.DocumentFilename) == true)
        {
            MessageBox.Show("地图文档只读,无法保存");
        }
        else
        {
            string fileSavePath = @"E:\World\newworld1.mxd";
            try
            {
                mapDocument.Save(mapDocument.UsesRelativePaths, true);
                MessageBox.Show("地图文档保存成功");
            }
            catch (Exception e)
            {
                MessageBox.Show("地图文档保存失败" + e.ToString());
            }
        }
    }
}
```

图层操作是 GIS 中常用的功能,涉及图层的添加、删除、移动等,下面的示例演示了,图层的添加、删除、移动功能。

添加图层可以是*.lyr 格式的图层文件,也可以是*.shp 文件,使用 MapControl 自带的 AddLayerFromFile 方法,提供 lyr 图层文件的路径,即可加载到地图控件中。

```csharp
//添加图层文件
private void addLayerFile()
{
    System.Windows.Forms.OpenFileDialog openFileDialog;
    openFileDialog = new OpenFileDialog();
    openFileDialog.Title = "打开图层文件";
    openFileDialog.Filter = "map documents(*.lyr)|*.lyr";
    if (openFileDialog.ShowDialog() == DialogResult.OK)
    {
        string filePath = openFileDialog.FileName;
        try
        {
            axMapControl1.AddLayerFromFile(filePath);
        }
        catch (Exception e)
        {
            MessageBox.Show("添加图层失败" + e.ToString());
        }
    }
}
```

MapControl 还提供了 AddShapeFile 方法，通过输入 shp 文件的路径目录和文件名来加载到地图中。

```csharp
//添加 SHp 文件
//ArcGIS 10 版本中 addshapefile 方法执行程序中，应在程序启动时添加如下语句
// ESRI.ArcGIS.RuntimeManager.BindLicense(ESRI.ArcGIS.ProductCode.Engine);
//或者在程序界面拖曳一个 LicenseControl 控件
private void addShapeFile()
{
    System.Windows.Forms.OpenFileDialog openFileDialog;
    openFileDialog = new OpenFileDialog();
    openFileDialog.Title = "打开图层文件";
    openFileDialog.Filter = "map documents(*.shp)|*.shp";
    if(openFileDialog.ShowDialog()==DialogResult.OK)
    {
        FileInfo fileInfo = new FileInfo(openFileDialog.FileName);
        String path= fileInfo.Directory.ToString();
        String fileName = fileInfo.Name.Substring(0, fileInfo.Name.IndexOf("."));
        try
        {
            axMapControl1.AddShapeFile(path, fileName);
        }
        catch (Exception e)
        {
            MessageBox.Show("添加图层失败" + e.ToString());
        }
    }
}
```

MapControl 提供的 DeleteLayer 方法用于删除图层，通过输入指定的图层索引即可，代码示例如下：

```csharp
axMapControl1.DeleteLayer(i);
```

MoveLayerTo 方法，则用于移动图层在地图中的叠加顺序，代码示例如下：

```csharp
//将最下层代码移动到最上层
axMapControl1.MoveLayerTo(axMapControl1.LayerCount - 1, 0);
```

该方法第一个参数是待移动图层，第二个参数是待移动图层移动后在地图中的图层索引位置，地图中其他图层则自动后移或前移。

图形绘制是 GIS 系统中的一个很重要、很常用的功能，MapControl 提供了常用线、圆、矩形、多边形等形式的绘制，例如，绘制圆形，则在 axMapControl1_OnMouseDown 事件中设置绘制方法，具体如下所示：

```csharp
     geometry = this.axMapControl1.TrackCircle();
//绘制线、圆、矩形
private void drawMapShape(IGeometry geometry )
{
    IRgbColor rgbColor;
    rgbColor = new RgbColorClass();
    rgbColor.Red = 255;
    rgbColor.Green = 255;
    rgbColor.Blue = 0;
    object symbol = null;
    if (geometry.GeometryType == esriGeometryType.esriGeometryPolyline ||
        geometry.GeometryType == esriGeometryType.esriGeometryLine)
    {
```

```csharp
            ISimpleLineSymbol simpleLineSymbol;
            simpleLineSymbol = new SimpleLineSymbolClass();
            simpleLineSymbol.Color = rgbColor;
            simpleLineSymbol.Width = 5;
            symbol = simpleLineSymbol;
        }
        else
        {
            ISimpleFillSymbol simpleFillSymbol;
            simpleFillSymbol = new SimpleFillSymbolClass();
            simpleFillSymbol.Color = rgbColor;
            symbol = simpleFillSymbol;
        }
        axMapControl1.DrawShape(geometry, ref symbol);
    }
```

MapControl 提供的绘制方法，在地图的跟踪层上绘制所需的类型，通过 DrawShape 绘制出来，根据绘制线型或面型的不同，设置线型或面型符号，在地图上可以直观看到绘制效果。

GIS 系统中经常需要在地图上临时标注，此时可以使用 MapControl 提供的 DrawText 方法，在地图上指定位置显示所需的文字，代码示例如下：

```csharp
//绘制文本对象
private void drawMapText(IGeometry geometry)
{
    IRgbColor color = new RgbColorClass();
    color.Red = 255;
    color.Blue = 0;
    color.Green = 0;
    ITextSymbol txtSymbol = new TextSymbolClass();
    txtSymbol.Color = color;
    object symbol = txtSymbol;
    this.axMapControl1.DrawText(geometry, "测试 DRAW TEXT ", ref symbol);
}
```

几乎所有的 GIS 系统都提供鹰眼图功能，使用鹰眼图可以很直观地看到主视图中的地图范围在整个地图范围内的位置，犹如鸟瞰一样，在 Engine 中非常方便实现，首先在地图上叠加一个 MapControl 控件，要实现这一功能，主要是保持两个控件显示的数据一致，以及在鹰眼控件的显示方框中让两个控件的数据共享。

在 axMapControl 控件的 OnExtentUpdated 事件中，添加对鹰眼图的控制，主要控制鹰眼图的比例尺、鹰眼图的中心点等，以保证鹰眼图和地图窗口中地图保持同步，代码如下所示：

```csharp
//设置鹰眼图像的显示范围中心点等…
private void setLoadEagle()
{
    axMapControl2.MapScale = this.axMapControl1.MapScale * 2.0;
    IPoint point = new PointClass();
    point.X = (this.axMapControl1.Extent.XMax + this.axMapControl1.Extent.XMin) / 2;
    point.Y = (this.axMapControl1.Extent.YMax + this.axMapControl1.Extent.YMin) / 2;
    axMapControl2.ShowScrollbars = false;
    axMapControl2.CenterAt(point);
    axMapControl2.Refresh();
}
```

任何一个系统都需要提供数据查询的功能，GIS 中常用的就是空间查询和属性查询，Engine 中提供了丰富的功能支持查询操作，空间查询主要用到 spatialFilter 接口，属性查询用到 queryFilter 接口，这两个接口用于设置查询的空间对象、查询的条件、返回值的内容等。FeatureCursor 接口是一个游标，提供了只向后顺序移动的查询结果游标，通过该接口的 NextFeature 方法，可以顺序获取查询结果集中的每个 Feature 对象。

MapControl 提供了 selectFeature 方法，用于对查询结果的设置选择状态，具体实现程序如下：

```csharp
//名称查询
private void searchByName()
{
    string searchName = this.textBox1.Text.Trim();
    if (null != searchName && searchName.Length > 1)
    {
        ILayer layer = axMapControl1.Map.get_Layer(1);
        IFeatureLayer featureLayer = layer as IFeatureLayer;
        IFeatureClass featureClass = featureLayer.FeatureClass;
        IQueryFilter queryFilter = new QueryFilterClass();
        IFeatureCursor featureCursor;
        IFeature feature = null;
        queryFilter.WhereClause = "continent like '%" + searchName + "%'";
        featureCursor = featureClass.Search(queryFilter, true);
        feature = featureCursor.NextFeature();
        if (feature != null)
        {
            axMapControl1.Map.SelectFeature(axMapControl1.get_Layer(1), feature);
            axMapControl1.Refresh(esriViewDrawPhase.esriViewGeoSelection, null, null);
        }
    }
}
```

2.4.2 布局控件与地图控件关联

在 GIS 系统中经常使用到地图的制图，在地图上标记指北针、图例、比例尺等，进行地图输出，Engine 提供了 PageLayoutControl 控件，该控件可以添加图元要素等进行地图整饰，下面代码演示了布局控件与地图控件关联，要实现该功能的核心接口是 IObjectCopy，该接口提供了 Copy 方法用于地图的复制，Overwrite 方法用于地图写入 PageLayoutControl 控件的视图中，具体实现程序如下：

```csharp
private void copyToPageLayout()
{
    IObjectCopy objectCopy = new ObjectCopyClass();
    object copyFromMap = axMapControl1.Map;
    object copyMap = objectCopy.Copy(copyFromMap);
    object copyToMap = axPageLayoutControl1.ActiveView.FocusMap;
    objectCopy.Overwrite(copyMap, ref copyToMap);
}
private void axMapControl1_OnMapReplaced(object sender, IMapControlEvents2_OnMapReplacedEvent e)
{
    copyToPageLayout();
}
private void axMapControl1_OnAfterScreenDraw(object sender, IMapControlEvents2_ OnAfterScreenDrawEvent e)
{
    IActiveView activeView = (IActiveView)axPageLayoutControl1.ActiveView.FocusMap;
    IDisplayTransformation displayTransformation = activeView.ScreenDisplay. DisplayTransformation;
    displayTransformation.VisibleBounds = axMapControl1.Extent;
    axPageLayoutControl1.ActiveView.Refresh();
    copyToPageLayout();
}
```

2.4.3 布局控件中属性设置与绘制元素

绘制元素常需要对元素进行符号样式设置，Engine 提供的 SymbologyControl 控件用于加载、设置图层符号，下面代码片段示例演示了从安装路径下读取图层的样式文件并添加到样式类中：

```
Microsoft.Win32.RegistryKey regKey = Microsoft.Win32.Registry.LocalMachine.OpenSubKey
("SOFT-WARE\\ESRI\\CoreRuntime",true);
axSymbologyControl1.LoadStyleFile(regKey.GetValue("InstallDir"+"\\Styles\\ESRI.Server-
Style")
axSymbologyControl1.GetStyleClass(esriSymbologyStyleClass.esriStyleClassBackgrounds).
Update();
axSymbologyControl1.GetStyleClass(esriSymbologyStyleClass.esriStyleClassBorders).Update();
axSymbologyControl1.GetStyleClass(esriSymbologyStyleClass.esriStyleClassShadows).Update();
```

SymbologyControl 提供了样式设置属性 StyleClass，通过该属性可以设置边框、阴影、背景、网格等，下面的代码片段演示这些内容：

```
//设置边框
axSymbologyControl1.StyleClass = esriSymbologyStyleClass.esriStyleClassBorders;
//设置阴影
axSymbologyControl1.StyleClass = esriSymbologyStyleClass.esriStyleClassShadows;
//设置背景
axSymbologyControl1.StyleClass = esriSymbologyStyleClass.esriStyleClassBackgrounds;
//设置网格
            IMap map;
            IActiveView activeView;
            activeView = axPageLayoutControl1.PageLayout as IActiveView;
            map = activeView.FocusMap;

            IMapGrid mapGrid;
            IMeasuredGrid measuredGrid;
            measuredGrid = new MeasuredGridClass();
            mapGrid = measuredGrid as IMapGrid;
            measuredGrid.FixedOrigin = true;
            measuredGrid.Units = map.MapUnits;
            measuredGrid.XIntervalSize = 10;
            measuredGrid.YIntervalSize = 10;
            measuredGrid.XOrigin = -180;
            measuredGrid.YOrigin = -90;

            IProjectedGrid projectedGrid;
            projectedGrid = measuredGrid as IProjectedGrid;
            projectedGrid.SpatialReference = map.SpatialReference;
            mapGrid.Name = "Measured Grid";
            IMapFrame mapFrame;
            IGraphicsContainer graphicsContainer;
            graphicsContainer = activeView as IGraphicsContainer;
            mapFrame = graphicsContainer.FindFrame(map) as IMapFrame;
            IMapGrids mapGrids = mapFrame as IMapGrids;
            mapGrids.AddMapGrid(mapGrid);
            activeView.PartialRefresh(esriViewDrawPhase.esriViewBackground, null, null);
```

下面代码片段演示如何添加图例：

```
//添加图例
private void button6_Click(object sender, EventArgs e)
{
    UID uid;
    IEnvelope envelope;
    IMapSurround mapSurround;
    IGraphicsContainer graphicsContainer;
    IMapFrame mapFrame;
```

```csharp
    IMapSurroundFrame mapSurroundFrame;
    IElement element;
    ITrackCancel trackCancel;

    uid = new UIDClass();
    uid.Value = "esriCarto.legend";
    envelope = new EnvelopeClass();
    envelope.PutCoords(1,1,2,2);
    graphicsContainer = axPageLayoutControl1.PageLayout as IGraphicsContainer ;
    mapFrame = graphicsContainer.FindFrame(axPageLayoutControl1.ActiveView.FocusMap) as IMapFrame ;

    mapSurroundFrame = mapFrame.CreateSurroundFrame(uid, null );
    mapSurroundFrame.MapSurround.Name = "图a?例¤y";
    element = mapSurroundFrame as IElement;
    element.Geometry = envelope;
    element.Activate(axPageLayoutControl1.ActiveView.ScreenDisplay);
    trackCancel = new CancelTrackerClass();
    element.Draw(axPageLayoutControl1.ActiveView.ScreenDisplay, trackCancel);
    graphicsContainer.AddElement(element, 0);
    axPageLayoutControl1.Refresh();
}
```

下面代码片段演示如何添加文字比例尺:

```csharp
//文字比例尺
private void button7_Click(object sender, EventArgs e)
{
    UID uid;
    IEnvelope envelope;
    IMapSurround mapSurround;
    IGraphicsContainer graphicsContainer;
    IMapFrame mapFrame;
    IMapSurroundFrame mapSurroundFrame;
    IElement element;
    ITrackCancel trackCancel;
    uid = new UIDClass();
    uid.Value = "esriCarto.ScaleText";
    envelope = new EnvelopeClass();
    envelope.PutCoords(1, 1, 2, 2);
    graphicsContainer = axPageLayoutControl1.PageLayout as IGraphicsContainer;
    mapFrame = graphicsContainer.FindFrame(axPageLayoutControl1.ActiveView.FocusMap) as IMapFrame;
    mapSurroundFrame = mapFrame.CreateSurroundFrame(uid, null);
    element = mapSurroundFrame as IElement;
    element.Geometry = envelope;
    element.Activate(axPageLayoutControl1.ActiveView.ScreenDisplay);
    trackCancel = new CancelTrackerClass();
    element.Draw(axPageLayoutControl1.ActiveView.ScreenDisplay, trackCancel);
    graphicsContainer.AddElement(element, 0);
    axPageLayoutControl1.Refresh();

}
```

下面代码片段演示如何添加图形比例尺:

```csharp
//图形比例尺
private void button8_Click(object sender, EventArgs e)
{
    UID uid;
    IEnvelope envelope;
    IMapSurround mapSurround;
    IGraphicsContainer graphicsContainer;
    IMapFrame mapFrame;
```

```
        IMapSurroundFrame mapSurroundFrame;
        IElement element;
        ITrackCancel trackCancel;
        uid = new UIDClass();
        uid.Value = "esriCarto.ScaleLine";
        envelope = new EnvelopeClass();
        envelope.PutCoords(1, 1, 10, 2);
        graphicsContainer = axPageLayoutControl1.PageLayout as IGraphicsContainer;
        mapFrame = graphicsContainer.FindFrame(axPageLayoutControl1.ActiveView.FocusMap) as
IMapFrame;
        mapSurroundFrame = mapFrame.CreateSurroundFrame(uid, null);
        element = mapSurroundFrame as IElement;
        element.Geometry = envelope;
        element.Activate(axPageLayoutControl1.ActiveView.ScreenDisplay);
        trackCancel = new CancelTrackerClass();
        element.Draw(axPageLayoutControl1.ActiveView.ScreenDisplay, trackCancel);
        graphicsContainer.AddElement(element, 0);
        axPageLayoutControl1.Refresh();
}
```

2.5 本章小结

本章主要介绍了 ArcEngine 提供的常用几个控件以及控件实现的接口，并提供了一个示例演示控件的使用、地图加载、图层管理、图形绘制、空间查询，读者可以运行本书配套的示例代码运行学习，布局控件与地图控件的关联、布局控件中添加边框、背景、图例、比例尺等，本书给出了简单的代码片段，读者可以根据该代码片段直接添加进系统中，本章例子的功能在以后章节的实例程序中都将用到，初学者应理解这些程序。

第 3 章 几何对象和空间坐标系

3.1 Geometry 对象

Geometry 是一个要素的基本组成部分，它确定了要素在地球上的位置。用户可以通过空间过滤器（SpatialFilter）对这些要素进行空间查询操作，如查询与某个几何对象包含其他的几何形体等，以及返回需要的几何对象。这个工作对于 GIS 用户而言使用非常普遍。GIS 也可以对几何形体对象进行空间运算，如缓冲区分、相交、合并等，有些空间运算还会涉及拓扑关系等方面的内容，这些操作一般都在两个 Geometry 对象之间进行。

Geometry 是 Arc Engine 中使用最为广泛的对象集之一，用户在新建、删除、编辑或进行地理分析的时候，就是在处理一个包含几何形体的矢量对象。在 Geometry 模型中，几何形体对象分为两个层次，一个是构成要素形状的几何图形，另一个是组成这些形状的构件。前者包括 Point、Multipoint、Envelope、Polyline、Polygon 等。Geometry 的主要对象模型如图 3-1 所示。本章着重围绕这些主要对象进行介绍。

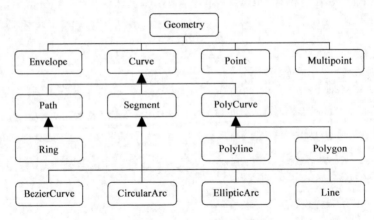

▲图 3-1 Geometry 主要对象模型

Point 对象是一个 0 维的几何图形，具有 x、y 坐标值，以及可选的属性，如高程（Z 值）、测量值（M）和 ID 号等，可用于描述需要精确定位的对象。

Multipoint 点集对象是无序点的群集，用于表示具有相同属性设置的同一组点。如一家公司不同的营业场所就可以使用点集来表示。

Envelope 包络线是一个矩形，用于表示要素的空间范围。它覆盖了几何对象的最小坐标和最大坐标，同时也记录了几何形体对象的 Z 值和 M 值的变化范围。所有的几何形体对象都拥有一个包络线，连其自己本身也有。

Polyline 多义线是一个有序路径（Path）的集合，这些路径既可以是连续的，也可以是离散的。这些对象可用于表示具有线状特征的对象，用户可以用单路径构成的多义线来表示简单线，也可以用具有多个路径的多义线来表示复杂线类型。

Polygon 多边形是环（Ring）的集合，环是一种封闭的路径。Polygon 可以由一个或者多个环组成，甚至环内嵌套环，但是内、外环之间不能重叠，它通常用来描述面状特性的要素。

Geometry 类是所有几何形体对象的父类，是一个抽象类，IGeometry 接口定义的属性和方法为所有的几何对象所拥有。如 Dimension 用于查询几何形体对象的纬度；Envelope 用于返回几何对象的包络线；GeometryType 用于返回对象的几何类型；IsEmpty 用于查看一个对象是否为空；SetEmpty 方法用于将一个几何对象设置为空；Project 方法用于设置一个几何对象的空间参考属性，用户可以产生或引用系统定义的空间参考。

Geometry 类型的集合接口主要有 IGeometryCollection、ISegmentCollection、IpointCollection 等。

IGeometryCollection 接口被多种几何对象继承实现，如 Polygons、Polylines、Multipoints、GeometryBags 等。该接口提供了方法，可以添加、改变和移除一个几何对象的组成元素，即子对象，如 Polygons 对象的组成元素 Ring 可以使用该接口的方法添加、改变和移除。几何对象都是有序的子对象的集合，每个子对象都有索引值，这个索引值确定了它们在组成方向上的排列顺序，可以通过 IGeometryCollection 的 Geometry 属性和索引值获取几何对象的子对象，GeometryCount 则用于返回这些子对象的数目。

IGeometryCollection 提供了一些方法可进行集合操作，如 AddGeometries、AddGeometry、AddGeometryCollection、InsertGeometries、InsertGeometriesCollection、SetGeometries 等。下面的代码演示了如何使用这些方法。

新建一个 VS2010 工程，命名为"GeometrySample"，在 Form1 窗体上添加一个 ToolbarControl 控件，一个 MapControl 控件，一个 menuStrip 控件，然后在 ToolbarControl 控件的属性里设置"伙伴控件"为 MapControl，添加全图、漫游、放大、缩小等工具，如图 3-2 所示。

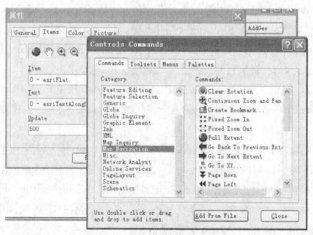

▲图 3-2 设置 ToolbarControl 控件的 4 个工具

在菜单控件中添加 IGeometryCollection 一级菜单，添加 AddGeometry、AddGeometryCollection、InsertGeometriesCollection、SetGeometries 二级菜单窗体整体效果如图 3-3 所示。

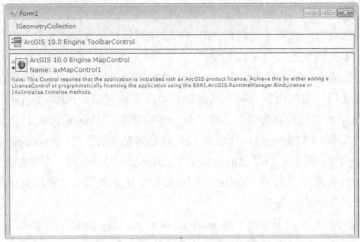

▲图3-3 窗体整体效果

本示例使用的Mxd文件在World目录下的Geometry.mxd文件中。本示例通过自动生成多点，并将多点通过IGeometryCollection创建Feature对象，并将对象添加在图层上进行显示查看代码执行的效果，示例中的 addFeature 方法是示例执行的关键，在该方法中使用了空间编辑对象，涉及IDataset 数据集、IWorkspace 工作空间、IWordspaceEdit 编辑工作空间。这几个接口在后面章节中会详细介绍。

Engine 的编辑机制是，在编辑工作空间中进行数据的编辑操作的同时可以在编辑工作空间中进行数据的版本管理。工作空间是由数据集组成，每个实体图层对象都有一个实体类（FeatureClass），该实体类由图层中所有Feature组成。

在开始数据编辑前，需要先获取图层的实体类（FeatureClass），实体图层的 FeatureClass 属性提供了该实体类的返回，获取实体类后将实体类转为 IDataset 集合，该接口提供了工作空间的获取，通过接口的 Workspace 属性可以获取到实体类的工作空间。

编辑工作空间（IWorkspaceEdit）继承自工作空间（IWorkspace），编辑工作空间提供了两对方法：StartEditing（开始编辑）和 StopEditing（结束编辑）、StartEditOperation（开始编辑操作）和 StopEditOperation（结束编辑操作），其中 StartEditOperation 和 StopEditOperation 必须在 StartEditing 方法之后，StopEditing 方法之前执行形成一个闭合的嵌套结构。

IFeatureClass 接口，提供了创建 FeatureBuffer 方法：CreateFeatureBuffer。通过该方法，在实体类中创建一个空 FeatureBuffer 对象，FeatureBuffer 对象的 Shape 属性用于获取或设置实体对象，如：点、线、面。

IFeatureClass 接口提供了丰富的实体编辑、查询方法，Search 方法用于查询实体类中的实体，Insert 方法用于创建实体指针（IFeatureCursor），该实体指针提供了 InsertFeatureBuffer 方法用于将 FeatureBuffer 插入到实体类中，InsertFeatureBuffer 方法将 FeatureBuffer 插入到实体类中，此时 FeatureBuffer 还处于内存中，并未真正保存在 shp 文件或数据库中，必须调用实体指针的 Flush 方法，该方法将 FeatureBuffer 真正写入 shp 文件或数据库中。

详细代码片段如下：

```
private void addFeature(string layerName,IGeometry geometry)
{
    int i = 0;
    ILayer layer = null;
    for (i = 0; i < axMapControl1.LayerCount; i++)
    {
```

```csharp
        layer = axMapControl1.Map.get_Layer(i);
        if (layer.Name.ToLower() == layerName)
        {
            break;
        }
    }
    IFeatureLayer featureLayer = layer as IFeatureLayer;
    IFeatureClass featureClass = featureLayer.FeatureClass;
    IDataset dataset = (IDataset)featureClass;
    IWorkspace workspace = dataset.Workspace;
    IWorkspaceEdit workspaceEdit = (IWorkspaceEdit)workspace;
    workspaceEdit.StartEditing(true);
    workspaceEdit.StartEditOperation();
    IFeatureBuffer featureBuffer = featureClass.CreateFeatureBuffer();
    IFeatureCursor featureCursor;
    featureCursor = featureClass.Search(null, true);
    IFeature feature;
    feature = featureCursor.NextFeature();
    while (feature!=null)
    {
        feature.Delete();
        feature = featureCursor.NextFeature();
    }
    featureCursor = featureClass.Insert(true);
    featureBuffer.Shape = geometry;
    object featureOID = featureCursor.InsertFeature(featureBuffer);
    featureCursor.Flush();
    workspaceEdit.StopEditOperation();
    workspaceEdit.StopEditing(true);
    System.Runtime.InteropServices.Marshal.ReleaseComObject(featureCursor);
}
```

IGeometryCollection 接口提供了 AddGeometry 方法,通过该方法将单个点对象加入到多点对象集合中,然后将多点对象作为一个整体,插入到多点对象类型图层中,作为一个实体对象。

```csharp
private void toolStripMenuItem1_Click(object sender, EventArgs e)
{
    IGeometryCollection geometryCollection = new MultipointClass();
    IMultipoint multipoint;
    object missing = Type.Missing;
    IPoint point;
    for (int i = 0; i < 10; i++)
    {
        point = new PointClass();
        point.PutCoords(i * 2, i * 2);

        geometryCollection.AddGeometry(point as IGeometry, ref missing, ref missing);
    }
    multipoint = geometryCollection as IMultipoint;
    addFeature("multipoint", multipoint as IGeometry);
    this.axMapControl1.Extent = multipoint.Envelope;
    this.axMapControl1.Refresh();
}
```

IGeometryCollection 接口提供了 AddGeometryCollection 方法,该方法将一个 IGeometryCollection 子类型对象集合作为一个实体对象插入到对象集合中(解释起来有点绕口,看下面的代码就明白了),等待插入对象集自动添加到被插入对象集队列的后面,具体实现如下:

```csharp
private void addGeometryCollectionToolStripMenuItem_Click(object sender, EventArgs e)
{
    IGeometryCollection geometryCollection1 = new MultipointClass();
```

```csharp
IGeometryCollection geometryCollection2 = new MultipointClass();
IMultipoint multipoint;
object missing = Type.Missing;
IPoint point;
for (int i = 0; i < 10; i++)
{
    point = new PointClass();
    point.PutCoords(i * 2, i * 2);
    geometryCollection1.AddGeometry(point as IGeometry, ref missing, ref missing);
}
geometryCollection2.AddGeometryCollection(geometryCollection1);
multipoint = geometryCollection2 as IMultipoint;
addFeature("multipoint", multipoint as IGeometry);
this.axMapControl1.Extent = multipoint.Envelope;
this.axMapControl1.Refresh();
}
```

InsertGeometryCollection 方法则是将一个对象集插入到另一个对象集合的指定索引位置，与 AddGeometryCollection 不同的是，AddGeometryCollection 没有指定索引位置，InsertGeometryCollection 将对象集合插入到指定的索引位置，如果该索引位置不是对象集合的索引最大值，则该索引值后面的索引自动后移，具体实现如下：

```csharp
private void insertGeometriesCollectionToolStripMenuItem_Click(object sender, EventArgs e)
{
    IGeometryCollection geometryCollection1 = new MultipointClass();
    IGeometryCollection geometryCollection2 = new MultipointClass();
    IGeometryCollection geometryCollection3 = new MultipointClass();
    IGeometryCollection geometryCollection4 = new MultipointClass();
    IMultipoint multipoint;
    object missing = Type.Missing;
    IPoint point;
    for (int i = 0; i < 10; i++)
    {
        point = new PointClass();
        point.PutCoords(i * 2, i);
        geometryCollection1.AddGeometry(point as IGeometry, ref missing, ref missing);
    }
    for (int i = 0; i < 10; i++)
    {
        point = new PointClass();
        point.PutCoords(i, i);
        geometryCollection2.AddGeometry(point as IGeometry, ref missing, ref missing);
    }
    for (int i = 0; i < 10; i++)
    {
        point = new PointClass();
        point.PutCoords(i, i * 2);
        geometryCollection3.AddGeometry(point as IGeometry, ref missing, ref missing);
    }
    geometryCollection1.InsertGeometryCollection(1, geometryCollection2);
    geometryCollection1.InsertGeometryCollection(1, geometryCollection3);

    multipoint = geometryCollection1 as IMultipoint;
    addFeature("multipoint", multipoint as IGeometry);
    this.axMapControl1.Extent = multipoint.Envelope;
    this.axMapControl1.Refresh();
}
```

GeometryBrige 是支持 IGeometry 接口的集合对象引用的集合，任何对象都可以通过 IGeometryCollection 接口添加到 GeometryBrige 中，但是在使用拓扑操作的时候，不同的几何类型可能会出现不兼容的情况，在向 GeometryBrige 中添加几何对象的时候，需要对 GeometryBag 对象指定控件参考，添加到其中的几何对象拥有和 GeometryBag 对象一样的空间参考。

GeometryBrige 接口提供了一个 SetGeometries 方法，该方法用于将整个 IGeometryCollection 几何中的对象用一个几何对象数据替换。下面的代码示例通过创建一个几何对象数组，使用 SetGeometries 方法，将 IGeometryCollection 中的对象替换：

```csharp
private void setGeometriesToolStripMenuItem_Click(object sender, EventArgs e)
{
    IGeometryCollection geometryCollection1 = new MultipointClass();
    IGeometryBridge geometryBridge = new GeometryEnvironmentClass();
    IGeometry[] geometryArray = new IGeometry[10];
    IMultipoint multipoint;
    object missing = Type.Missing;
    IPoint point;
    for (int i = 0; i < 10; i++)
    {
        point = new PointClass();
        point.PutCoords(i * 2, i * 2);
        geometryArray[i] = point as IGeometry;
    }
    geometryBridge.SetGeometries(geometryCollection1, ref geometryArray);
    multipoint = geometryCollection1 as IMultipoint;
    addFeature("multipoint", multipoint as IGeometry);
    this.axMapControl1.Extent = multipoint.Envelope;
    this.axMapControl1.Refresh();
}
```

ISegmentCollection 接口是 Segment 集合对象，由 Path、Polygon、Polyline 和 Ring 4 个类实现。使用这个接口可以处理 Segment 集合对象中的每个组成元素，该接口与 IGeometryCollection 类似，可以添加、改变和移除一个 Segment，每个 Segment 都有一个索引值。该接口也提供了一些方法，如 EnumSegments 用于返回一个 SegmentCollection 对象中的 Segment 对象，以枚举值返回；AddSegment 用于添加 Segment 对象；AddSegmentCollection 方法则可用于添加另一个 SegmentCollection 对象中的 Segment，即合并两个 SegmentCollection 对象，还可通过 IGeometryBridge 的 QuerySegment 方法对 Segment 对象进行查询。下面的代码将演示如何使用这些方法。

在 GeometrySample 窗体菜单项上添加 AddSegment、QuerySegment、SetSegment 3 个菜单项。

ISegmentCollection 提供的 AddSegment 方法，用于将 Segment 对象添加到 SegmentCollection 集合中，下面的示例以 Polyline 为例。添加 Segment 对象也可使用 IGeometryBrige 的 AddSegments 方法添加 Segment 数组，下面的代码示例演示如何使用 AddSegment 方法：

```csharp
private void addSegmentColllectionToolStripMenuItem_Click(object sender, EventArgs e)
{
    ISegment[] segmentArray = new ISegment[10];
    IPolyline polyline = new PolylineClass();
    ISegmentCollection segmentCollection = new PolylineClass();
    for (int i = 0; i < 10; i++)
    {
        ILine line = new LineClass();
        IPoint fromPoint = new PointClass();
        fromPoint.PutCoords(i * 10, i * 10);
        IPoint toPoint = new PointClass();
        toPoint.PutCoords(i * 15, i * 15);
        line.PutCoords(fromPoint, toPoint);
        segmentArray[i] = line as ISegment;
        segmentCollection.AddSegment(line as ISegment,Type.Missing,Type.Missing);
```

```csharp
}
//也可通过 IGeometryBridge 对象的 AddSegments 方法进行整个 Segment 数据的添加
//IGeometryBridge geometryBridge = new GeometryEnvironmentClass();
//geometryBridge.AddSegments(segmentCollection, ref segmentArray);
polyline = segmentCollection as IPolyline;
addFeature("polyline", polyline as IGeometry);
this.axMapControl1.Extent = polyline.Envelope;
this.axMapControl1.Refresh();
}
```

IGeometryCollection 接口提供 QueryGeometrys 方法，用于查询对象集合中的元素，但是，ISegmentCollection 接口没有提供该方法，不过可以通过 IGeometryBridge 接口提供的 QuerySegments 方法进行查询，查询结果返回一个 Segment 数组，下面的代码演示如何进行 QuerySegments 查询：

```csharp
private void querySegmentsToolStripMenuItem_Click(object sender, EventArgs e)
{
    ISegment[] segmentArray = new ISegment[10];
    for (int i = 0; i < 10; i++)
    {
        ILine line = new LineClass();
        IPoint fromPoint = new PointClass();
        fromPoint.PutCoords(i * 10, i * 10);
        IPoint toPoint = new PointClass();
        toPoint.PutCoords(i * 15, i * 15);
        line.PutCoords(fromPoint, toPoint);
        segmentArray[i] = line as ISegment;
    }
    ISegmentCollection segmentCollection = new PolylineClass();
    IGeometryBridge geometryBridge = new GeometryEnvironmentClass();
    geometryBridge.AddSegments(segmentCollection, ref segmentArray);

    int index = 0;
    ISegment[] outputSegmentArray = new ISegment[segmentCollection.SegmentCount - index];
    for (int i = 0; i < outputSegmentArray.Length; i++)
    {
        outputSegmentArray[i] = new LineClass();
    }
    geometryBridge.QuerySegments(segmentCollection, index, ref outputSegmentArray);
    String report = "";
    for (int i = 0; i < outputSegmentArray.Length; i++)
    {
        ISegment currentSegment = outputSegmentArray[i];
        ILine currentLine = currentSegment as ILine;
        report = report + "index = " + i + " , FromPoint X = " + currentLine.FromPoint.X +
            " , FromPoint Y = " + currentLine.FromPoint.X;
        report = report + " , ToPoint X = " + currentLine.ToPoint.X + " , ToPoint Y = " +
            currentLine.ToPoint.X + "\n";
    }
    System.Windows.Forms.MessageBox.Show(report);
}
```

ISegmentCollection 接口提供的 SetSegmentCollection 方法用于将一个 ISegmentCollection 设置成一个新的 ISegmentCollection 对象，和 IGeometryCollection 接口不一样，没有提供 SetSegments 之类的方法，可以通过 IGeometryBridge 接口的 SetSegments 方法，将一个 Segment 数组替换成 SegmentCollection 对象，也可通过 ISegmentCollection 接口提供的 AddSegment 方法逐个将 Segment 添加到 SegmentCollection 中。下面的代码演示使用 IGeometryBridge 的 SetSegments 方法：

```csharp
private void setSegmentsToolStripMenuItem_Click(object sender, EventArgs e)
{
    IPolyline polyline = new PolylineClass();
    ISegmentCollection segmentCollection = new PolylineClass();
    IGeometryBridge geometryBridge = new GeometryEnvironmentClass();
    ISegment[] insertSegmentArray = new ISegment[5];
    for (int i = 0; i < 5; i++)
    {
        ILine insertLine = new LineClass();
        IPoint insertFromPoint = new PointClass();
        insertFromPoint.PutCoords(i, 1);
        IPoint insertToPoint = new PointClass();
        insertToPoint.PutCoords(i * 10, 1);
        insertLine.PutCoords(insertFromPoint, insertToPoint);
        insertSegmentArray[i] = insertLine as ISegment;
    }
    geometryBridge.SetSegments(segmentCollection, ref insertSegmentArray);
    polyline = segmentCollection as IPolyline;
    addFeature("polyline", polyline as IGeometry);
    this.axMapControl1.Extent = polyline.Envelope;
    this.axMapControl1.Refresh();
}
```

IPointCollection 接口是点集合对象。任何一个几何对象都是由点构成，因此任何一个几何对象都有一个点集。该接口可被多种几何对象类实现，如 Multipoint、Path、Ring、Polyline 和 Polygon 等。这些几何类型都是 PointCollection 对象，可以通过该接口定义的方法获取、添加、插入、查询和移除集合中的某个顶点。和 IGeometryCollection 一样，该接口也提供了类似的方法进行集合操作。下面的代码将演示如何使用 AddPointCollection、QueryPoints、UpdatePoint 3 个方法。

在 GeometrySample 工程的菜单上添加 AddPointCollection、QueryPoints、UpdatePoint 3 个菜单项。

IPointCollection 接口与 IGeometryCollection 一样，提供了 AddPointCollection 方法，将待添加 IPointCollection 对象作为一个整体添加到目标 IPointCollection 中。下面的示例代码用 IGeometryBridge 接口的 SetPoints 方法创建一个 IPointCollection 对象作为待添加 IPointCollection 对象，添加到目标 IPointCollection 中：

```csharp
private void addPointCollectionToolStripMenuItem_Click(object sender, EventArgs e)
{
    IPointCollection4 pointCollection = new MultipointClass();
    IPointCollection pointCollection2 = new MultipointClass();
    IGeometryBridge geometryBridge = new GeometryEnvironmentClass();
    IPoint[] points = new PointClass[10];
    IMultipoint multipoint;
    object missing = Type.Missing;
    IPoint point;
    for (int i = 0; i < 10; i++)
    {
        point = new PointClass();
        point.PutCoords(i * 5, i);
        points[i] = point;
    }
    geometryBridge.SetPoints(pointCollection, ref points);
    pointCollection2.AddPointCollection(pointCollection);
    multipoint = pointCollection2 as IMultipoint;
    addFeature("multipoint", multipoint as IGeometry);
    this.axMapControl1.Extent = multipoint.Envelope;
    this.axMapControl1.Refresh();
}
```

IPointCollection 接口的 QueryPoint 方法根据索引位置为查询点，下面的代码演示如何使用该方法，IGeometryBridge 接口提供了类似的方法 QueryPoints，该方法返回的是一个点的数组：

```csharp
private void queryPointsToolStripMenuItem_Click(object sender, EventArgs e)
{
    IPoint point1 = new PointClass();
    point1.PutCoords(10, 10);
    IPoint point2 = new PointClass();
    point2.PutCoords(20, 20);
    IPoint[] inputPointArray = new IPoint[2];
    inputPointArray[0] = point1;
    inputPointArray[1] = point2;
    IPointCollection4 pointCollection = new MultipointClass();
    IGeometryBridge geometryBridge = new GeometryEnvironmentClass();
    geometryBridge.AddPoints(pointCollection, ref inputPointArray);

    int index = 0;
    IPoint[] outputPointArray = new IPoint[2];
    for (int i = 0; i < outputPointArray.Length; i++)
    {
        outputPointArray[i] = new PointClass();
    }
    pointCollection.QueryPoint(0, outputPointArray[0]);
    //geometryBridge.QueryPoints(pointCollection, index, ref outputPointArray);
    for (int i = 0; i < outputPointArray.Length; i++)
    {
        IPoint currentPoint = outputPointArray[i];
        if (currentPoint.IsEmpty==true)
        {
            System.Windows.Forms.MessageBox.Show("Current point = null");
        }
        else
        {
            System.Windows.Forms.MessageBox.Show("X = " + currentPoint.X + ", Y = " +
                currentPoint.Y);
        }
    }
}
```

IPointCollection 接口的 UpdatePoint 方法根据索引位置替换该索引位置的点，下面的代码演示如何使用该方法。

```csharp
private void updatePointToolStripMenuItem_Click(object sender, EventArgs e)
{
    IMultipoint multipoint;
    object missing = Type.Missing;
    IPoint point1 = new PointClass();
    point1.PutCoords(10, 10);
    IPoint point2 = new PointClass();
    point2.PutCoords(20, 20);
    IPointCollection pointCollection = new MultipointClass();
    pointCollection.AddPoint(point1, ref missing, ref missing);
    pointCollection.AddPoint(point2, ref missing, ref missing);
    point1 = new PointClass();
    point1.PutCoords(40, 10);
    pointCollection.UpdatePoint(1, point1);
    multipoint = pointCollection as IMultipoint;
    addFeature("multipoint", multipoint as IGeometry);
    System.Windows.Forms.MessageBox.Show("X = " + pointCollection.get_Point(1).X +
        ", Y = " + pointCollection.get_Point(1).Y);
```

```
this.axMapControl1.Extent = multipoint.Envelope;
this.axMapControl1.Refresh();
}
```

3.2 Envelope 对象

Envelope 也称包络线，是一个矩形区域，是每个几何形体的最小外接矩形。每个 Geometry 都拥有一个 Envelope，包括 Envelope 自身。IEnvelope 是包络线对象的主要接口，定义了 XMax、XMin、YMax、YMin、Height 和 Width 等属性，用于获取或设置一个存在的包络线对象的空间坐标。IEnvelope 接口提供了一些方法，如 Expand、offset、CenterAt、PutCoords 等。Expand 方法用于按比例缩放包络线的范围，产生一个新的包络对象；offset 是一个偏移方法，通过一个给定的（X，Y）移动包络线；CenterAt 方法则通过改变包络线的中心点来移动包络线；PutCoords 方法是通过指定的坐标点来构造包络线。

IEnvelope 接口还提供了两个拓扑运算方法 Intersect 和 Union。Intersect 用来计算两个包络线相交，返回相交部分作为结果；Union 则是合并两个包络线，并以两个包络线的最小外接矩形作为合并结果，如图 3-4 所示。

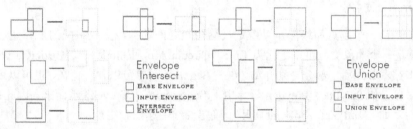

▲图 3-4 IEnvelope 接口

下面的代码演示了如何使用 Envelope。

```
IEnvelope envelope1 = new EnvelopeClass();
IEnvelope envelope2 = new EnvelopeClass();
envelope1.PutCoords(100, 100, 200, 200);
envelope2.PutCoords(150, 150, 250, 250);
envelope1.Intersect(envelope2);
```

envelope1 为两个 Envelope 相交的交集。

```
IEnvelope envelope1 = new EnvelopeClass();
IEnvelope envelope2 = new EnvelopeClass();
envelope1.PutCoords(100, 100, 200, 200);
envelope2.PutCoords(150, 150, 250, 250);
envelope1.Union(envelope2);
```

两个 Envelope 合并为一个 Envelope。

3.3 Curve 对象

Curve 是几何体中重要的类型，除了点、点集、包络线外，可以将其他的几何形体都看做是曲线（Curve），如 Line、Polyline、Path、CircularArc、Polygon 等，这些对象都实现了 ICurve 接口。该接口提供了操作这些对象的属性和方法，但不能用于产生一个新的曲线对象。例如，Length 属性用于返回一个曲线对象的长度；FromPoint、ToPoint 属性用于获取或设置曲线的起点和终点；ReverseOrientation 方法用于改变一个曲线的节点次序，即将起点和终点对换，通过此方法可以改变曲线的方向；IsClosed 属性则用于说明曲线的起点和终点是否在同一个位置，即可以检查曲线是否闭合，如图 3-5 所示。

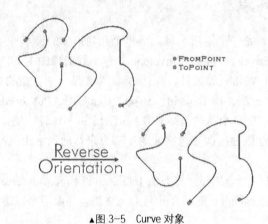

▲图 3-5　Curve 对象

3.3.1　Segment 对象

Segment 是由一个起点和终点，以及定义这两点之间的曲线函数组成的一维几何形体对象。Segment 对象是一个抽象类，可以泛化为 Line、CircularArc、EllipticArc、BezierCurve 4 种类型，这 4 种类型将在本节的后一部分分别介绍。这些 Segment 对象可以独立存在，也可以用于构造其他的几何形体对象，如 Polyline、Polygon 等。由于 Segment 是从 Curve 继承而来，因此也拥有 Curve 的方法和属性，如 FromPoint、ToPoint 可用于确定曲线的起点和终点。

ISegment 是 Segment 对象的主要接口，该接口提供了两个方法：SplitAtDistance、SplitDivideLength，将一个 Segment 对象分割成了多个 Segment 对象。在分割 Segment 前会自动复制一份副本，然后对副本进行分割，分割操作对源对象不产生影响。SplitAtDistance 方法是通过一个给定的长度或比率，在 Segment 对象的起点到终点方向上，在给定长度距离的位置上产生一个分割点。SplitDivideLength 是将一个 Segment 对象分割成多个新的 Segment，如图 3-6 所示。

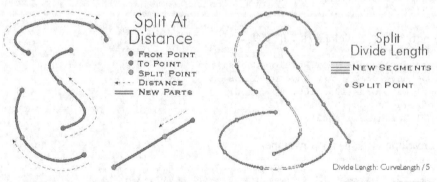

▲图 3-6　Segment 对象

下面的代码将演示如何使用该对象的 SplitAtDistance 方法，该方法通过指定距离，将一个 Segment 打断成两个 Segment 对象，下面的示例代码中也给出了 IGeometryBridge 接口的 SplitDivideLength 方法的示例，该方法将一个 Segment 对象按距离打断成多个 Segment 对象：

```
IPoint fromPoint = new PointClass();
fromPoint.PutCoords(0, 0);
IPoint toPoint = new PointClass();
toPoint.PutCoords(100, 0);
ILine line = new LineClass();
line.FromPoint = fromPoint;
line.ToPoint = toPoint;
double offset = 0;
double length = 10;
bool asRatio = false;
int numberOfSplittedSegments;
ISegment[] splittedSegments = new ISegment[10];
for (int i = 0; i < 10; i++)
{
    splittedSegments[i] = new PolylineClass() as ISegment;
}
ISegment segment = line as ISegment;
segment.SplitAtDistance(length, asRatio, out splittedSegments[0],
    out splittedSegments[1]);
//IGeometryBridge2 geometryBridge = new GeometryEnvironmentClass();
//geometryBridge.SplitDivideLength(segment, offset, length, asRatio,
//    out numberOfSplittedSegments, ref splittedSegments);
```

3.3.1.1　CircularArc 对象

CircularArc 对象是一个圆弧，是圆的一部分，如图 3-7 所示。

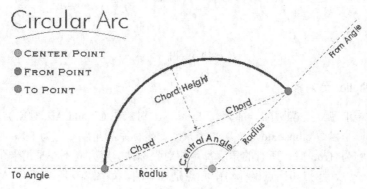

▲图 3-7　CircularArc 对象

CircularArc 是一个组件类。ICircularArc 是 CircularArc 的主要接口，提供了 FromAngle、ToAngle、CentralAngle、CenterPoint、ChordHeight 等属性，通过这些属性可以改变这个圆弧的形状。同时也提供了类型检查属性，如 IsLine、IsPoint、IsMinor、IsCounterClockwise 等。如果 IsLine 为真，则圆弧的半径无限大，即为直线；如果 IsPoint 为真，则圆弧的半径为 0，即为点；如果 IsMinor 为真，则 CentralAngle 小于半圆，即为劣弧；如果 IsCounterClockwise 为真，则 CentralAngle 为正值。通过接口提供的 Complement 方法，可以封闭该圆弧对象，产生一个圆对象。

CircularArc 对象可以使用 IConstructCircularArc 接口提供的方法进行构造，该接口提供了 34 种构造方法，如 ConstructCircle 构造器，该构造器通过指定一个 CenterPoint（圆心）和一个 Radius（半径）来构造；ConstructArcDistance 构造器，该构造器通过指定一个圆心、起始点、圆弧方向和圆弧来构造；ConstructChordDistance 构造器，该构造器通过指定一个起点、圆弧的弦长度、圆弧的方向和中心点来构造；ConstructFilletPoint 构造器，该构造器基于两个 Segment 对象以及内切弧，在两个 Segment 上的点可以产生两条线段和圆弧的内切线。在"帮助"中提供了这些方法的详细说明，读者可以查看"帮助"中的具体条目。

3.3.1.2 Line 对象

Line 是最简单的 Segment 对象，由起点和终点构成一条直线段，用于构造 Polyline、Polygon、Ring 和 Path 等对象。ILine 是 Line 对象的主要接口，定义了一系列用于构造和设置线段对象的属性和方法。如 QueryCoords 属性用于返回 Line 的起点和终点，PutCoords 属性用于设置 Line 的起点和终点坐标，Angle 属性用于返回 Line 与 x 轴的夹角。

除了 ILine 外，Line 还继承自 IConstructLine，该接口提供了两个方法：ConstructAngleBisector 和 ConstructExtended。ConstructAngleBisector 方法通过指定的 3 个点对象和新构造线的长度来构造，原理是：通过 3 个点构造一个夹角，在夹角的角平分线上根据线的长度构造一条新的线，如图 3-8 所示。

ConstructExtended 通过扩展一个已经存在的线段对象来产生一个新的线段。扩展类型将在后面进行详细的介绍。

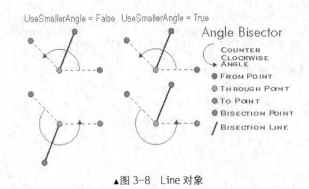

▲图 3-8 Line 对象

3.3.1.3 EllipticArc 对象

EllipticArc（椭圆弧）是椭圆的一部分。EllipticArc 对象与 CircularArc 对象类似。在 EllipticArc 方法中需要用到一个参数 ellipseStd，该参数用来改变坐标系和角度，它为 false 时，系统是标准的笛卡尔坐标系；它为 True 时，所有的角度都是相对坐标，0 度方向不在是笛卡尔直角坐标系，而是和椭圆的长轴一致，起点和终点坐标与中心点是相对的，如图 3-9 所示。

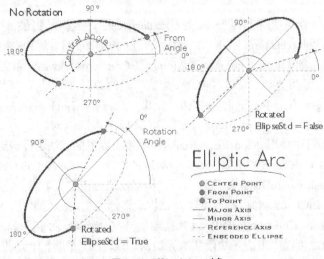

▲图 3-9 EllipticArc 对象

IEllipticArc 接口是 EllipticArc 的主要接口,提供了一些方法用于操作椭圆弧。如 GetAxes 方法用于得到椭圆的长半轴和短半轴以及比例,PutAxes 方法用于改变一个已经存在的椭圆弧对象的半轴长度。另外接口还提供了几个属性用于检查椭圆弧的情况,如 IsCircular 用于检查长半轴和短半轴的长度是否相等,即检查是否是圆;IsLine 如果为 True,则短半轴为 0,为一直线;IsPoint 如果为 True,则长半轴和短半轴都是 0,为一个点对象;ISCounterClockwise 如果为 True,则 CentralAngle 为正值;IsMinor 如果为 True,则弧的长度小于椭圆的一半,为劣弧。IEllipticArc 接口还提供了 Complement、PutCoord、QueryCoords、QueryCoordsByAngle 等方法,这些方法的使用同 ICircularArc 类似。

EllipticArc 可以用 IConstructEllipticArc 构造器来构造,该构造器提供了 ConstructEnvelope、ConstructquarterEllipse、ConstructTwoPointsEnvelope、ConstructUpToFivePoints 4 种方法。ConstructEnvelope 方法,通过一个给定的包络线对象来产生一个内置的椭圆对象,这个椭圆对象是逆时针方向,如图 3-10 所示。

ConstructQuarterEllipse 方法,通过起点、终点和方向来构造一个椭圆弧,如图 3-11(1)所示。

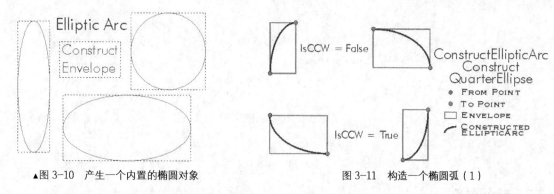

▲图 3-10 产生一个内置的椭圆对象　　　　图 3-11 构造一个椭圆弧(1)

ConstructTwoPointsEnvelope 方法,通过起点、终点、包络线和方向来构造一个椭圆弧,如图 3-12(2)所示。

ConstructUpToFivePoints 方法,通过起点、终点、一个弧上的任意点以及两个椭圆对象上的点,来构造一个椭圆弧,如图 3-13(3)所示。

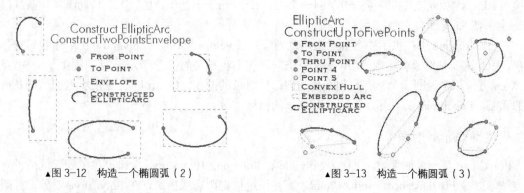

▲图 3-12 构造一个椭圆弧(2)　　　　▲图 3-13 构造一个椭圆弧(3)

3.3.1.4 BezierCurve 对象

BezierCurve 对象定义一条两个节点之间的弧线。IBezierCurve 接口定义了设置和查询曲线属性的方法,通过 PutCoord 方法可以为曲线增加一个顶点。除了 PutCoord 方法外,Arc Engine 还提供了 IConstructBezierCurve 接口定义更高级方法,来实现 BezierCurve 曲线的构建。

3.3.2　Path 对象

Path 是连续多个 Segment 对象的集合,每个 Segment 通过首尾相连构成一条路径。IPath 是 Path 对象的主要接口,它定义了设置一个路径对象的多个方法,如 Generalizes 方法用于抽象化一个平滑的路径对象,即将一个平滑的曲线变成几条相连的线段;Smooth 方法则用于将一个非平滑的路径对象平滑;SmoothLocal 方法则只用于将某个 Segment 连接点处平滑。

Path 对象可以由两种途径来构造,其一,使用 ISegmentCollection 接口,通过接口的 AddSegment 方法为 Path 对象添加子元素;其二,使用 IConstructPath 接口,该接口提供了 ConstructRigidStretch 方法,用于旋转或缩放一个已经存在的路径对象的形状,或改变路径上顶点的位置,如图 3-14 所示。

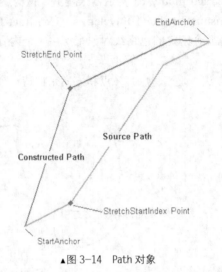

▲图 3-14　Path 对象

3.3.3　Ring 对象

Ring 对象是一种封闭的路径对象,它的起点和终点是同一个点。Ring 对象具有"内部"和"外部"属性,Polygon 是由 Ring 产生。由于 Ring 是封闭的 Path,因此其组成是一系列首尾相连的同方向的 Segment 对象,并且不能自相交。IRing 接口定义了 Ring 对象的处理方法,如 Close 方法用于检查 Ring 对象的起点和终点是否是同一个点,如果 Ring 不是闭合的,则会自动添加一条线段到 Ring 上,使 Ring 闭合。

在 Arc Engine 中提供了 3 种面状类型的对象:Envelope、Ring、Polygon。由于面状对象都有面积,因此这 3 种对象都实现了 IArea 接口来获取与面积相关的信息。IArea 接口提供了一些方法,如 Area 方法用于返回一个具有封闭特性几何形体对象的面积,Centroid 方法用于返回这些几何形体的重心,LabelPoint 方法用于返回几何形体的标注点等。

3.3.4　PolyCurve 对象

PolyCurve 是一个抽象类,泛化为 Polyline 和 Polygon。IPolyCurve 接口提供了处理这两种对象的方法,如 SplitAtDistance、SplitAtPoint,都是通过指定距离或指定点在 PolyCurve 对象上打断;Generalize 方法则用于对 PolyCurve 对象进行概化整形。关于 Polyline 和 Polygon 类,将在后面两节介绍。

> 提示　点对象的代码片段,代码直接复制到一个按钮事件中即可执行,AddFeature 函数,可自行练习编写,该函数的作用将实体添加到地图上。

3.4 点对象

点（Point）代表了一个 0 维的具有 X、Y 坐标的几何对象。点是没有任何形状的，可用于描述点类型的要素，而且 Geometry 中的任何类型都是用点来产生的。

构成几何形状的顶点存在着 3 种可以选择的属性，即 Z、M 和 ID。Z 值在大多数情况下都可用于表示一个点的 Z 坐标，还可以将 Z 值作为一个点的辅助值来使用；M 即度量值，可以一个路径对象的线性度量，用于交通工程中一条公路的不同点的位置；ID 值即为一个点的唯一标识值。

点集（Multipoint）是具有相同属性的点的集合，用于构成高级几何对象、几何对象动态模拟等。

IPoint 接口定义了 Point 对象的属性和方法，点（Point）可以使用 PutCoords 方法创建，也可以使用 IConstructPoint 接口来创建。IConstructPoint 接口提供了 10 种创建点的方法，下面分别介绍。

（1）ConstructAlong 沿线创建法：基于一个曲线上的起始点对象，通过给定距离、比例和扩展类型，沿着曲线创建一个新点。如果距离比曲线的长度长，那么点将沿着它的切线生成。如图 3-15 所示。

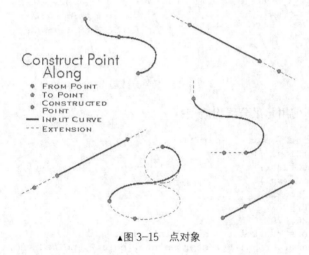

▲图 3-15　点对象

下面代码片段演示使用该方法创建两个点：

```
IFeature polylineFeature=null;
ICurve curve = polylineFeature.Shape as ICurve;
IConstructPoint constructPoint = new PointClass();
constructPoint.ConstructAlong(curve, esriSegmentExtension.esriNoExtension, 10, true);
IPoint point1 = constructPoint as IPoint;
constructPoint = new PointClass();
constructPoint.ConstructAlong(curve, esriSegmentExtension.esriNoExtension, 20, false);
IPoint point2 = constructPoint as IPoint;
```

其扩展参数如下。

EsriNoExtension：没有扩展。在这种方式下构建点将始终在输入曲线上。如果输入距离小于 0，则构建点在曲线的起点；如果输入距离大于曲线的长度，则构建点在曲线的终点。

EsriExtendTangentAtFrom：起点切线。在这种方式下，如果输入距离小于 0，则构建点在曲线起点的切线且与曲线反方向的距离位置上。

EsriExtendTangentAtTo：终点切线。在这种方式下，如果输入距离大于曲线长度，则构建点放在曲线终点的切线且与曲线同方向的距离位置上。

EsriExtendEmbeddedAtFrom：与起点切线类似，但使用嵌入几何不是切线。

EsriExtendEmbeddedAtTo：与终点切线类似，但使用嵌入几何不是切线。

EsriExtendEmbedded：在这种方式下，根据输入距离自动使用 esriExtendEmbeddedAtFrom 或 esriExtendEmbeddedAtTo 创建构建点。

EsriExtendTangents：在这种方式下，根据输入距离自动使用 esriExtendTangentAtFrom 或 esriExtendTangentAtTo 创建构建点。

（2）ConstructAngleBisector 角平分线构建法：这种方法使用 3 个点，即起点（FromPoint）、经由点（Through Point）和终止点（ToPoint）。该方法通过平分 3 点形成的夹角，并设置一个距离在平分线上寻找一个点，如图 3-16 所示。

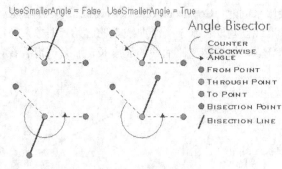

▲图 3-16　角平分线构建

下面代码片段演示使用角平分线构造点：

```
IPoint point1, point2, point3, point4;
point1 = new PointClass();
point2 = new PointClass();
point3 = new PointClass();
point1.PutCoords(0, 0);
point2.PutCoords(0, 20);
point3.PutCoords(20, 20);
IConstructPoint constructPoint = new PointClass();
constructPoint.ConstructAngleBisector(point1, point2, point3, 50, true);
point4 = constructPoint as IPoint;
addFeature("point", point1);
addFeature("point", point2);
addFeature("point", point3);
addFeature("point", point4);
axMapControl1.Refresh();
```

（3）ConstructAngleDistance 构造角度距离法：通过一个给定点和一个相对点的绝对角度和距离构建一个点，如图 3-17 所示，下面代码演示该方法使用：

```
IPoint point1, point2;
point1 = new PointClass();
point1.PutCoords(0, 0);
double distance = 20;
double angle = 60;
IConstructPoint constructPoint = new PointClass();
double angleRad = angle * 2 * Math.PI / 360;
constructPoint.ConstructAngleDistance(point1, angleRad, distance);
point2 = constructPoint as IPoint;
addFeature("point", point1);
addFeature("point", point2);
axMapControl1.Refresh();
```

（4）ConstructAngleIntersection 构造角度交点法：通过给定的两个点和两个角度产生两条射线，然后在两条射线的交点处产生一个交点，如图 3-18 所示。

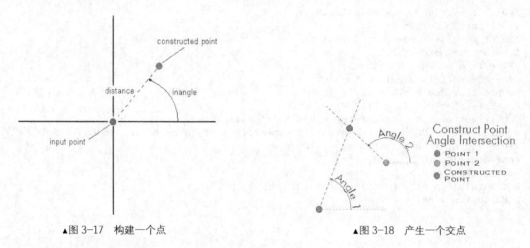

▲图 3-17　构建一个点　　　　　　　　　　▲图 3-18　产生一个交点

下面代码演示构造角度交点法创建点：

```
IPoint point1, point2;
point1 = new PointClass();
point2 = new PointClass();
point1.PutCoords(0, 0);
point2.PutCoords(10, 0);
double angleRad1 = 45 * 2 * Math.PI / 360;
double angleRad2 = 60 * 2 * Math.PI / 360;
IConstructPoint constructPoint = new PointClass();
constructPoint.ConstructAngleIntersection(point1, angleRad1, point2, angleRad2);

addFeature("point", point1);
addFeature("point", point2);
addFeature("point", constructPoint as IPoint);
axMapControl1.Refresh();
```

（5）ConstructDeflection 构造偏转角度法：通过给定一条基准线、一个偏转角度和一个距离，沿着线的偏转角度的射线方向以给定距离位置构建一个点，如图 3-19 所示。下面代码片段演示构造偏转角度法生成点：

```
IPoint fromPoint = new PointClass();
fromPoint.PutCoords(0, 0);
IPoint toPoint = new PointClass();
toPoint.PutCoords(1, 1);
ILine line = new LineClass();
line.PutCoords(fromPoint, toPoint);
double distance = 1.414;
double angle = Math.PI / 4;
IConstructPoint constructPoint = new PointClass();
constructPoint.ConstructDeflection(line, distance, angle);
IPoint point = constructPoint as IPoint;
addFeature("point", fromPoint);
addFeature("point", toPoint);
addFeature("point", point);
axMapControl1.Refresh();
```

（6）ConstructDeflectionIntersection 构造偏转角交点法：通过指定一条线段作为基线，从该线段的起点和终点的偏转角度引出两个射线，然后在两条射线的交点处构建一个点。该方法的

bRightSide 指定偏转角在线段的左边或右边，如图 3-20 所示，下面代码片段演示构造偏转角交点法创建点。

```
IPoint fromPoint = new PointClass();
fromPoint.PutCoords(0, 0);
IPoint toPoint = new PointClass();
toPoint.PutCoords(1, 1);
ILine line = new LineClass();
line.PutCoords(fromPoint, toPoint);
double startAngle = Math.PI / 4;
double endAngle = Math.PI / 4;
IConstructPoint constructPoint = new PointClass();
constructPoint.ConstructDeflectionIntersection(line, startAngle, endAngle,false);
IPoint point = constructPoint as IPoint;
addFeature("point", fromPoint);
addFeature("point", toPoint);
addFeature("point", point);
axMapControl1.Refresh();
```

▲图 3-19　构建一个点　　　　　　▲图 3-20　构造偏转角交点

（7）ConstructOffset 构造偏移点法：该方法通过指定一条曲线，沿着曲线的距离或比例进行水平偏移。可以设置参数 Offset 的值，如果值为正，则该点向右水平偏移；如果值为负，则向左偏移。如图 3-21 所示。

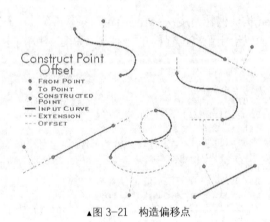

▲图 3-21　构造偏移点

下面代码片段演示构造偏移点创建点：

```
IPoint[] points = new IPoint[4];
for (int i = 0; i < 4; i++)
{
    points[i] = new PointClass();
}
```

```
points[0].PutCoords(0, 0);
points[1].PutCoords(10, 0);
points[2].PutCoords(20, 0);
points[3].PutCoords(30, 0);
IPointCollection polyline = new Polyline();
IGeometryBridge geometryBridge = new GeometryEnvironmentClass();
geometryBridge.AddPoints(polyline as IPointCollection4, ref points);
IConstructPoint constructPointRight = new PointClass();
IConstructPoint constructPointLeft = new PointClass();
constructPointRight.ConstructOffset(polyline as ICurve,
    esriSegmentExtension.esriNoExtension, 15, false, 5);
IPoint outPutPoint1 = constructPointRight as IPoint;
constructPointLeft.ConstructOffset(polyline as ICurve,
    esriSegmentExtension.esriNoExtension, 1, false, -5);
IPoint outPutPoint2 = constructPointLeft as IPoint;
addFeature("point", points[0]);
addFeature("point", points[1]);
addFeature("point", points[2]);
addFeature("point", points[3]);
addFeature("point", outPutPoint1);
addFeature("point", outPutPoint2);
axMapControl1.Refresh();
```

其扩展参数如下。

EsriNoExtension：没有扩展。在这种方式下构建点将始终在输入曲线上。如果输入距离小于0，则构建点在曲线的起点；如果输入距离大于曲线的长度，则构建点在曲线的终点。

EsriExtendTangentAtFrom：起点切线。在这种方式下，如果输入距离小于0，则构建点在曲线起点的切线且与曲线反方向的距离位置上。

EsriExtendTangentAtTo：终点切线。在这种方式下，如果输入距离大于曲线长度，则构建点放在曲线终点的切线且与曲线同方向的距离位置上。

EsriExtendEmbeddedAtFrom：与起点切线类似，但使用嵌入几何不是切线。

EsriExtendEmbeddedAtTo：与终点切线类似，但使用嵌入几何不是切线。

EsriExtendEmbedded：在这种方式下，根据输入距离自动使用 esriExtendEmbeddedAtFrom 或 esriExtendEmbeddedAtTo 创建构建点。

EsriExtendTangents：在这种方式下，根据输入距离自动使用 esriExtendTangentAtFrom 或 esriExtendTangentAtTo 创建构建点。

EsriExtendAtFrom：在这种方式下，包括了 esriExtendTangentAtFrom 和 esriExtendEmbeddedAtFrom。

EsriExtendAtT：在这种方式下，包括了 esriExtendTangentAtTo 和 esriExtendEmbeddedAtTo。

EsriExtendEmbeddedAtTo，在这种方式下，包括了 esriExtendTangentAtTo 和 EsriExtendEmbeddedAtTo，如图3-22所示。

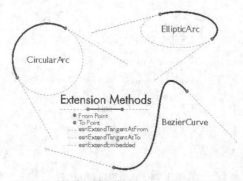

▲图 3-22　构造偏移点参数 esriExtendAtTo 的示例

（8）ConstructParallel 构造平行线上点法：该方法是给定一条路径 Path、一个参考点 Point 及一个距离或比率，在平行线上构建一个点，如图 3-23 所示。

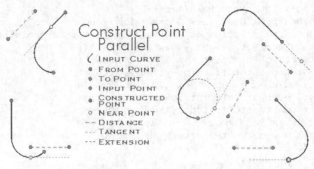

▲图 3-23　构造平行线上点法

下面代码片段演示构造平行线上点法：

```
IPoint[] points = new IPoint[2];
for (int i = 0; i < 2; i++)
{
    points[i] = new PointClass();
}
points[0].PutCoords(0, 0);
points[1].PutCoords(20, 0);
ISegment segment;
ILine line = new LineClass();
line.FromPoint = points[0];
line.ToPoint = points[1];
segment = line as ISegment;
IPoint fromPoint = new PointClass();
fromPoint.X = points[0].X + 10;
fromPoint.Y = points[0].Y + 5;
IConstructPoint constructPoint = new PointClass();
constructPoint.ConstructParallel(segment, esriSegmentExtension.esriNoExtension,
    fromPoint, segment.Length);
addFeature("point", points[0]);
addFeature("point", points[1]);
addFeature("point", constructPoint as IPoint);
axMapControl1.Refresh();
```

其扩展参数同 ConstructOffset。

（9）ConstructPerpendicular 构造垂直线上点法：通过给定一条路径 Path、一个参考点 Point 及一个距离，在参考点与路径的垂直方向上，在给定距离位置构建一个点，如图 3-24 所示。

▲图 3-24　构造垂直线上点法

下面代码片段演示构造垂直线上点法：

```
IPoint[] points = new IPoint[2];
for (int i = 0; i < 2; i++)
{
    points[i] = new PointClass();
}
points[0].PutCoords(0, 0);
points[1].PutCoords(20, 0);
ISegment segment;
ILine line = new LineClass();
line.FromPoint = points[0];
line.ToPoint = points[1];
segment = line as ISegment;
IPoint fromPoint = new PointClass();
fromPoint.X = points[0].X + 10;
fromPoint.Y = points[0].Y + 10;
IConstructPoint constructPoint = new PointClass();
constructPoint.ConstructPerpendicular(segment, esriSegmentExtension.esriNoExtension,
    fromPoint, 3,false);
addFeature("point", points[0]);
addFeature("point", points[1]);
addFeature("point", constructPoint as IPoint);
axMapControl1.Refresh();
```

（10）ConstructThreePointResection 后方交会定点法：该方法即测量学中的后方交会定点法，即构建点对 3 个已知点进行观测，得到两个角度，然后确定构建点的位置，如图 3-25 所示。

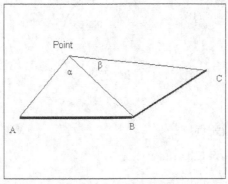

▲图 3-25　后方交会定点法

下面的代码片段演示后方交会定点法创建点：

```
IPoint[] points = new IPoint[3];
for (int i = 0; i < 3; i++)
{
    points[i] = new PointClass();
}
IConstructPoint constructPoint = new PointClass();
points[0].PutCoords(0, 10);
points[1].PutCoords(20, 20);
points[2].PutCoords(0, 0);
double angle1 = Math.PI / 4;
double angle2 = Math.PI / 4;
double angle3 = 0;
try
{
    constructPoint.ConstructThreePointResection(points[0],angle1,
        points[1],angle2,points[2], out angle3);
```

```
        addFeature("point", points[0]);
        addFeature("point", points[1]);
        addFeature("point", points[2]);
        addFeature("point", constructPoint as IPoint);
    }
    catch (Exception ex)
    {
    }
    axMapControl1.Refresh();
```

Multipoint 对象是一个具有相同结构的点的集合,通过使用 Add 方法来添加一个点到集合中。这个过程需要使用 IPointCollection 接口来完成,在 IConstructMultipoint 接口中定义了"ConstructArcPoints"、"ConstructDivideEqual"、"ConstructDivideLength"、"ConstructIntersection"、"ConstructTangent" 5 种构造方法。

(1) ConstructArcPoints 构造圆弧点:通过给定的一段圆弧,返回该圆弧的起点、终点、圆心和切线的交点,如图 3-26 所示。

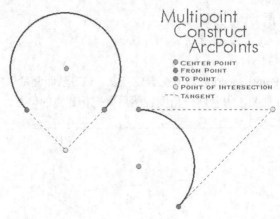

▲图 3-26 构造圆弧点

下面代码片段演示构造圆弧点:

```
IPoint centerPoint = new PointClass();
centerPoint.PutCoords(10, 0);
IPoint fromPoint = new PointClass();
fromPoint.PutCoords(0, 0);
IPoint toPoint = new PointClass();
toPoint.PutCoords(0, 20);
IConstructCircularArc circularArcConstruction = new CircularArcClass();
circularArcConstruction.ConstructThreePoints(fromPoint, centerPoint, toPoint, false);
IConstructMultipoint constructMultipoint = new MultipointClass();
constructMultipoint.ConstructArcPoints(circularArcConstruction as ICircularArc);
IPointCollection pointCollection = constructMultipoint as IPointCollection;
for (int i = 0; i < pointCollection.PointCount; i++)
{
    addFeature("point", pointCollection.get_Point(i));
}
axMapControl1.Refresh();
```

(2) ConstructDivideEqual 构造等分点法:通过一条曲线和需要返回的点数来产生一个点集合,如图 3-27 所示,下面代码片段演示构造等分点法:

```
IPoint centerPoint = new PointClass();
centerPoint.PutCoords(10, 10);
```

```
IPoint fromPoint = new PointClass();
fromPoint.PutCoords(0, 0);
IPoint toPoint = new PointClass();
toPoint.PutCoords(10, 20);
ICircularArc circularArc = new CircularArcClass();
circularArc.PutCoords(centerPoint, fromPoint, toPoint,
    esriArcOrientation.esriArcClockwise);
IConstructMultipoint constructMultipoint = new MultipointClass();
constructMultipoint.ConstructDivideEqual(circularArc as ICurve,
    (int) circularArc.Length/5);
IPointCollection pointCollection = constructMultipoint as IPointCollection;
for (int i = 0; i < pointCollection.PointCount; i++)
{
    addFeature("point", pointCollection.get_Point(i));
}
axMapControl1.Refresh();
```

（3）ConstructDivideLength 构造等长点法：通过给定的一条曲线对象和已经定义的间距长度，返回所有处于这条曲线上的点。下面代码片段演示构造等长点法：

```
IPoint centerPoint = new PointClass();
centerPoint.PutCoords(10, 0);
IPoint fromPoint = new PointClass();
fromPoint.PutCoords(0, 0);
IPoint toPoint = new PointClass();
toPoint.PutCoords(0, 20);
ICircularArc circularArc = new CircularArcClass();
circularArc.PutCoords(centerPoint, fromPoint, toPoint,
    esriArcOrientation.esriArcClockwise);
IConstructMultipoint constructMultipoint = new MultipointClass();
constructMultipoint.ConstructDivideLength(circularArc as ICurve,10);
IPointCollection pointCollection = constructMultipoint as IPointCollection;
for (int i = 0; i < pointCollection.PointCount; i++)
{
    addFeature("point", pointCollection.get_Point(i));
}
axMapControl1.Refresh();
```

（4）ConstructIntersection 构造交点法：通过给定的两条曲线对象，以及这两条曲线的延长线、切线，计算这些线的交点，如图 3-28 所示。

▲图 3-27 构造等分点

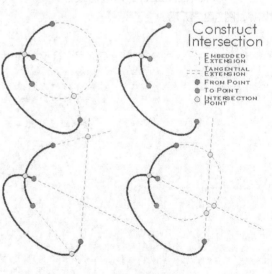

▲图 3-28 构造交点法

下面代码片段演示构造交点法：

```
IPoint[] points = new IPoint[4];
for (int i = 0; i < 4; i++)
{
    points[i] = new PointClass();
}
points[0].PutCoords(15, 10);
points[1].PutCoords(20, 60);
points[2].PutCoords(40, 60);
points[3].PutCoords(45, 10);
//构造Bezier曲线
IBezierCurveGEN bezierCurve = new BezierCurveClass();
bezierCurve.PutCoords(ref points);
IPoint centerPoint = new PointClass();
centerPoint.PutCoords(30, 30);
IPoint fromPoint = new PointClass();
fromPoint.PutCoords(10, 10);
IPoint toPoint = new PointClass();
toPoint.PutCoords(50, 10);
//构造圆弧
IConstructCircularArc constructCircularArc = new CircularArcClass();
constructCircularArc.ConstructThreePoints(fromPoint, centerPoint, toPoint, false);
object param0;
object param1;
object isTangentPoint;
IConstructMultipoint constructMultipoint = new MultipointClass();
constructMultipoint.ConstructIntersection(constructCircularArc as ISegment,
    esriSegmentExtension.esriNoExtension, bezierCurve as ISegment,
    esriSegmentExtension.esriNoExtension, out param0, out param1,
    out isTangentPoint);
IMultipoint multipoint = constructMultipoint as IMultipoint;
IPointCollection pointCollection = multipoint as IPointCollection;
for (int i = 0; i < pointCollection.PointCount; i++)
{
    addFeature("point", pointCollection.get_Point(i));
}
axMapControl1.Extent = multipoint.Envelope;
axMapControl1.Refresh();
```

（5）**ConstructTangent** 构造切线点：通过一条曲线和一个指定的点，产生指定的点到曲线的切线，如图 3-29 所示。

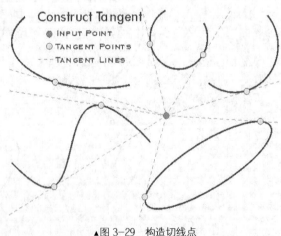

▲图 3-29 构造切线点

下面代码片段演示构造切线点:

```
IPoint[] points = new IPoint[4];
for (int i = 0; i < 4; i++)
{
    points[i] = new PointClass();
}
points[0].PutCoords(15, 10);
points[1].PutCoords(20, 60);
points[2].PutCoords(40, 60);
points[3].PutCoords(45, 10);
//构造¬Bezier 曲²线?
IBezierCurveGEN bezierCurve = new BezierCurveClass();
bezierCurve.PutCoords(ref points);
IConstructMultipoint constructMultipoint = new MultipointClass();
constructMultipoint.ConstructTangent(bezierCurve as ICurve, points[1]);
IMultipoint multipoint = constructMultipoint as IMultipoint;
IPointCollection pointCollection = multipoint as IPointCollection;
for (int i = 0; i < pointCollection.PointCount; i++)
{
    addFeature("point", pointCollection.get_Point(i));
}
axMapControl1.Extent = multipoint.Envelope;
axMapControl1.Refresh();
```

3.5 线对象

Polyline(多义线)对象是相连或不相连的路径对象的有序集合,如图 3-30 所示。组成 Polyline 的 Path 对象都是有效的,Path 不会重合、相交或自相交。多个 Path 对象可以连接于一个节点,也可以是分离的,长度为 0 的 Path 对象是不被允许的。

IPolyline 是 Polyline 类的主要接口,它定义了两个主要的方法,一个方法是使用一个路径对象给一个存在的 Polyline 整形;另一个方法是 SimplifyNetwork,该方法主要用于清除 0 长度的 Segment,并重新组织 Polyline 中的 Segment。

Polyline 构造可以通过 IGeometryCollection 接口添加路径对象来产生,可以参阅 3.1 节的代码。

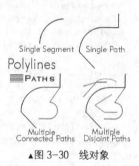

▲图 3-30 线对象

3.6 面对象

Polygon 对象是一个有序 Ring 对象的集合,如图 3-31 所示。

构造 Polygon 对象必须保证每个构成 Ring 都是有效的,Ring 之间的边界不能重合,外部 Ring 方向是顺时针,内部 Ring 方式是逆时针,不存在面积为 0 的 Ring。IPolygon 接口是 Polygon 类的主要接口,定义了一系列的属性和方法来控制 Ring。如 ExteriorRingCount 属性用于返回一个多边形的全部外部环的数目,InteriorRingCount 属性用于返回一个多边形内部环的数目,QueryExteriorRings 属性用于查询全部的外部 Ring,QueryInteriorRings 属性则用于查询全部的内部 Ring。Polygon

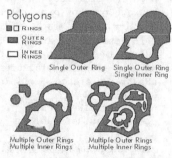

▲图 3-31 面对象

对象也可以使用 IGeometryCollection 接口的 AddSegment 方法添加 Ring 来产生，可以参阅 3.1 节的代码。

3.7 空间坐标系及变换

空间数据都有一个坐标系（地理坐标系统或投影坐标系统），它定义了空间数据在地球上的位置。一幅地图上显示的空间数据地理坐标系是一致的，否则会造成数据无法正确拼合。

地理坐标系统也称为真实世界的坐标系，是确定地物在地球上位置的坐标系，以经纬度作为地图的存储单位。

投影坐标系统是将三纬地理坐标系统上的经纬网投影到二纬平面地图上使用的坐标系统，因此地理信息系统必然要考虑地图投影，地图投影的使用保证了空间信息在地域上的联系和完整性，在各类地理信息系统的建设过程中，都要考虑选择何种地图投影系统。

Arc Engine 提供了 3 种组件：ProjectedCoordinateSystem、GeographicCoordinateSystem 和 SpatialReferenceEnvironmentClass，这些组件可用于自定义坐标系统。下面的代码演示了如何使用这些组件。

```
//改变一个图层的空间参考
IFeatureLayer player ;
player = pMap.get_Layer(0) as IFeatureLayer;
IFeatureClass pFeatureClass;
pFeatureClass = player.FeatureClass;
IGeoDataset pGeoDataset;
pGeoDataset = pFeatureClass as IGeoDataset;
IGeoDatasetSchemaEdit pGeoDatasetEdit;
pGeoDatasetEdit = pGeoDataset as IGeoDatasetSchemaEdit;
if (pGeoDatasetEdit.CanAlterSpatialReference == true)
{
ISpatialReferenceFactory2 pSpatRefFact;
pSpatRefFact = new SpatialReferenceEnvironmentClass();
IGeographicCoordinateSystem pGeoSys;
pGeoSys =
pSpatRefFact.CreateGeographicCoordinateSystem(4214);
pGeoDatasetEdit.AlterSpatialReference(pGeoSys);
}
pActiveView.Refresh();
//得到一个图层的空间参考
IFeatureLayer pLayer;
pLayer = pMap.get_Layer(0)as IFeatureLayer;
IGeoDataset pGeoDataset;
ISpatialReference pSpatialReference;
pGeoDataset = pLayer as IGeoDataset;
pSpatialReference = pGeoDataset.SpatialReference;
MessageBox.Show(pSpatialReference.Name);
//设置一个地图的空间参考，使用一个空间参考对话框
IProjectedCoordinateSystem pSpatialReference;
ISpatialReferenceDialog pDialog ;
pDialog = new SpatialReferenceDialogClass();
pSpatialReference = pDialog.DoModalCreate(true, false,false,0) as IprojectedCoordinateSystem;
pMap.SpatialReference = pSpatialReference;
pActiveView.Refresh();
```

Arc Engine 的空间参考对象模型中有 3 个组件类：GeographicCoordinateSystem（地理坐标系统）、ProjectedCoordinateSystem（投影坐标系统）和 UnknownCoordinateSystem（未知坐标系统），这 3

个组件类都实现了 ISpatialRefrence 接口，该接口提供了操作方法和属性来设置一个数据集空间参考属性，如空间域和坐标精度等。如 Changed 是这个接口中最重要的一个方法，用于检查一个坐标系统中的参数是否发生了变化；GetDomain 和 SetDomain 方法分别用于获取和设置一个坐标系统的域范围等。

IGeographicCoordinateSystem 是 GeographicCoordinateSystem 类接口，提供了 CoordinateUnit（坐标系的角度单位）、Datum（椭球体）和 PrimeMeridian（本初子午线）等属性。

IProjectedCoordinateSystem 是 ProjectedCoordinateSystem 类的接口，该接口提供了新建一个投影坐标系统的方法。在一个新的投影坐标系统中，需要设置 projection（投影方式）、GeographicCoordinateSystem、CoordinateUnit 和 Parameters 等。

3.8 本章小结

本章介绍了基于 ArcEngine 的几何对象模型和空间参考内容，ArcEngine 的几何对象是一种逐级组合的关系，通过线段组合为路径，路径组合为多义线，环组合为多边形。本章详细介绍了这些几何对象的组件类以及组件类实现的接口，特别是对象的创建，并给出了各种对象创建的构造方法实例。

第 4 章　地图组成及图层控制

4.1 地图对象

地图（Map）是 ArcEngine 的主要组成部分。Map 对象既是数据的管理容器，同时也是数据显示的主要载体。Map 对象的主要接口有 IMap、IGraphicsContainer、IActiveView、IActiveViewEvents、IMapBookmark 和 ITableCollection 等。Map 对象共有 35 个接口，更多的接口请查看"帮助"文件。本节介绍几个主要的接口。

4.1.1 IMap 接口

IMap 接口主要用于管理 Map 对象中的 layer 对象、要素选择集、MapSourround 对象和标注等。

Map 对象通过图层的方式管理地理数据。在 IMap 接口中定义了大量的方法来操作其中的图层对象，Map 对象是整个 GIS 系统的最重要对象，本节将一些重要、常用的属性、方法从帮助文件中翻译出来，供各位参考。

- ActiveGraphicsLayer：地图的活动图形图层，如果没有则会自动创建一个内存图形图层。
- AddLayer：向地图中添加单个图层。
- AddLayers：向地图中添加多个图层。
- AddMapSurround：向地图中添加辅助图形元素（如：比例尺、指北针等）。
- AnnotationEngine：地图使用的注记。
- BasicGraphicsLayer：基本图形图层。
- ClearLayer：从地图中移除所有图层。
- ClearMapSurrounds：从地图中移除所有辅助图形元素。
- ClearSelection：清除地图选择集。
- ComputeDistance：计算地图上两个点的距离。
- CreateMapSurround：创建并初始化一个辅助图形元素。
- DelayDrawing：暂停绘制。
- DelayEvents：该方法在地图选择改变时，设置为 True 时，地图选择改变事件将触发每一个功能添加或删除，如果设置为 false，则在地图选择操作完成后，发送一个通知。
- DeleteLayer：从 MAP 中删除图层。
- DeleteMapSurround：从地图上删除辅助图形元素。
- DistanceUnits：地图的距离单位。
- FeatureSelection：地图中的选择的要素。
- IsFramed：指示地图是否在一个框架内而不是在整个窗口中绘制。
- LayerCount：地图中图层的数目。

- Layers：地图中的图层集合。
- MapScale：数字形式的地图比例尺。
- MapSurroundCount：地图上辅助图形元素。
- MapUnits：地图单位。
- MoveLayer：将一个图层移动到另一个位置。
- Name：地图名称。
- RecalcFullExtent：重新计算全幅范围。
- ReferenceScale：分数形式的地图参考比例尺。
- SelectByShape：用一个几何对象或范围在地图中选择要素。
- SelectFeature：在地图上选择一个指定的实体对象。
- SelectionCount：选择的要素数目。
- SetPageSize：设置地图的页面大小。
- SpatialReference：地图的空间参考。
- SpatialReferenceLocked：锁定空间参考改变。

在第 2 章的例子代码中演示了如何加载一个图层文件（lyr 和 SHP）两种格式，本示例代码片段演示 AddLayer 方法加载一个实体图层到地图上。示例中用到 IWorkspaceFactory 接口，该接口是工作空间工厂，通过该工厂打开一个工作空间，WorkspaceFactory 是 GeoDatabase 的入口，它是一个抽象类，派生了许多子类，如 AccessWorkspaceFactory、ShapefileWorkspaceFactory、FileGDBWorkspaceFactoryClass、SdeWorkspaceFatoryClass 等，这些对象都实现了 IWorkspaceFactory 接口，不同类型的文件需要不同的工作空间工厂对象来打开一个工作空间，WorkspaceFactory 负责管理 Workspace 允许用户通过一系列的连接属性连接 Workspace，所有的工作空间对象都可以通过该接口产生：

```
//添加图层
    private void button1_Click(object sender, EventArgs e)
    {
        System.Windows.Forms.OpenFileDialog openFileDialog;
        openFileDialog = new OpenFileDialog();
        openFileDialog.Title = "打开 SHP 文件";
        openFileDialog.Filter = "shp layer(*.shp)|*.shp";
        if (openFileDialog.ShowDialog() == DialogResult.OK)
        {
            string file = openFileDialog.FileName;
            string filePath, fileName;
            int index = 0;
            index = file.LastIndexOf("\\");
            filePath = file.Substring(0, index);
            fileName = file.Substring(index + 1, file.Length - index - 1);
            IWorkspaceFactory workspaceFactory;
            IFeatureWorkspace featureWorkspace;
            IFeatureLayer featureLayer;
            workspaceFactory = new ShapefileWorkspaceFactoryClass();
            featureWorkspace = workspaceFactory.OpenFromFile(filePath, 0) as IFeatureWorkspace;
            featureLayer = new FeatureLayerClass();
            featureLayer.FeatureClass = featureWorkspace.OpenFeatureClass(fileName);
            featureLayer.Name = featureLayer.FeatureClass.AliasName;
            this.axMapControl1.AddLayer(featureLayer as ILayer);
            this.axMapControl1.Refresh();
        }
    }
```

第 2 章示例代码中实现了在地图上选择对象并高亮显示，采用了 Map 类的 SelectFeature 方法。

该方法通过指定要要找的实体对象，在地图上高亮显示，Map 类还提供了 SelectByShape 方法可以实现同样功能，该方法第一个参数指定选择选择实体的对象，通过该对象在地图上选择实体，SelectByShape 方法的第二个参数指定一个 SelectionEnvironmentClass 对象，通过该对象可以设置高亮的颜色、查询范围等，程序在选择实体后，通常需要对视图进行局部刷新，否则看不到高亮效果，部分刷新使用 IActiveView 接口提供的 PartialRefresh 方法，该方法可以设置刷新的类型，如刷新全部、刷新背景灯：

```
private void axMapControl1_OnMouseDown(object sender, IMapControlEvents2_OnMouseDownEvent e)
{
    IMap pMap;
    IActiveView pActiveView;
    pMap = axMapControl1.Map;
    pActiveView = pMap as IActiveView;
    IEnvelope pEnv;
    pEnv = axMapControl1.TrackRectangle();
    ISelectionEnvironment pSelectionEnv;
    pSelectionEnv = new SelectionEnvironmentClass();
    pSelectionEnv.DefaultColor = getRGB(110, 120, 210);
    pMap.SelectByShape(pEnv, pSelectionEnv, false);
    pActiveView.PartialRefresh(esriViewDrawPhase.esriViewGeoSelection, null, null);
}
private IRgbColor getRGB(int r,int g,int b)
{
    IRgbColor pColor;
    pColor = new RgbColorClass();
    pColor.Red = r;
    pColor.Green = g;
    pColor.Blue = b;
    return pColor;
}
```

4.1.2　IGraphicsContainer 接口

Map 对象通过 IGraphicsContainer 接口来管理图形元素（包括图形元素和框架元素）。接口提供了添加、更新、删除元素的方法：AddElement、UpdateElement、DeleteElement，以及选择元素的方法，如 LocateElements 方法。使用一个点来选择元素，它需要传入一个点对象和一个容差值；LocateElementsByEnvelope 方法，通过在 Map 上拖曳一个矩形区域，然后根据矩形区域选择区域内的元素。

通过 IGraphicsContainer 接口提供的 AddElement 方法添加图形元素，该方法第一个参数是一个 IElement 类型对象，MapFreame、ILineElement、IPolygonElement、IPictureElement 等类型的图形元素都继承自 IElement，因此，图形元素和框架元素都可以通过该方法添加到 GraphicsContainer 中：

```
private void button2_Click(object sender, EventArgs e)
{
    IGraphicsContainer graphicsContainer;
    IMap map = this.axMapControl1.Map;
    ILineElement lineElement = new LineElementClass();
    IElement element;
    IPolyline polyline = new PolylineClass();
    IPoint point = new PointClass();
    point.PutCoords(1, 5);
    polyline.FromPoint = point;
```

```
        point.PutCoords(80, 5);
        polyline.ToPoint = point;
        element = lineElement as IElement;
        element.Geometry = polyline as IGeometry;
        graphicsContainer = map as IGraphicsContainer;
        graphicsContainer.AddElement(element, 0);
        this.axMapControl1.ActiveView.PartialRefresh(esriViewDrawPhase.esriViewGraphics,
null, null);
    }
```

GraphicsContainer 对象是一个顺序索引指针，Reset 方法用于将指针指向第一个元素，Next 方法用于顺序取的下一个元素。UpdateElement 方法用于更新元素。

```
    private void button3_Click(object sender, EventArgs e)
    {
        IGraphicsContainer graphicsContainer;
        IPolyline polyline = new PolylineClass();
        IPoint point = new PointClass();
        point.PutCoords(1, 5);
        polyline.FromPoint = point;
        point.PutCoords(80, 20);
        polyline.ToPoint = point;
        IElement el;
        graphicsContainer = this.axMapControl1.Map as IGraphicsContainer;
        graphicsContainer.Reset();
        el = graphicsContainer.Next();
        if (el != null)
        {
            el.Geometry = polyline;
            graphicsContainer.UpdateElement(el);
        }
        this.axMapControl1.ActiveView.PartialRefresh(esriViewDrawPhase.esriViewGraphics,
null, null);
    }
```

4.1.3　IActiveView 接口

IActiveView 接口是 Map 对象最主要、最常用接口之一，该接口定义了 Map 对象的数据显示功能。通过该接口，可以在 Map 上绘制图形、改变视图范围、获取 ScreenDisplay 对象的指针、显示或隐藏标尺和滚动条，也可以刷新视图。

Arc Engine 中的 PageLayout 和 Map 对象都实现这个接口，分别代表了两种不同的视图：布局视图和数据视图。在任何一个时刻只能有一个视图处于活动状态。两种视图之间的关系，如图 4-1 所示。

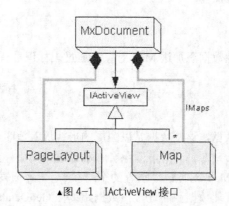

▲图 4-1　IActiveView 接口

IActiveView 接口提供了几个重要的方法和属性,如 Extent 属性,该属性用于返回 Map 对象当前视图的范围,是一个 Envelope 对象;FullExtent 属性,用于返回视图的全图范围;ScreenDisplay 属性,用于指向一个 ScreenDisplay 对象,每个视图都有一个 ScreenDisplay 对象用于控制视图的图形绘制工作。ScreenDisplay 拥有任意数量的缓存(Cache),缓存是保存在内存中的相关位图,当图形需要重绘时,可以不用读取数据库,而直接从内存中读取缓存,这样可以提高重绘的速度。在 Map 中可以产生 3 种类型的缓存:与所有图层相关的缓存、注记、图形或要素选择集,当重绘事件发生时,可以按地理数据、选择集、标注的顺序进行绘制。PartialRefresh 方法则是常用的一种刷新方法,该方法通过指定 esriViewDrawPhase 参数,可以实现不同方式的局部刷新 PartialRefresh 方法共有 3 个参数,第一个参数指定刷新对象类型,第二个参数用于指定刷新的对象:

```
//用于刷新指定图层
axMapControl1.ActiveView.PartialRefresh(esriViewDrawPhase.esriViewGeography, pLayer, null);
//用于刷新所有图层
axMapControl1.ActiveView.PartialRefresh(esriViewDrawPhase.esriViewGeography, null, null);
//用于刷新所选择对象
axMapControl1.ActiveView.PartialRefresh(esriViewDrawPhase.esriViewGeoSelection, null, null);
//用于刷新图形元素
axMapControl1.ActiveView.PartialRefresh(esriViewDrawPhase.esriViewGraphics, null, null);
//用于刷新指定的图形元素
axMapControl1.ActiveView.PartialRefresh(esriViewDrawPhase.esriViewGraphics,pElement,null);
//用于刷新所选择的图形元素
axMapControl1.ActiveView.PartialRefresh(esriViewDrawPhase.esriViewGraphicSelection, null, null);
//不刷新元素
axMapControl1.ActiveView.PartialRefresh(esriViewDrawPhase.esriViewNone, null, null);
```

4.1.4　IActiveViewEvents 接口

该接口是地图对象缺省的外向接口,使 Map 对象可以监听某些与活动视图相关的事件并做出相应的反应,如 AfterDraw、SelectionChanged 等。在对地图事件监听中,一般会用到 MapControl 控件的 IMapControlEvents2 和 Map 的 IActiveViewEvents 接口,这两个接口中方法大多相同,但 IMapControlEvents2 是对地图控件的事件接口,监听依据控件绑定,IActiveViewEvents 是对地图对象进行绑定,当 MapControl 中的地图对象更换后,对之前添加的地图事件监听就不再有效了。

4.1.5　IMapBookmark 接口

Map 对象可以管理所有的空间书签对象,使用该接口可以得到一个已经存在的空间书签,也可以进行产生和删除空间书签操作。

4.1.6　ITableCollection 接口

该接口用于将关系型属性数据添加进 Map 对象,通过表标识号获取相应的表。

4.2　图层对象

Map 对象用来装载地理数据,这些数据是以图层的形式放入地图对象的。Layer 作为装载的单位,当第 1 个图层装载进 Map 对象时,Map 对象会自动设置空间坐标系为当前图层的空间坐标系,以后再装载的图层都将使用 Map 对象已经设置的空间参考。Layer 对象本身没有装载数据,其作用是获得数据的引用,用于管理数据源的连接,数据始终在 GeoDatabase 或地图文件中。

4.2.1 ILayer 接口

ILayer 是所有图层类都实现的接口,该接口定义了所有图层的公共方法和属性,所有的图层接口都继承自该接口,下面将一些常用的属性和方法翻译出来。

- Cached:指出图层是否需要显示缓存。
- Draw:按照 drawphase 向指定的 display 绘制图层。
- MaximumScale:图层显示的最大比例尺。
- MinimunScale:图层显示的最小比例尺。
- Name:用于返回图层名称。
- ShowTips:属性用于指示当鼠标光标放在图层中某个要素上的时候,是否会出现 Tip。
- TipText:属性用于确定图层提示显示的区域。
- SpatialReference:属性用于设置图层的空间参考。
- Valid:属性用于指示当前图层是否有效。
- Visible:属性用于指示当前图层是否可视。

4.2.2 要素图层

要素图层是 GIS 中最常使用的数据类型之一,用于表示离散的矢量对象的信息。

4.2.2.1 IFeatureLayer 接口

IFeaturelayer 接口用于管理要素图层的数据源,即要素类(FeatureClass)。DataSourceType 属性用于返回图层的数据源类型。Search 方法通过两个参数,一个是过滤器,它是一个 IQueryFilter 类型的对象;一个是布尔值,用于说明返回的要素游标是否循环,返回一个 ICursor 类型的对象。下面的程序是演示如向通过该接口返回一个 ICursor 类型的对象:

```
private void searchFeatures(string sqlfilter, IFeatureLayer pFeatureLayer)
{
    IFeatureLayer pFeatLyr;
    pFeatLyr = pFeatureLayer;
    IQueryFilter pFilter;
    pFilter = new QueryFilterClass();
    //设置查询条件
    pFilter.WhereClause = sqlfilter;
    IFeatureCursor pFeatCursor;
    pFeatCursor = pFeatLyr.Search(pFilter, true);
    IFeature pFeat;
    pFeat = pFeatCursor.NextFeature();
    while (pFeat != null)
    {
        pFeat = pFeatCursor.NextFeature();
        if (pFeat != null)
        {
            ISimpleFillSymbol pFillsyl;
            pFillsyl = new SimpleFillSymbolClass();
            pFillsyl.Color = getRGB(220, 100, 50);
            object oFillsyl;
            oFillsyl = pFillsyl;
            IPolygon pPolygon;
            pPolygon = pFeat.Shape as IPolygon;
            //闪烁实体
            axMapControl1.FlashShape(pPolygon, 15, 20, pFillsyl);
            axMapControl1.DrawShape(pPolygon, ref oFillsyl);
        }
    }
}
```

4.2.2.2 IGeoFeatureLayer 接口

IGeoFeatureLayer 接口继承了 ILayer 和 IFeatureLayer 两个接口，用于控制要素图层中与地理相关的内容，如标注等，CadFeatureLayer、FeatureLayer、GdbRasterCatalogLayer 3 个图层类实现了该接口。该接口提供了一个 SearchDisplayFeature 方法，使用该方法只显示符合查询要求的要素，其他的要素都会消失掉。Renderer 属性用于设置图层的着色对象；DisplayAnnotation 属性用于设置要素图层是否出现标注，设置为 True 时，用户可以在这个要素图层上依据要素类的某个字段进行标注。

4.2.2.3 IGeoDataset 接口

IGeoDataset 接口仅有两个属性，用于管理地理要素集，其中 Extent 属性用于返回当前数据集的范围，SpatialReference 属性用于获取这个数据集的空间参考。Arc Engine 中有 40 多个类实现了 IGeoDataset 接口。

4.2.2.4 IFeatureSelection 接口

IFeatureSelection 接口负责管理一个图层中要素的选择集的方法和属性。Arc Engine 中有多个类实现了 IFeatureSelection 接口，如 CadFeatureLayer、CoverageAnnotationLayer、FeatureLayer、GdbRasterCatalogLayer 等。

IFeatureSelection 接口的 Add 方法用于将本图层的一个要素添加到图层的选择集中，SelectFeatures 方法用于使用一个过滤器把符合要求的要素放入图层的选择集中，Clear 方法用于把图层的所有选择集清空。下面的代码使用了选择集选择。

```csharp
private void searchSelection(string sqlfilter, IFeatureLayer pFeatureLayer)
{
    IQueryFilter pFilter;
    pFilter = new QueryFilterClass();
    //设置查询条件 t
    pFilter.WhereClause = sqlfilter;
    IFeatureSelection pFeatureSelection;
    pFeatureSelection = pFeatureLayer as IFeatureSelection;
    pFeatureSelection.SelectFeatures(pFilter,
    esriSelectionResultEnum.esriSelectionResultNew, true);
    IRgbColor pColor;
    pColor = new RgbColorClass();
    pColor.Red = 220;
    pColor.Green = 112;
    pColor.Blue = 60;
    pFeatureSelection.SelectionColor = pColor;
    axMapControl1.CtlRefresh(esriViewDrawPhase.esriViewGeoSelection, null, null);
    ISelectionSet pFeatSet;
    pFeatSet = pFeatureSelection.SelectionSet;
    IFeatureCursor pFeatCursor;
    ICursor pCursor;
    pFeatSet.Search(null, true, out pCursor);
    pFeatCursor = pCursor as IFeatureCursor;
    IFeature pFeat;
    pFeat = pFeatCursor.NextFeature();
    while (pFeat != null)
    {
        if (pFeat != null)
        {
            ISimpleFillSymbol pFillsyl2;
            pFillsyl2 = new SimpleFillSymbolClass();
```

```
            pFillsyl2.Color = getRGB(220, 60, 60);
            axMapControl1.FlashShape(pFeat.Shape, 15, 20, pFillsyl2);
        }
        pFeat = pFeatCursor.NextFeature();
    }
}
```

选择集有两种：Feature Selections 和 Element Selections，这两者都实现了 ISelection 接口，在返回值时分别返回 IenumFeature 和 IEnumElement。PageLayout 对象和 Map 对象都有自己的选择集对象，PageLayout 只有元素选择集，Map 则拥有元素选择集和要素选择集。

4.2.2.5　IFeatureLayerDefinition 接口

该接口定义了 CreateSelectionLayer 方法，该方法用一个图层选择集中的要素创建一个新的图层。该方法要求有 4 个参数，其中 LayerName 定义了新图层的名称，joinTableNames 是一个与当前图层关联的表的名称，DefinitionExpression 用于设定一个选择过滤，将要素选择集中符合条件的要素放入新的图层。下面的代码详细介绍了如何使用该方法。

```
private void CreateSelLayer()
{
    IFeatureLayer pFeatLyr;
    IMappMap = axMapControl1.Map;
    IActiveView pActiveView = pMap as IActiveView;
    pFeatLyr = pMap.get_Layer(1) as IFeatureLayer;
    IQueryFilter pQueryFilter ;
    pQueryFilter = new QueryFilterClass();
    pQueryFilter.WhereClause = "area > 100000";
    IFeatureSelection pFeatSel;
    pFeatSel = pFeatLyr as IFeatureSelection;
    pFeatSel.SelectFeatures(pQueryFilter,
    esriSelectionResultEnum.esriSelectionResultNew, false);
    IFeatureLayerDefinition pFeatLyrDef;
    pFeatLyrDef = pFeatLyr as IFeatureLayerDefinition;
    pFeatLyrDef.DefinitionExpression = "pop1999 > 1000000";
    IFeatureLayer pnewfeat;
    pnewfeat = pFeatLyrDef.CreateSelectionLayer("new" + pFeatLyr.Name , true, "",
    "");
    pFeatSel.Clear();
    pMap.AddLayer(pnewfeat);
    pActiveView.Refresh();
}
```

4.2.2.6　ILayerFields 接口

该接口提供了 Field、FieldCount、FieldInfo 和 FindField 4 个属性和方法，FindField 方法用于查询字段，FieldInfo 用于获取字段的信息，FieldCount 用于获取字段数目。

4.2.2.7　IIdentify 接口

该接口只定义了一个方法 Identify，用于获取图层的单个要素，返回一个 IArray 数组对象。下面的代码演示了如何使用该方法。

```
private void axMapControl1_OnMouseDown(object sender,IMapControlEvents2_OnMouseDownEvent e)
{
    IIdentify pIdentify;
    IPoint pPoint;
    IArray pIDArray;
    IFeatureIdentifyObj pFeatIdObj;
```

```csharp
        IIdentifyObj pIdObj;
        IMap pMap = axMapControl1.Map;
        pIdentify = pMap.get_Layer(1) as IIdentify;
        pPoint = new PointClass();
        pPoint.PutCoords(e.mapX, e.mapY);
        pPoint.SpatialReference = pMap.SpatialReference;
        pIDArray = pIdentify.Identify(pPoint);
        if (pIDArray != null)
        {
            pFeatIdObj = pIDArray.get_Element(0) as IFeatureIdentifyObj;
            pIdObj = pFeatIdObj as IIdentifyObj;
            pIdObj.Flash(axMapControl1.ActiveView.ScreenDisplay);
            MessageBox.Show("Layer: " + pIdObj.Layer.Name +
            System.Environment.NewLine + "Feature: " + pIdObj.Name);
        }
        else
            MessageBox.Show("没有找到要素.");
    }
```

4.2.3 CAD 文件

Map 对象用于载入多种格式的地理和非地理数据，其中包括 CAD 的 DWG 文件。DWG 文件是两种不同形式的混合体：一是要素图层，保存矢量数据，使用 FeatureLayer 来管理；二是栅格图像，作为地图背景使用，可以使用 cadLayer 对象来管理。当 DWG 文件载入 Map 对象时，Arc engine 将要素图层的元素分为 4 种类型，即点、线、多边形和标注，并在要素的属性中建立图层标识字段，而不按照 DWG 文件本身的图层号进行分层，CAD 文件中的文字可以使用 CadAnnotationLayer 对象来操作。

Engine 中对 CAD 数据的操作专门提供了一套相应的接口和类，如 CadWorkspaceFactryClass 类提供了打开 CAD 文件目录，和 FeatureLayer 数据的工作空间工厂一样，该工作空间工厂提供了 OpenCadDrawingDataset 方法用于打开该目录下的 CAD 文件。

```csharp
private void AddCADLayer()
{
    IWorkspaceFactory pCadWorkspaceFactory;
    pCadWorkspaceFactory = new CadWorkspaceFactoryClass();
    IWorkspace pWorkspace;
    pWorkspace =
    pCadWorkspaceFactory.OpenFromFile(@"C:\arcgis\ArcTutor\Editor\ExerciseData\EditingCAD",
    0);
    ICadDrawingWorkspace pCadDrawingWorkspace ;
    pCadDrawingWorkspace = pWorkspace as ICadDrawingWorkspace;
    ICadDrawingDataset pCadDataset;
    pCadDataset=
    pCadDrawingWorkspace.OpenCadDrawingDataset("PARCELS.DWG");
    ICadLayer pCadLayer ;
    pCadLayer = new CadLayerClass();
    pCadLayer.CadDrawingDataset = pCadDataset;
    axMapControl1.AddLayer(pCadLayer,0);
}
private void AddCADFeatures()
{
    IWorkspaceFactory pCadWorkspacefactory ;
    pCadWorkspacefactory = new CadWorkspaceFactoryClass();
    IFeatureWorkspace pWorkspace;
    String filePath="c:\\aaa\\cad";
    pWorkspace =pCadWorkspacefactory.OpenFromFile(filePath,0) as IFeatureWorkspace;
    IFeatureDataset pFeatDataset;
```

```
        pFeatDataset = pWorkspace.OpenFeatureDataset("PARCELS.DWG");
        IFeatureClassContainer pFeatClassContainer;
        pFeatClassContainer = pFeatDataset as IFeatureClassContainer;
        IFeatureClass pFeatClass;
        IFeatureLayer pFeatLayer;
        int i;
        for( i = 0; i<= pFeatClassContainer.ClassCount - 1;i++)
        {
            pFeatClass = pFeatClassContainer.get_Class(i);
            if (pFeatClass.FeatureType ==esriFeatureType.esriFTCoverageAnnotation)
            {
                pFeatLayer = new CadAnnotationLayerClass();
            }
            else
            {
                pFeatLayer = new FeatureLayerClass();
            }
            pFeatLayer.Name = pFeatClass.AliasName;
            pFeatLayer.FeatureClass = pFeatClass;
            axMapControl1.AddLayer(pFeatLayer,0);
        }
    }
```

4.2.4 TIN 图层

TIN 表示连续的表面，如地表高程或温度梯度。TinLayer 接口定义了 TIN 如何在图层上显示的方法和属性，如 Datasetet 用于显示 Tin 图层的数据源；DisplayField 是 TinLayer 图层的主要显示字段；AddRenderer 方法用于加入一个着色对象；ClearRenderers 方法用于清除图层中所有的着色对象；GetRenderer 方法用于通过索引值，获取着色对象；RendererCount 方法用于返回着色对象的数目。

4.2.5 GraphicsLayer

GraphicsLayer 是一个抽象类，当在 MapControl 控件上绘制图形时，这些绘制图形所在的层就是 GraphicsLayer 对象。该类泛化出 CompositeGraphicsLayer、FDOGraphicsLayer、GlobeGraphicsLayer、GraphicsLayer3D、GraphicsSubLayer 等。每个 Map 对象都管理着一个 CompositeGraphicsLayer 对象，该对象是一个图形图层集合，该集合中有一个默认的图形图层——基本图形图层，可以通过 BasicGraphicsLayer 直接获取。该图层是默认的，不能在 CompositeGraphicsLayer 中删除，CompositeGraphicsLayer 对象提供了集合中添加、删除、查找图层的方法。

FDOGraphicsLayer 是一个与要素相关的标注图层，可以有两种方法添加标注，一种是使用 Label 方法，通过鼠标光标选取要素，逐个给这些要素添加标注文本，文本放置在 FDOGraphicsLayer 上；另一种是使用 Annotation 方法，它可以对要素图层进行自动注记，并且这些注记的文本信息都可以保存在数据库中。

4.3 屏幕显示对象

MapControl 中的每个视图都有一个 ScreenDisplay 对象，用于控制视图中的图形绘制。ScreenDisplay 是一个与窗体相联系的显示设备，除了管理窗体屏幕的显示属性外，也管理缓存和视图屏幕的变化等。很多对象都是由 ScreenDisplay 去管理与它们相关的可视化窗体，通过 IActiveView 接口的 ScreenDisplay 属性，获取 ScreenDisplay 对象。

ScreenDisplay 对象都拥有 DisplayTransformation 对象，用于设备单位和地图单位的转换。ScreenDisplay 实现了 Idisplay 接口，可以通过该接口的 DisplayTransformation 属性获取该对象。每

个 DisplayTransformation 都与一个 Map 相关，它拥有地图的空间参考属性。

ScreenDisplay 实现了 IDisplay、IDraw、IscreenDisplay 和 IScreenCacheManager 等 10 个接口。IScreenDisplay 接口是最主要的接口，提供了缓存操作、漫游、旋转、绘制等方法和属性，如 AddCache 方法用于将一个新的缓存对象添加进 ScreenDisplay 对象，CacheCount 属性用于返回缓存数目。此外，还提供了绘制高级几何体的方法，如 DrawPoint、DrawPolyline、DrawPolygon、DrawRectangel 和 DrawText 等，这些几何体在绘制前都必须使用 StartDrawing 方法开始绘制，绘制完毕需要使用 FinishDrawing 方法结束。ScreenDisplay 也提供了多种移动地图的方法，如 TrackPan 方法支持在控件的 MouseDown 事件中移动地图，也可以在 MouseDown、MouseMove 和 MouseUp 等事件中分别使用 panStart、panMoveto 和 panStop 等方法。ScreenDisplay 还提供了和 Pan 操作类似的旋转方法。下面的代码演示了如何在 ScreenDisplay 中绘制线。

```
private void axMapControl1_OnMouseDown(object sender, ESRI.ArcGIS.Controls.
IMapControlEvents2_OnMouseDownEvent e)
{
    IActiveView pActiveView = axMapControl1.ActiveView;
    IScreenDisplay screenDisplay = pActiveView.ScreenDisplay;
    ISimpleLineSymbol lineSymbol = new SimpleLineSymbolClass();
    IRgbColor rgbColor = new RgbColorClass();
    rgbColor.Red = 255;
    lineSymbol.Color = rgbColor;
    IPolyline pLine = axMapControl1.TrackLine() as IPolyline;

    screenDisplay.StartDrawing(screenDisplay.hDC,(short)esriScreenCache.esriNoScreenCache);
    screenDisplay.SetSymbol((ISymbol)lineSymbol);
    screenDisplay.DrawPolyline(pLine);
    screenDisplay.FinishDrawing();
}
```

4.4 页面布局对象

在第 2 章曾经介绍过页面布局对象。PageLayout 和 Map 对象一样，都是图形元素的容器，但是 PageLayout 除了保存图形元素外，还可以保存 MapFrame 的框架元素（Frame Element）。

PageLayout 类主要实现了 IPageLayout 接口，该接口定义了用于修改页面版式（layout）的方法和属性，其中包括图形的位置属性、标尺和对齐网格的设置，以及确定页面显示在屏幕上的方法。通过该接口可以管理 RulerSettings、SnapGrid、SnapGuides 和 Page 等对象。

- Page：属性用于获得放在 PageLayout 对象中的 page 对象。
- RulerSettings：用于获得 PageLayout 控制的标尺对象。
- SnapGrid 和 VerticalSnapGrid：用于控制 PageLayout 对象中显示的网格对象。

接口还提供了一些属性用于设置显示尺寸、比例、范围宽度等。

- ZoomToWhole：用于设置 PageLayout 以最大尺寸显示。
- ZoomPercent、ZoomToPercent：用于设置按照输入的比例显示。
- ZoomToWith：用于让视图的显示范围与控件的宽度匹配。

PageLayout 对象不但实现了 IGraphicsContainer 接口用来管理元素，同时还实现了 IGraphicsContainerSelect 接口用于管理被选择的元素。

- SelectAllElements：方法用于选择所有的元素。
- SelectionBounds：属性用于获得选择的包络线。
- UnselectAllElements：用于清空所有的选择元素。
- ElementSelectionCount：属性用于获得选择的元素数目。

4.5 地图排版

4.5.1 Page 对象

在 PageLayout 对象创建后，会自动产生一个 Page 对象来管理视图中的页面，Page 对象只作为一个装载地图数据的容器，不提供查询、分析的功能。Page 类的主要接口是 IPage，用于管理 Page 的颜色、尺寸、方向、版式单位、边框类型、打印区域等属性，如 Background 属性用于设置 Page 背景样式，BackgroundColor 用于设置背景的颜色，Border 属性用于设置 Page 的边框，Units 用于获取 Page 所使用的单位。此外 Page 对象还继承了 IPageEvents 接口，拥有 PageColorChanged、PageMarginsChanged、PageSizeChanged 和 PageUnitsChanged 4 个事件，当 Color、Margin、Size 和 Unit 等发生变化的时候触发。

Arc Engine 提供了 esriPageFromID 枚举值设置 Page 对象的尺寸，如设置 Page 的尺寸为 esriPageFromSameAsPrinter，则无论打印机的页面设置如何变化，版式页面的尺寸始终会和打印机保持一致，使用 esriPageFromID 比用 PutCustomSize 设置 Page 尺寸的速度更快。

4.5.2 SnapGrid 对象

SnapGrid 是 PageLayout 上用于摆放元素而设置的辅助点，这些点便于用户对齐元素，通过 IPageLayout 接口的 SnapGrid 属性来获得 SnapGrid 对象。SnapGrid 类实现了 ISnapGrid 接口，用于设置 SnapGrid 的各种属性，如 HorizontalSpacing 和 VerticalSpacing 属性用于设置网点之间的水平距离和垂直距离，IsVisible 用于确定这些网点是否处于可见状态，Draw 方法用于将 SnapGrid 对象绘制在 Page 上。下面的代码演示了如何使用 SnapGrid 对象的方法和属性。

```
private void SetSnapGrid()
{
    ISnapGrid pSnapGrid;
    IPageLayout pPageLayout = axPageLayoutControl1.PageLayout;
    pSnapGrid = pPageLayout.SnapGrid;
    pSnapGrid.VerticalSpacing = 2;
    pSnapGrid.HorizontalSpacing = 1;
    pSnapGrid.IsVisible = true;
    axPageLayoutControl1.CtlRefresh((esriViewDrawPhase)65535, Type.Missing, Type.Missing);
}
```

4.5.3 SnapGuides 对象

SnapGuides 是绘制在 PageLayout 上的辅助线，分为水平辅助线和垂直辅助线两种，通过 IPageLayout 接口的 HorizontalSnapGuides 或 VerticalSnapGuides 获得，每个 SnapGuides 都管理着一个 Guide 集合。

SnapGuides 类实现了 ISnapGuides 接口，定义了 SnapGuides 类的方法和属性，如 AddGuides 方法用于将一个 Guide 放在指定的位置上，AreVisible 属性用于设置 SnapGuides 是否可见，Draw 方法用于绘制辅助线，DrawHighlight 用于绘制高亮辅助线，GuideCount 属性用于返回 SnapGuides 对象中 Guide 的数目，RemoveAllGuides 和 RemoveGuide 方法分别用于清除所有的 Guide 和按照索引值清除 Guide。下面的代码演示了如何使用 SnapGuides 对象的主要方法和属性。

```
private void AddGuide()
{
    ISnapGuides pSnapGuides;
    IPageLayout pPageLayout = axPageLayoutControl1.PageLayout;
    pSnapGuides = pPageLayout.HorizontalSnapGuides;
    pSnapGuides.AddGuide(5);
```

```
    pSnapGuides.AddGuide(7);
    pSnapGuides.AddGuide(9);
    pSnapGuides.AddGuide(11);
    pSnapGuides = pPageLayout.VerticalSnapGuides;
    pSnapGuides.AddGuide(2);
    pSnapGuides.AddGuide(4);
    pSnapGuides.AddGuide(6);
    pSnapGuides.AddGuide(8);
    pSnapGuides.AreVisible = true;
    axPageLayoutControl1.CtlRefresh((esriViewDrawPhase)65535, Type.Missing,
    Type.Missing);
}
```

4.5.4 RulerSettings 对象

标尺对象也是辅助元素在 PageLayout 上的设置，可以通过 IPageLayout 的 RulerSettings 获得当前 PageLayout 相关的标尺。RulerSettings 类实现了 IRulerSettings 接口，该接口只有一个 SmallestDivision 属性，用于设置最小的区分值。

4.6 Element 对象

Element 对象就是地图上不保存到数据库中的元素，像 MapControl 中的 DrawShape 和 DrawText 方法，绘制的图形都是 Element 对象，这些对象都是内存中的对象，当 MapControl 显示地图发生变化时，这些对象就会消失。Element 对象分为两大部分：图形元素（Graphic Element）和框架元素（Frame Element）。

图形元素包括 GroupElement、MarkerElement、LineElement、TextElement、DataGraphElement、PictureElement 和 FillShapElement 等对象，Arc Engine 提供了 23 种图形元素，这些对象都是作为图形的形式而存在，在视图上是可见的。框架元素包括 FrameElement、MapFrame、MapSurroundFrame、OleFrame 和 TableFrame 等，它们都是作为不可见的容器而存在的。

IElement 是所有图形元素和框架元素类都实现的接口，提供了 Geometry 属性和查询、绘制元素的方法。

4.6.1 图形元素

4.6.1.1 LineElement 和 MarkerElement 对象

LineElement 和 MarkerElement 是最简单的图形元素，在数据视图（data view）或者布局视图（Pagelayout view）上表现为线和点的形式。ILineElement 接口只提供了一个属性 Symbol，用于设置线的样式。LineElement 的 Geometry 只能使用 Line 或者 Polyline。IMarkerElement 接口也只提供一个属性 Symbol，用于设置点的样式。下面的代码演示了如何添加这两种元素。

```
private void axMapControl1_OnMouseDown(object sender,
IMapControlEvents2_OnMouseDownEvent e)
{
    IMap pMap;
    IActiveView pActiveView;
    pMap = axMapControl1.Map;
    pActiveView = pMap as IActiveView;
    IPoint pPt;
    pPt = new PointClass();
    pPt.PutCoords(e.mapX, e.mapY);
    IMarkerElement pMarkerElement;
    pMarkerElement = new MarkerElementClass();
```

```
ISimpleMarkerSymbol pMarkerSymbol;
pMarkerSymbol = new SimpleMarkerSymbolClass();
pMarkerSymbol.Color = getRGB(11, 200, 145);
pMarkerSymbol.Size = 2;
pMarkerSymbol.Style = esriSimpleMarkerStyle.esriSMSDiamond;
IElement pElement;
pElement = pMarkerElement as IElement;
pElement.Geometry = pPt;
pMarkerElement.Symbol = pMarkerSymbol;
IGraphicsContainer pGraphicsContainer;
pGraphicsContainer = pMap as IGraphicsContainer;
pGraphicsContainer.AddElement(pMarkerElement as IElement, 0);
pActiveView.PartialRefresh(esriViewDrawPhase.esriViewGraphics, null, null);
}
```

4.6.1.2 TextElement 对象

地图为了显示图形的属性信息，一般都采用文字标注来完成。地图标注有两种形式：一种是保存在地理数据库中以标注类的形式存在，另一种是使用文字元素。TextElement 对象实现了 ITextElement 接口，该接口提供了 3 个属性：ScaleText（文字尺寸）、Text（字符）和 Symbol（文字符号）。下面的程序演示如何使用 TextElement 对象：

```
private void axMapControl1_OnMouseDown(object sender,
IMapControlEvents2_OnMouseDownEvent e)
{
    IMap pMap;
    IActiveView pActiveView;
    pMap = axMapControl.Map;
    pActiveView = pMap as IActiveView;
    ITextElement pTextEle;
    IElement pEles;
    pTextEle = new TextElementClass();
    pTextEle.Text = "ArcObjects";
    pEles = pTextEle as IElement;
    IPoint pPoint;
    pPoint = new PointClass();
    pPoint.PutCoords(e.mapX, e.mapY);
    pEles.Geometry = pPoint;
    IGraphicsContainer pGraphicsContainer;
    pGraphicsContainer = pMap as IGraphicsContainer;
    pGraphicsContainer.AddElement(pEles, 0);
    pActiveView.PartialRefresh(esriViewDrawPhase.esriViewGraphics, null, null);
}
```

4.6.1.3 GroupElement 对象

GroupElement 对象用于将多个元素编为一组作为一个整体使用，该类实现了 IGroupElement 接口，该接口定义了 6 个属性和方法，如 AddElement 方法用于将一个元素添加到 GroupElement 对象中，ClearElements 用于将 GroupElement 中的元素清空，DeleteElement 用于删除 GroupElement 中的某个元素，ElementCount 用于返回 GroupElement 中元素的数目。下面的代码演示了如何使用该类。

```
private void GroupElement(IGraphicsContainer pGraphicsContainer)
{
    IGroupElement pGroupEle;
    pGroupEle = new GroupElementClass();
    IElement pEle;
    pGraphicsContainer.Reset();
    pEle = pGraphicsContainer.Next();
    while (pEle != null)
    {
```

```
        pGroupEle.AddElement(pEle);
        pEle = pGraphicsContainer.Next();
    }
    MessageBox.Show(pGroupEle.ElementCount.ToString());
}
```

4.6.1.4 FillShapeElement 对象

FillShapeElement 是一个抽象类，该类泛化出 CircleElement、EllipseElement、PolygonElement 和 RectangleElement 等。该类实现了 IFillShapeElement 接口，通过接口提供的 Symbol 属性可以设置元素的样式，Symbol 必须是 IFillsymbol 对象。下面的代码演示了如何使用该对象。

```
private void axMapControl1_OnMouseDown(object sender,
IMapControlEvents2_OnMouseDownEvent e)
{
    IMap pMap;
    IActiveView pActiveView;
    pMap = axMapControl1.Map;
    pActiveView = pMap as IActiveView;
    IPolygon pPolygon;
    pPolygon = axMapControl1.TrackPolygon() as IPolygon;
    ISimpleFillSymbol pSimpleFillsym;
    pSimpleFillsym = new SimpleFillSymbolClass();
    pSimpleFillsym.Style = esriSimpleFillStyle.esriSFSDiagonalCross;
    pSimpleFillsym.Color = getRGB(102, 200, 103);
    IFillShapeElement pPolygonEle;
    pPolygonEle = new PolygonElementClass();
    pPolygonEle.Symbol = pSimpleFillsym;
    IElement pEle;
    pEle = pPolygonEle as IElement;
    pEle.Geometry = pPolygon;
    IGraphicsContainer pGraphicsContainer;
    pGraphicsContainer = pMap as IGraphicsContainer;
    pGraphicsContainer.AddElement(pEle, 0);
    pActiveView.PartialRefresh(esriViewDrawPhase.esriViewGraphics, null, null);
}
```

4.6.1.5 图片元素对象

图片元素对象有 7 类：BmpPictureElement、EmfPictureElement、GifPictureElement、Jp2PictureElement、JpgPictureElement、PngPictureElement 和 TifPictureElement，这 7 类对象都实现了 IPictureElement 接口。该接口提供了 6 个属性和方法，如 Filter 属性提供 OpenFileDialog 使用的过滤器，MaintainAspectRatio 属性用于调整尺寸是否保持其长宽比例，PictureDescription 用于添加图片的附加描述信息，SavePictureInDocument 属性用于确定图片是否保存到 MXD 文件，ImportPictureFromFile 方法用于取得图片文件。下面的程序演示如何使用图片元素对象：

```
private void axPageLayoutControl1_OnMouseDown(object sender,
ESRI.ArcGIS.PageLayoutControl.IPageLayoutControlEvents_OnMouseDownEvent e)
{
    IPageLayout pPageLayout;
    pPageLayout = axPageLayoutControl1.PageLayout;
    IGraphicsContainer pGraphicsContainer;
    pGraphicsContainer = pPageLayout as IGraphicsContainer;
    IActiveView pActiveView;
    pActiveView = pPageLayout as IActiveView;
    IPictureElement pBmpPicEle;
    pBmpPicEle = new BmpPictureElementClass();
    pBmpPicEle.ImportPictureFromFile(@"C:\My Documents\sim.bmp");
    pBmpPicEle.MaintainAspectRatio = true;
    IEnvelope pEnv;
```

```
            pEnv = axPageLayoutControl1.TrackRectangle();
            IElement pEle;
            pEle = pBmpPicEle as IElement;
            pEle.Geometry = pEnv;
            pGraphicsContainer.AddElement(pEle, 0);
            pActiveView.PartialRefresh(esriViewDrawPhase.esriViewGraphics, null, null);
}
```

4.6.2 框架元素

框架元素主要有两个对象：MapFrame（地图框架）和 MapSurroundFrame（地图修饰框架）。MapFrame 对象是 Map 的容器，用于管理 Map 对象；MapSurroundFrame 对象用于管理 MapSurround 对象，该对象包括比例尺、比例文本、指北针等。每个 MapSurroundFrame 都与一个 MapFrame 相联系，如果一个 MapFrame 被删除，与其相联系的 MapSurroundFrame 对象都会被删除。框架元素是一种包含其他地图元素的容器，所有的框架元素都实现 IFrameElement 接口。

4.6.2.1 MapFrame 对象

MapFrame 对象是供 PageLayout 对象使用，用于管理 Map 对象。MapFrame 实现了 IMapFrame 接口，该接口提供了属性和方法对 Map 对象进行控制，如 Map 属性用于获得这个地图框架内的地图对象，MapBounds 属性用于返回地图对象的范围，MapScale 属性用来确定地图显示的比例，CreateSurroundFrame 方法用于返回一个 MapSurroundFrame 对象。同时，MapFrame 对象还实现了 IMapGrids 接口。

4.6.2.2 MapSurroundFrame 对象

MapSurroundFrame 是一种用于管理 MapSurround 对象的框架元素。MapSurround 会自动与某个地图对象关联，随着地图视图的变化而变化。MapSurroundFrame 实现了 IMapSurroundFrame 接口。

4.7 MapGrid 对象模型

Arc Engine 中提供了 MapGrids 对象，是用于辅助显示地图的地图网格，在地图的边缘显示经纬度。地图网格由 GridLine（网格线）、GridLabel（网格标注）和 GridBorder（网格边框）等 3 部分组成，如图 4-2 所示。

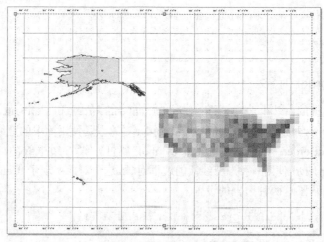

▲图 4-2 MapGrid 对象模型

4.7.1 MapGrid 对象

MapGrid 对象是布局视图中的一种参考点或线，辅助用户在地图中进行要素的定位。MapGrid 类是抽象类，泛化出 5 个子类：CustomOverlayGrid、Graticule、IndexGrid、MeasuredGrid 和 MgrsGrid。其中有 4 个类型的 MapGrid 继承自 IMapGrid 接口，分别是 IMeasuredGrid、IGraticule、IndexGrid 和 ICustomGridOverlay。

IMapGrid 是所有类型的地图网格类都实现的接口，用于设置 MapGrid 对象的一般属性和方法，如 Border 属性用于设置地图网格的边框，LabelFormat 属性用于设置地图网格上的标签格式，Linesymbol 用于设置网格的样式。此外还定义了多个设置 tick 对象的属性，所谓 tick 就是小记号点，如网线之间的交点、网线与边的交点等，如 QuerySubTickVisibility、QueryTickVisibility、SubTickCount 等。

CustomOverlayGrid 是用户定义的网格对象。

Graticule 是使用经纬度来划分地图的地图网格对象，它实现了两个接口：IGraticule 和 IMeasuredGrid。由于 Graticule 对象是使用经纬网，因而需要设置空间参考属性。

MeasuredGrid 也是使用经纬度作为地图网格来划分地图的，和 Graticule 对象的不同之处在于它的空间参考属性可以与 MapFrame 对象一致，也可以不一致。该对象实现了 IMeasuredGrid 和 IProjectedGrid 接口。

IndexGrid 是使用索引值的方式来划分地图的区域的对象，适合小区域内地块的划分。IndexGrid 类实现 IIndexGrid 接口，提供了 ColumnCount、QueryCellExtent、RowCount、XLabel 和 Ylabel 等，XLabel、YLabel 属性用于设置网格 X、Y 轴上的标签，ColumnCount、RowCount 属性分别用于设置网格划分的列、行的数目，QueryCellExtent 用于通过给定的行和列查找对应的网格。

下面的程序演示如何使用索引网格对象：

```
private IIndexGrid CreateIndexGrid()
{
    IIndexGrid pIndexGrid;
    pIndexGrid = new IndexGridClass();
    pIndexGrid.ColumnCount = 5;
    pIndexGrid.RowCount = 5;
    int i;
    for (i = 0; i <= pIndexGrid.ColumnCount - 1; i++)
    {
        pIndexGrid.set_XLabel(i, (i + 1).ToString());
    }
    for (i = 0; i <= pIndexGrid.RowCount - 1; i++)
    {
        pIndexGrid.set_YLabel(i, i.ToString() + "A");
    }
    return pIndexGrid;
}
```

4.7.2 MapGridBorder 对象

地图网格是有边框的，边框分为两类：SimpleMapGridBorder 和 CalibratedMapGridBorder。边框都实现了 IMapGridBorder 接口，SimpleMapGridBorder 对象只是使用简单的直线来作为地图的边框，通过 ISimpleMapGridBorder 接口提供的 LineSymbol 属性，设置边框线的样式、宽度和颜色。CalibratedMapGridBorder 是使用一种渐变线段的边框对象，这个对象实现了 ICalibratedMapGridBorder 接口。下面的代码演示了这两种边框的设置。

```
private ISimpleMapGridBorder CreateSimpleMapGridBorder()
{
    ISimpleMapGridBorder pSimpleMapGridBorder;
    pSimpleMapGridBorder = new SimpleMapGridBorderClass();
```

```
ISimpleLineSymbol pLineSymbol;
pLineSymbol = new SimpleLineSymbolClass();
pLineSymbol.Style = esriSimpleLineStyle.esriSLSSolid;
pLineSymbol.Color = getRGB(0, 0, 0);
pLineSymbol.Width = 2;
pSimpleMapGridBorder.LineSymbol = pLineSymbol;
return pSimpleMapGridBorder;
}
```

4.8 MapSurround 对象

MapSurround 是一个修饰地图的辅助图形元素对象，该对象会根据 Map 对象的变化而做出反应。如地图视图范围改变后，ScaleBar 对象会自动调整比例尺，ScaleBarText 也会相应地改变它的比例值。

4.8.1 图例对象

图例（Legend）是一个与 Map 对象中图层的着色操作（Renderer）相关的对象，通过着色对象在地图上产生专题图，每个着色对象可以拥有多个 LegendGroup（图例组），LegendGroup 都有一个或多个 LegendClass（着色类），LegendClass 是一个使用自身符号和标签制作的图例分类。

Legend 类主要实现 ILegend 接口，提供了修改属性和获取它的组成对象的属性和方法，如 AddItem 用于添加图例；ClearItems 用于清除图例；AutoAdd 为 true 时，Map 对象加入新图层后，与 Map 关联的图例对象会响应 Map 图层变化，增加一个条目；AutoVisibility 属性用于保持图例条目的顺序与 Map 中图层的顺序一致。

Legend 还继承了 IReadingDirection 接口，提供了一个属性 RightToLeft，该属性的作用是将图例中的描述字符从 Patch 的右边移到左边（Patch 是图例前面用于显示着色符号的小图片）。

可以将 Legend 看做 Map 中各个图层图例项的集合，通过 ILegendItem 接口提供的方法和属性来控制这些图例项。LegendItem（图例项）对象是一个抽象类，泛化出 4 个子类：HorizontalBarLegendItem、HorizontalLegendItem、NestedLegendItem 和 VerticalLegendItem。LegendItem 默认实现了 ILegendItem 接口，定义了 LegendItem 的一般属性，如是否显示标题、标签、高度、宽度等。泛化出的 4 个子类均有各自的接口，都提供了定制不同类型图例条目的属性和方法。

LegendClassFormat 对象用于控制单个 LegendItem 的外观，如 DescriptionSymbol、Patch 和 LinePatch 等，LegendFormat 对象用于控制一个 Legend 的属性。Patch 可以通过 IPatch 接口定义的属性和方法来创建。下面的代码是新建 patch 的例子。

```
private void CreatePatch(IGeometry pGeo)
{
    IPageLayout pPageLayout;
    pPageLayout = axPageLayoutControl1.PageLayout;
    IGraphicsContainer pGraphicsContainer;
    pGraphicsContainer = pPageLayout as IGraphicsContainer;
    IActiveView pActiveView;
    pActiveView = pPageLayout as IActiveView;
    IMap pMap;
    pMap = pActiveView.FocusMap;
    ILegend pLegend;
    pLegend = pMap.get_MapSurround(0) as ILegend;
    ILegendItem pLegendItem;
    pLegendItem = pLegend.get_Item(0);
    ILegendFormat pLegendFormat;
    pLegendFormat = pLegend.Format;
```

```
    IPatch pPatch;
    pPatch = new AreaPatchClass();
    pPatch.Geometry = pGeo;
    pLegendFormat.DefaultAreaPatch = pPatch as IAreaPatch;
    pLegend.Refresh();
    pActiveView.PartialRefresh(esriViewDrawPhase.esriViewGraphics, null, null);
}
```

4.8.2 指北针对象

MarkerNorthArrow（指北针）是用于指示地图空间方位的图形，该对象继承自 NorthArrow 抽象类，是一个 MapSurround 对象。该对象主要实现了两个接口：IMarkerNorthArrow 和 INorthArrow，INorthArrow 接口用于设置指北针对象的一般属性，如颜色、尺寸、位置等，IMarkerNorthArrow 接口定义了一个属性 MarkerSymbol，用于设置指北针的符号。

4.8.3 比例尺对象

地图上的图形与现实空间地物之间存在着一定的比例关系，通过比例尺来说明地图上的单位长度，代表现实世界的实际长度。ScaleBar（比例尺）对象也是一种 MapSurround，该类泛化出多个子类，如 ScaleLine、AlternatingScaleBar、DoubleAlternatingScaleBar、HollowScaleBar、SingleDivisionScaleBar 和 SteppedScaleLine 等。这些类都实现了 IScaleBar 和 IScaleMarks 接口。IScaleBar 管理比例尺的颜色、高度、比例尺对象上 Label 属性等，如 LabelSymbol 用于设置比例尺中的标识字符符号，LabelPosition 用于设置位置。IScaleMarks 接口用于管理比例尺相关的单个标记（mark）的属性，如高度、符号和位置等，如图 4-3 所示。

SingleDivisionScaleBar 类实现了 ISingleFillScaleBar 接口，如图 4-4 所示。

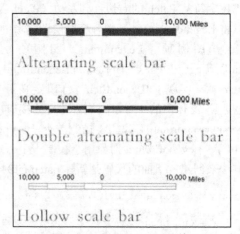

▲图 4-3　比例尺对象　　　　　　▲图 4-4　IsingleFiuscoleBar 接口

AlternatingScaleBar、DoubleAlternatingScaleBar、HollowScaleBar 等类实现了 IDoubleFillScaleBar 接口，提供了两个属性：FillSymbol1 和 FillSymbol2，这是比例尺的两种符号，如图 4-5 所示。

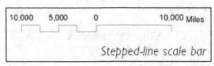

▲图 4-5　比例尺的两种符号

ScaleLine 和 SteppedScaleLine 类实现了 IScaleLine 接口。下面的代码演示了如何添加一个比例尺对象。

```csharp
private void AddSacleBar(IEnvelope pEnv, int strBarType)
{
    IScaleBar pScaleBar;
    IMapFrame pMapFrame;
    IMapSurroundFrame pMapSurroundFrame;
    IMapSurround pMapSurround;
    IElementProperties pElementPro;
    UID pUID = new UIDClass();
    pUID.Value = " esriCarto.scalebar";
    IPageLayout pPageLayout;
    pPageLayout = axPageLayoutControl1.PageLayout;
    IGraphicsContainer pGraphicsContainer;
    pGraphicsContainer = pPageLayout as IGraphicsContainer;
    IActiveView pActiveView;
    pActiveView = pGraphicsContainer as IActiveView;
    IMap pMap;
    pMap = pActiveView.FocusMap;
    pMapFrame = pGraphicsContainer.FindFrame(pMap) as IMapFrame;
    pMapSurroundFrame = pMapFrame.CreateSurroundFrame(pUID, null);
    switch (strBarType)
    {
        case 0:
            pScaleBar = new AlternatingScaleBarClass();
            break;
        case 1:
            pScaleBar = new DoubleAlternatingScaleBarClass();
            break;
        case 2:
            pScaleBar = new HollowScaleBarClass();
            break;
        case 3:
            pScaleBar = new ScaleLineClass();
            break;
        case 4:
            pScaleBar = new SingleDivisionScaleBarClass();
            break;
        case 5:
            pScaleBar = new SteppedScaleLineClass();
            break;
        default:
            pScaleBar = new ScaleLineClass();
            break;
    }
    pScaleBar.Division = 4;
    pScaleBar.Divisions = 4;
    pScaleBar.LabelGap = 4;
    pScaleBar.LabelPosition = esriVertPosEnum.esriAbove;
    pScaleBar.Map = pMap;
    pScaleBar.Name = "myScaleBar";
    pScaleBar.Subdivisions = 2;
    pScaleBar.UnitLabel = "";
    pScaleBar.UnitLabelGap = 4;
    pScaleBar.UnitLabelPosition = esriScaleBarPos.esriScaleBarAbove;
    pScaleBar.Units = esriUnits.esriKilometers;
    pMapSurround = pScaleBar;
    pMapSurroundFrame.MapSurround = pMapSurround;
    pElementPro = pMapSurroundFrame as IElementProperties;
    pElementPro.Name = "my scalebar";
    axPageLayoutControl1.AddElement(pMapSurroundFrame as IElement,
```

```
    pEnv, Type.Missing, Type.Missing, 0);
    pActiveView.PartialRefresh(esriViewDrawPhase.esriViewGraphics, null, null);
}
```

4.8.4 比例文本对象

上一小节介绍了比例尺对象，Arc Engine 还提供了 ScaleText（比例尺文本）对象，用于显示明确的比例值。ScaleText 是一个文本元素，显示的比例值随地图比例尺的变化而变化。该类实现了 IScaleText 接口，定义了文本的属性和方法，如 Symbol、style、text 等，其中 text 属性用来获取比例文本的字符。

4.9 Style 对象

上一节介绍的指北针、比例尺等统称为样式，一个样式是由多个样式条目（Style Gallery Item）组成，这些样式条目提供了得到单个地图元素或符号的方法，相似的条目被组织成样式类（Style Gallery Class）。如 Marker Symbols 就是一个样式类，右边的符号和元素是这个样式类的条目，样式类中的条目可以依据类型进行分组（Categories），如 Marker Symbols 样式类可分为 Asterisk 和 Circle 等多个组，如图 4-6 所示。

▲图 4-6　Style 对象

StyleGallery 是一个与文档对象相关的 Style 的集合对象，该类实现了 IStyleGallery 接口，提供了操作样式类、样式条目的方法和属性。如 Class 属性通过指定的索引值，获得 Style 文件中的某个 StyleGalleryClass 对象；ClassCount 属性用于返回样式类数目；Categories 通过指定的类名返回样式类；Items 用于获取样式条目集对象。StyleGallery 还实现了 IStyleGalleryStorage 接口，提供了 Stylegallery 对象中获取一个 Style 文件的方法，如 DefaultStylePath 用于返回 Style 文件的缺省目录，TargerFile 属性用于新建一个 Style 文件，CanUpdate 属性允许更改一个 Style 文件。

StyleGalleryItem 对象表示一个具体的样式条目，包含一个地图元素和符号，以及一些相关的信息。该类实现了 IStyleGalleryItem 接口，如 Category 属性用于确定条目在样式类中的类别，Item 属性用于获取样式条目中的符号或元素。下面的代码是一个获得样式条目的例子。

```
private void StyleGalleryDemo()
{
    IStyleGallery pStyleGallery;
    pStyleGallery = new StyleGalleryClass();
```

```csharp
            pStyleGallery.LoadStyle(@"C:\Program Files\ArcGIS\Styles\esri.style", " ");
            for (int i = 0; i <= pStyleGallery.ClassCount - 1; i++)
            {
                MessageBox.Show(pStyleGallery.get_Class(i).Name);
            }
            IEnumStyleGalleryItem pEnumStyleGalleryItem;
            IStyleGalleryItem pStyleGalleryItem;
            pEnumStyleGalleryItem = pStyleGallery.get_Items("Scale Bars", @"C:\Program Files\ArcGIS\Styles\esri.style", "hollowscalebar");
            pEnumStyleGalleryItem.Reset();
            pStyleGalleryItem = pEnumStyleGalleryItem.Next();
            while (pStyleGalleryItem != null)
            {
                MessageBox.Show(pStyleGalleryItem.Name);
                pStyleGalleryItem = pEnumStyleGalleryItem.Next();
            }
        }
        private void AddStyleItem()
        {
            IStyleGallery pStyleGallery;
            pStyleGallery = new StyleGalleryClass();
            IRgbColor pRgbColor;
            pRgbColor = new RgbColorClass();
            pRgbColor.Red = 255;
            pRgbColor.Green = 0;
            pRgbColor.Blue = 0;
            IStyleGalleryItem pStyleItem;
            pStyleItem = new StyleGalleryItemClass();
            pStyleItem.Name = " Red";
            pStyleItem.Category = " Defalut";
            pStyleItem.Item = pRgbColor;
            IStyleGalleryStorage pStyleStorage;
            pStyleStorage = pStyleGallery as IStyleGalleryStorage;
            pStyleStorage.TargetFile = @"D:\temp\test.style";
            pStyleGallery.AddItem(pStyleItem);
        }
```

4.10 添加、删除图层数据

ArcEngine 支持多种文件类型，矢量数据如 ArcInfo Coverage 、ESRI Shapefile，以及 AutoCAD 的 DXF、DWG 文件等，栅格数据如 BMP、GRID 等。

4.10.1 矢量数据的添加

矢量数据通常是 SHP 文件，添加此类文件一般需要用到 Map、Dataset、FeatureLayer 和 FeatureClass 等对象。这些对象一般用到的主要接口有 IMap、IActiveView、IDataset、IfeatureLayer 和 IFeatureClass 等。

Map 组件类主要是地图图层数据显示和操作的一个平台、一个 Map 对象，可以包含各种图层。该组件类主要实现 IMap 接口，主要用于控制地图数据和相关元素，通过该接口可以添加、删除图层，访问各种数据源和 Map 的各种特性，以及通过各种方式选择要素。如 ActiveGraphicsLayer 属性用于获取 Map 中处于激活状态的图层，Layer 属性用于获取索引指定的图层，LayerCount 用于获取 Map 中的图层数量，AddLayer 方法用于向 Map 中添加图层，ClearSelection 方法用于清除所有选中的对象的选中状态。

Dataset 抽象类表示数据集，是 Workspace 中数据的集合，一个 Workspace 可以包含一个或者多个 Dataset，一个 Dataset 可以包含其他的 Dataset。该抽象类不能创建对象，可以通过该类的派生类：GeoDataset、Table、FeatureDataset、RasterDataset 等来创建具体类型的数据集。该抽象类主要实现了 IDataset 接口，主要用来管理数据集和提供数据集本身的相关属性信息，该接口同时也被 Workspace 类、FeatureLayer 组件类实现。该接口提供了对数据集操作的属性和方法，如 Name 属性用于获取数据集的名称，Category 用于获取数据集的类型，Workspace 用于获取数据集所在的工作空间，CanCopy 方法用于检测数据集是否可以被复制，CanDelete 用于检测数据集是否可以被删除，CanRename 用于检测数据集是否可以被重命名。

FeatureLayer 组件类是要素图层，是要素 Feature 的集合及其可视化表达，该类主要实现了 ILayer、IFeatureLayer 接口。

ILayer 接口提供了方法和属性，来决定图层的范围、最大和最小显示比例尺、空间参考、图层名以及显示方案等，如 Name 属性用来设置或获取图层名，SpatialReference 用来设置空间参考，MaximumScale 用来设置最大比例尺。

IFeatureLayer 接口是从 ILayer 接口继承下来的，该接口继承了 ILayer 接口所有的属性和方法，同时还提供了一些自己的属性和方法，比如数据源的类型、显示的字段、符号是否随比例尺变化，以及搜索等功能。如 DataSourceType 用于获取或设置数据源的类型；FeatureClass 用于获取图层的要素类；ScaleSymbols 用于获取或设置要素层中的符号是否随比例尺的变化而变化；Search 方法是该接口一个很重要也很常用的方法，该方法通过指定的查询过滤器在要素层中搜索符合给定查询条件的要素。

FeatureClass 要素类是 Feature 的集合，该要素类中所有的要素具有相同的属性字段，另外还有一个 shape 字段专门存放几何图形。该类主要实现 IFeatureClass 接口，提供了一些属性和方法，用于获取和设置要素类的属性，如 CreateFeature 用于创建一个新的实体，DeleteField 用于删除要素类的字段。

4.10.2 栅格数据的添加

栅格数据是 GIS 中重要的数据源之一，如卫星图像，扫描的地图、照片等。栅格数据常见有以下几种格式：BMP、TIFF、JPG、GRID 等。添加栅格数据主要使用 RasterLayer 组件类，以及 IMap、ILayer、IRasterLayer 接口等。RasterLayer 组件类实现了 ILayer、IRasterLayer、ITable、IDataset 等接口，主要用于栅格数据的打开和显示状态的设置等。

IRasterLayer 接口继承自 ILayer 接口，和 IFeatureLayer 一样，该接口还提供了一些自己的方法和属性，如 CreateFromRaster 方法用于从内存中已有的一个 Raster 对象创建一个图层，CreateFromDataset 方法用于从硬盘中的某个数据集中创建一个图层，Raster 用于获取 IRasterLayer 中的 Raster 对象，DisplayResolutionFactor 用于设置栅格数据的分辨率。

4.10.3 删除图层数据

删除图层即将图层对象从内存中删除，使 Map 对象中不再包含该图层。可以通过 IMap 接口提供的 DeleteLayer 方法来删除图层，如果需要删除全部图层，则可使用该接口的 ClearLayers 方法。

4.11 图层控制

地图是由地图图层通过一定的叠加顺序叠加起来的，每个图层表示一种类型的要素，这些图层一般分为点、线、面等，通过对图层的控制来展示多样的地图。

4.11.1 图层间关系的调整

所谓图层间关系的调整，即是通过指定或移动图层之间的叠加顺序，以使最关心的内容不会被其他图层遮盖。要调整图层之间的叠加顺序，可以使用 IMap 接口的 MoveLayer 函数实现。

4.11.2 图层显示状态的控制

图层显示状态是指图层是否可见。通过设置某些图层的可见性，一方面可以简化显示的内容，易于阅读；另一方面也能提高显示刷新的效率。ILayer 接口提供了 Visible 属性用于设置图层的可见性。

4.12 本章小结

本章是本书最重要的章节之一，用了大量篇幅介绍地图的各种对象以及对象的主要接口、各种类型图层对象、显示对象、地图排版、地图网格、图层样式，以及图层的添加、删除操作。

地图视图中的图形包括两种，一种是实体要素，另一种是图形要素，实体要素保存在空间数据库中，而图形要素则保存在文档文件中，作为实体要素展现的补充，视图中的这些图形都由相应的容器对象来装载，通过这些装载容器来管理这些图形要素。

本章涉及的接口非常多，这些接口也是 ArcGISEngine 开发的基础，只有熟练掌握这些接口的使用才能开发出基于 ArcGISEngine 的应用程序，本书的下一章节地图制图也是本书的重点章节，通过下一章节介绍制图内容，我们就可以控制地图的美观。

第 5 章 地图制图

5.1 地图标注

地图标注是地图的重要特性,是表示制图对象的名称或数量及质量特征的文字和数字等文字语言。用于说明制图对象的名称、种类、性质和数量等具体特征,不仅可以弥补地图符号的不足,丰富地图的内容,而且在某种程度上还可以起到符号的作用。使地图具有可读性、可翻译性以及成为一种信息传输工具,主要用来标识各种对象、指示对象的属性、表明对象间的关系和转译等。

Arc Engine 中的标注分为两种,一种是标注(Label),另一种是注记(Annotation)。Label 方式比较简单,如 ArcMap 中通过设置图层的 Label 属性,图层就以 Label 属性中设置的字段进行标注;Annotation 以更复杂的方法和属性对要素图层进行注记,这个过程是自动进行的,不需要用户干预,而且注记的内容还可以保存在地图数据库中。

AnnotationLayerPropertiesCollection 对象是一个要素图层的属性,是一个标注集对象的集合。标注集是与某个要素图层相关联的,用于描述要素图层如何被标注,可以通过 IGeoFeaturelayer 的 AnnotationProperties 属性获取。IAnnotationLayerPropertiesCollection 接口提供了对保存在集合中的 IannotationLayerProperties(LabelEngineLayerProperties、MaplexLabelEngineLayerProperties)对象进行操作,通过该接口,开发者可以对集合中的组件进行添加、删除、排序和查询等操作。LabelEngineLayerProperties 对象维持着一个要素图层的标注实例。IAnnotationLayerProperties 的 WhereClause 属性用于设置一个 SQL 语句,用于确定哪些要素可以被标注;AnnotationMaximunScale 和 AnnotationMinMunScale 用于设置文字标注的最大和最小范围。

LabelEngineLayerProperties 对象也实现了 ILabelEngineLayerProperties 接口,提供了用于控制标注过程中的主要属性,设置文字符号、标注文字排放等,如 BasicOverposterLayerProperties 属性用于设置标注文本如何被放置,以及处理文字之间的冲突;Expression 用于输入 VBScript 或 JavaScript 脚本。BasicOverposterLayerProperties 对象还实现了 IBasicOverposterLayerProperties 接口,提供了一些重要的属性,如 LineLabelPlacementPriorities 用于设置标注文本的摆设路径的权重,LineLabelPosition 用于设置标注文本的排放位置,PointPlacementPriorities 用于设置与一个点相关的标注路径的权重等。下面的程序演示如何使用注记:

```
private void Annotation(IGeoFeatureLayer pGeoFeatLyr, string annoField)
{
    IGeoFeatureLayer pGeoFeatLayer;
    pGeoFeatLayer = pGeoFeatLyr;
    IAnnotateLayerPropertiesCollection pAnnoProps;
    pAnnoProps = pGeoFeatLyr.AnnotationProperties;
    pAnnoProps.Clear();
    IAnnotateLayerProperties pAnnoLayerProps;
    ILineLabelPosition pPosition ;
```

```
    ILineLabelPlacementPriorities pPlacement;
    IBasicOverposterLayerProperties pBasic;
    ILabelEngineLayerProperties pLabelEngine ;
    ITextSymbol pTextSyl;
    pTextSyl = new TextSymbolClass();
    stdole.StdFont pFont;
    pFont = new stdole.StdFontClass() ;
    pFont.Name = "verdana";
    pFont.Size = 10;
    pTextSyl.Font = pFont as stdole.IFontDisp;
    pTextSyl.Color = HSVColor(250, 160, 200);
    pPosition = new LineLabelPositionClass();
    pPosition.Parallel = false;
    pPosition.Perpendicular = true;
    pPlacement = new LineLabelPlacementPrioritiesClass();
    pBasic = new BasicOverposterLayerPropertiesClass();
    pBasic.FeatureType =
    esriBasicOverposterFeatureType.esriOverposterPolyline;
    pBasic.LineLabelPlacementPriorities = pPlacement;
    pBasic.LineLabelPosition = pPosition;
    pLabelEngine = new LabelEngineLayerPropertiesClass();
    pLabelEngine.Symbol = pTextSyl;
    pLabelEngine.BasicOverposterLayerProperties = pBasic;
    pLabelEngine.Expression = annoField;
    pAnnoLayerProps = pLabelEngine as IAnnotateLayerProperties;
    pAnnoProps.Add(pAnnoLayerProps);
    pGeoFeatLyr.DisplayAnnotation = true;
    axMapControl1.CtlRefresh(esriViewDrawPhase.esriViewBackground,null,null);
}
```

5.2 符号及符号库

地图符号是表达空间数据的基本手段，是地图的语言单位，是可视化表达地理信息内容的基础工具，不仅能表示事物的空间位置、形状、质量和数量等特征，而且可以表示各事物之间的相互联系及区域总体特征。地图符号由形状不同、大小不一、色彩有别的图形和文字组成，既是地图的语言，也是一种图形语言。地图符号不仅具有确定的客观事物空间位置、分布特点，以及质量和数量特征的基本功能，而且具有相互联系和共同表达地理环境各要素总体的特殊功能。

Arc Engine 提供了丰富的符号组件来满足地图设计中的各种需求，这些组件包括 Renderer、Color、Symbol 3 大系列，地图符号化就是通过多个组件间的合作来完成的。

5.2.1 颜色对象

颜色（Color）是现实世界中最普遍的事物属性，人们使用不同的颜色模型来模拟现实中的颜色，例如最常见的 RGB 颜色模型，是通过红色、绿色和蓝色的不同值调配出来的，这些颜色的数值介于 0 和 255 之间。此外还有其他的颜色模型，如 CMYK 颜色模型主要用于印刷；HSV 颜色模型是由色调（Hue）、饱和度（Saturation）和（Value）值组成的，这个模型基于一个限制范围的颜色轮带；此外还有 GrayColor（灰度颜色）、CIELAB 颜色等。

RGB 模型：包含了可见光谱中的大部分颜色，通过设置 Red、Blue 和 Green 3 个属性值的比例和强度混合而成，如图 5-1 所示。

CMYK 模型：该模型以打印在纸上的油墨的光线吸收特性为基础，当白光照射到半透明的油墨上时，某些可见光波长被吸收，而其他的波长则被反射回眼睛。纯青色（C）、洋红色（M）和黄色（Y）色素在合成后可以吸收所有的光线并产生黑色，但由于油墨含有杂质，因此实际上这 3 种色素生成土灰色，必须与黑色（K）油墨合成后才能生成真正的黑色。这些油墨混合重现颜色的过程也称为四色印刷，如图 5-2 所示。

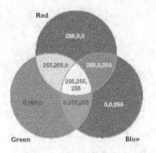
▲图 5-1 RGB 模型

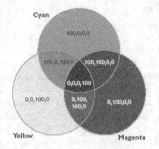
▲图 5-2 CMYK 模型

HSV 模型：该模型对应于圆柱坐标系的一个圆锥形子集，圆锥的顶面对应于 V=1，代表的是颜色较亮，色彩 H 由绕着 V 轴的旋转角给定，红色对应于 0°，绿色对应于 120°，蓝色对应于 240°。饱和度 S 取值在 0 与 1 之间，由圆心向圆周过渡，如图 5-3 所示。

Cray 模型：灰度模型是没有彩色的。灰度图像由 8 位信息组成，并使用 256 级的灰色来模拟颜色的层次，灰度图像的每个像素有一个 0（黑色）到 255（白色）之间的亮度值，如图 5-4 所示。

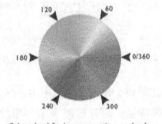
▲图 5-3 HSV 模型

▲图 5-4 Cray 模型

5.2.1.1 Color 对象

Color 对象是一个抽象类，该类泛化为 5 个类：CmykColor、GrayColor、HlsColor、HsvColor 和 RgbColor，这 5 个子类分别实现了接口 ICmykColor、IGrayColor、IHlsColor、IHsvColor 和 IrgbColor。该对象实现了 IColor 接口，提供了一些方法和属性，如 CMYK 、GetCIELAB、NullColor、RGB、SetCIELAB、Transparency 和 UseWindowsDithering 等。Arc Engine 中最常用的是 RGB 和 HSV 两种颜色模型。

下面的代码演示了如何使用颜色对象。

```
private IHsvColor HSVColor(int hue , int saturation , int val )
{
    IHsvColor pHsvColor;
    pHsvColor = new HsvColorClass();
    pHsvColor.Hue = hue;
    pHsvColor.Saturation = saturation;
    pHsvColor.Value = val;
    return pHsvColor;
}
    private IRgbColor getRGB(int r,int g,int b)
```

```
{
    IRgbColor pColor;
    pColor = new RgbColorClass();
    pColor.Red = r;
    pColor.Green = g;
    pColor.Blue = b;
    return pColor;
}
```

5.2.1.2　ColorRamp 对象

在制作专题图的过程中，需要的颜色常常不是一种，而是随机或有序产生的一个颜色带。ColorRamp 类的对象用于产生颜色带，这个类实现了 IColorRamp 接口，定义了一系列的颜色带公共方法，如 Size 用于指定产生多少种颜色，Colors 指示颜色带，是个实现 IEnumColor 接口的类对象，共有 4 个类继承了该接口，分别是：AlgorithmicColorRamp、MultiPartColorRamp、PresetColorRamp、RandomColorRamp。

AlgorithmicColorRamp 颜色带是通过指定起始颜色、终止颜色确定一个颜色带，起始、终止颜色使用 HSV 模型。该类还实现了两个接口：IColorRamp 和 IAlgorithmicColorRamp，后一接口继承了前一接口，因此一般使用 IAlgorithmicColorRamp 接口。

MultiPartColorRamp 是叠加产生颜色带，RandomColorRamp 产生随机颜色带，PresetColorRamp 使用 HSV 颜色模型来产生一串颜色预设的颜色模式，可存储 13 种颜色。下面的代码演示了如何创建这些颜色带。

```
private IColorRamp CreateAlgorithmicColorRamp()
{
    //创建一个新AlgorithmicColorRampClass对象
    IAlgorithmicColorRamp algColorRamp = new AlgorithmicColorRampClass();
    IRgbColor fromColor = new RgbColorClass();
    IRgbColor toColor = new RgbColorClass();
    //创建起始颜色对象
    fromColor.Red = 255;
    fromColor.Green = 0;
    fromColor.Blue = 0;
    //创建终止颜色对象
    toColor.Red = 0;
    toColor.Green = 255;
    toColor.Blue = 0;
    //设置AlgorithmicColorRampClass的起止颜色属性
    algColorRamp.ToColor = fromColor;
    algColorRamp.FromColor = toColor;
    //设置梯度类型
    algColorRamp.Algorithm = esriColorRampAlgorithm.esriCIELabAlgorithm;
    //设置颜色带颜色数量
    algColorRamp.Size = 10;
    //创建颜色带
    bool bture = true;
    algColorRamp.CreateRamp(out bture);
    return algColorRamp;

}
private IColorRamp CreateMultiPartColorRamp()
{
    IMultiPartColorRamp pMultiPartColorRamp = new MultiPartColorRampClass();
    bool bture=true ;
    //叠加颜色带
    pMultiPartColorRamp.AddRamp(CreateAlgorithmicColorRamp());
```

```csharp
        pMultiPartColorRamp.AddRamp(CreateRandomColorRamp());
    //设置颜色带颜色数量
    pMultiPartColorRamp.Size = 10;
    pMultiPartColorRamp.CreateRamp (out bture);
    return pMultiPartColorRamp;
}
private IColorRamp CreateRandomColorRamp()
{
    IRandomColorRamp pRandomColorRamp = new RandomColorRampClass();
    // 制作一系列介于橘黄色和蓝绿色之间的随机颜色
    pRandomColorRamp.StartHue = 40;
    pRandomColorRamp.EndHue = 120;
    pRandomColorRamp.MinValue = 65;
    pRandomColorRamp.MaxValue = 90;
    pRandomColorRamp.MinSaturation = 25;
    pRandomColorRamp.MaxSaturation = 45;
    pRandomColorRamp.Size = 20;
    pRandomColorRamp.Seed = 23;
    bool bture = true;
    pRandomColorRamp.CreateRamp(out bture);
    return pRandomColorRamp;
}
private IColorRamp CreatePresetColorRamp()
{
    IPresetColorRamp pPresetColorRamp = new PresetColorRampClass();
    IRgbColor rgbColor =new RgbColorClass ();
    //预置颜色值
    for (int i =0;i<10;i++)
    {
        rgbColor.Red =100+11*i;
        rgbColor.Green =100+5*i;
        rgbColor.Blue =0+8*i;
        pPresetColorRamp.set_PresetColor (i,rgbColor as IColor );
    }
    pPresetColorRamp.Size = 10;
    bool bture = true;
    pPresetColorRamp.CreateRamp(out bture);
    return pPresetColorRamp;
}

//查看AlgorithmicColorRamp 颜色带效果
private void button1_Click(object sender, EventArgs e)
{
    IEnumColors pEnumColors = null;
    IColor color;
    pEnumColors = CreateAlgorithmicColorRamp( ).Colors ;
    pEnumColors.Reset();
    this.pictureBox1.BackColor =
        ColorTranslator.FromOle(pEnumColors.Next().RGB);
    this.pictureBox2.BackColor =
        ColorTranslator.FromOle(pEnumColors.Next().RGB);
    this.pictureBox3.BackColor =
        ColorTranslator.FromOle(pEnumColors.Next().RGB);
    this.pictureBox4.BackColor =
        ColorTranslator.FromOle(pEnumColors.Next().RGB);
    this.pictureBox5.BackColor =
        ColorTranslator.FromOle(pEnumColors.Next().RGB);
    this.pictureBox6.BackColor =
        ColorTranslator.FromOle(pEnumColors.Next().RGB);
}
//查看MultiPartColorRamp 颜色带效果
private void button2_Click(object sender, EventArgs e)
```

```csharp
        {
            IEnumColors pEnumColors = null;
            IColor color;
            pEnumColors = CreateMultiPartColorRamp().Colors ;
            pEnumColors.Reset();
            this.pictureBox1.BackColor =
                ColorTranslator.FromOle(pEnumColors.Next().RGB);
            this.pictureBox2.BackColor =
                ColorTranslator.FromOle(pEnumColors.Next().RGB);
            this.pictureBox3.BackColor =
                ColorTranslator.FromOle(pEnumColors.Next().RGB);
            this.pictureBox4.BackColor =
                ColorTranslator.FromOle(pEnumColors.Next().RGB);
            this.pictureBox5.BackColor =
                ColorTranslator.FromOle(pEnumColors.Next().RGB);
            this.pictureBox6.BackColor =
                ColorTranslator.FromOle(pEnumColors.Next().RGB);
        }
        //查看RandomColorRamp颜色带效果
        private void button3_Click(object sender, EventArgs e)
        {
            IEnumColors pEnumColors = null;
            IColor color,fromColor,toColor;
            pEnumColors = CreateRandomColorRamp().Colors ;
            pEnumColors.Reset();
            this.pictureBox1.BackColor =
                ColorTranslator.FromOle(pEnumColors.Next().RGB);
            this.pictureBox2.BackColor =
                ColorTranslator.FromOle(pEnumColors.Next().RGB);
            this.pictureBox3.BackColor =
                ColorTranslator.FromOle(pEnumColors.Next().RGB);
            this.pictureBox4.BackColor =
                ColorTranslator.FromOle(pEnumColors.Next().RGB);
            this.pictureBox5.BackColor =
                ColorTranslator.FromOle(pEnumColors.Next().RGB);
            this.pictureBox6.BackColor =
                ColorTranslator.FromOle(pEnumColors.Next().RGB);
        }
        //查看PresetColorRamp颜色带效果
        private void button4_Click(object sender, EventArgs e)
        {
            IEnumColors pEnumColors = null;
            IColor color;
            pEnumColors = CreatePresetColorRamp().Colors ;
            pEnumColors.Reset();
            this.pictureBox1.BackColor =
                ColorTranslator.FromOle(pEnumColors.Next().RGB);
            this.pictureBox2.BackColor =
                ColorTranslator.FromOle(pEnumColors.Next().RGB);
            this.pictureBox3.BackColor =
                ColorTranslator.FromOle(pEnumColors.Next().RGB);
            this.pictureBox4.BackColor =
                ColorTranslator.FromOle(pEnumColors.Next().RGB);
            this.pictureBox5.BackColor =
                ColorTranslator.FromOle(pEnumColors.Next().RGB);
            this.pictureBox6.BackColor =
                ColorTranslator.FromOle(pEnumColors.Next().RGB);
        }
    }
}
```

5.2.2 Symbol 对象

地图用符号和标记来表示地理对象的某些描述信息，Symbol 就是用来在地图上修饰要素或者图形元素的对象。GIS 中的离散实体有 3 种：点、线、面实体，在 Arc Engine 中用 3 种符号来表示，分别是 MarkSymbol、LineSymbol 和 FillSymbol。此外还有 TextSymbol 用于文字标注，3D Chart 符号用来显示饼图等三维对象。

所有的符号都实现了 ISymbol 和 IMapLevel 接口，ISymbol 定义了一个符号对象的基本属性和方法，IMapLevel 定义 MapLevel 属性可以确定符号显示的层，和图层类似，用于确定符号的叠加顺序。

5.2.2.1 MarkerSymbol 对象

MarkerSymbol 对象是用于修饰点对象的符号，拥有 12 个子类：ArrowMarkerSymbol、BarChartSymbol、CharacterMarker3DSymbol、CharacterMarkerSymbol、Marker3DSymbol、MultiLayerMarkerSymbol、PictureMarkerSymbol、PieChartSymbol、SimpleMarker3DSymbol、SimpleMarkerSymbol、StackedChartSymbol 和 TextMarkerSymbol。所有的 MarkerSymbol 类都实现了 IMarkerSymbol 接口，定义了 Angle（角度）、Color（颜色）、Size（大小）、Xoffset（X 偏移量）和 Yoffset（Y 偏移量）。下面介绍常用的 5 种类型。

（1）SimpleMarkerSymbol：简单类型的点状符号。

SimpleMarkerSymbol 类实现了 IsimpleMarkerSymbol 和 IMarkerSymbol 接口，该类有 5 种类型的符号，如 esriSMSCircle、esriSMSSquare、esriSMSCross、esriSMSX 和 esriSMSDiamond。

（2）ArrowMarkerSymbol：箭头形式的符号。

ArrorMarkerSymbol 类的主要接口是 IarrorMarkerSymbol 和 IMarkerSymbol，该类的 Length 属性用于指定箭头的顶点到底边的距离，Width 用于指定箭头底边的宽度，Style 用于指定箭头的符号样式，该属性只有一个样式 esriAMSPlain。

（3）CharacterMarkerSymbol：字符形式的符号。

CharactermarkerSymbol 类实现了 ICharacterMarkerSymbol 接口，该接口继承自 ImarkerSymbol。CharacterMarkerSymbol 用于将一个点要素显示为字符状，而字符的字体来自系统中已经安装的字符集；而 Characterindex 属性用来确定显示的字符，字符来自 ASCII 码。

（4）PictureMarkerSymbol：以图片为背景的符号。

PictureMarkerSymbol 类实现了 IpictureMarkerSymbol 接口。IMarkerSymbol 对象用于把一个点对象的外形表示为一张位图图片，该对象设置点符号的方法有两种，一种是指定 picture 属性，另一种是使用 CreateMarkSymbolFromFile 方法，从硬盘上获取一张位图。

（5）MultiLayerMarkerSymbol：使用多个符号叠加生成新的符号。

MultiLayerMarkerSymbol 类提供了一种将多个符号进行叠加的方式，可生成全新的组合符号，组合的个数不受限制，具体实现如下：

```
//获取颜色对象
private IRgbColor getRGB(int r, int g, int b)
{
    IRgbColor pColor;
    pColor = new RgbColorClass();
    pColor.Red = r;
    pColor.Green = g;
    pColor.Blue = b;
    return pColor;
}
```

```csharp
//简单类型点状符号
private void button1_Click(object sender, EventArgs e)
{
    ISimpleMarkerSymbol iMarkerSymbol;
    ISymbol iSymbol;
    IRgbColor iRgbColor;
    iMarkerSymbol = new SimpleMarkerSymbol();
    iMarkerSymbol.Style = esriSimpleMarkerStyle.esriSMSCircle;
    iRgbColor = new RgbColor();
    iRgbColor = getRGB(100, 100, 100);
    iMarkerSymbol.Color = iRgbColor;
    iSymbol = (ISymbol)iMarkerSymbol;
    iSymbol.ROP2 = esriRasterOpCode.esriROPNotXOrPen;
    IPoint point1 = new PointClass();
    IPoint point2 = new PointClass();
    point1.PutCoords(5, 5);
    point2.PutCoords(5, 10);
    this.axMapControl1.FlashShape(point1 as IGeometry, 3, 200, iSymbol);
    this.axMapControl1.FlashShape(point2 );
}
//箭头形式符号
private void button2_Click(object sender, EventArgs e)
{
    IArrowMarkerSymbol arrowMarkerSymbol =new ArrowMarkerSymbolClass ();
    IRgbColor iRgbColor;
    iRgbColor = new RgbColor();
    iRgbColor = getRGB(100, 100, 100);
    arrowMarkerSymbol.Angle =90;
    arrowMarkerSymbol.Color =iRgbColor ;
    arrowMarkerSymbol.Length =30;
    arrowMarkerSymbol.Width = 20;
    arrowMarkerSymbol.XOffset = 0;
    arrowMarkerSymbol.YOffset = 0;
    arrowMarkerSymbol.Style = esriArrowMarkerStyle.esriAMSPlain;
    IPoint point1 = new PointClass();
    IPoint point2 = new PointClass();
    point1.PutCoords(5, 5);
    point2.PutCoords(5, 10);
    this.axMapControl1.FlashShape(point1 as IGeometry, 3, 500, arrowMarker Symbol);
    this.axMapControl1.FlashShape(point2 as IGeometry);
}
//字符形式符号
private void button3_Click(object sender, EventArgs e)
{
    ICharacterMarkerSymbol characterMarkerSymbol = new CharacterMarker Symbol();
    stdole.IFontDisp fontDisp = (stdole.IFontDisp)(new stdole.StdFontClass ());
    IRgbColor rgbColor = new RgbColor();
    rgbColor = getRGB(100, 100, 100);
    fontDisp.Name = "arial";
    fontDisp.Size = 12;
    fontDisp.Italic = true;
    characterMarkerSymbol.Angle = 0;
    characterMarkerSymbol.CharacterIndex = 97;
    characterMarkerSymbol.Color = rgbColor;
    characterMarkerSymbol.Font = fontDisp;
    characterMarkerSymbol.Size = 24;
    characterMarkerSymbol.XOffset = 0;
    characterMarkerSymbol.YOffset = 0;
    IPoint point1 = new PointClass();
    IPoint point2 = new PointClass();
    point1.PutCoords(5, 5);
    point2.PutCoords(5, 10);
```

```csharp
    this.axMapControl1.FlashShape(point1 as IGeometry, 3, 500, characterMarker Symbol);
    this.axMapControl1.FlashShape(point2 as IGeometry);
}
//图片形式符号
private void button4_Click(object sender, EventArgs e)
{
    IRgbColor rgbColor = new RgbColorClass();
    IPictureMarkerSymbol pictureMarkerSymbol = new PictureMarkerSymbolClass();
    string fileName = @"E:\qq.bmp";
    pictureMarkerSymbol.CreateMarkerSymbolFromFile(esriIPictureType.
    esriIPictureBitmap, fileName);
    pictureMarkerSymbol.Angle = 0;
    pictureMarkerSymbol.BitmapTransparencyColor = rgbColor;
    pictureMarkerSymbol.Size = 20;
    pictureMarkerSymbol.XOffset = 0;
    pictureMarkerSymbol.YOffset = 0;
    IPoint point1 = new PointClass();
    IPoint point2 = new PointClass();
    point1.PutCoords(5, 5);
    point2.PutCoords(5, 10);
    this.axMapControl1.FlashShape(point1 as IGeometry, 3, 200, pictureMarker Symbol);
    this.axMapControl1.FlashShape(point2 as IGeometry);
}
//多符号叠加
private void button5_Click(object sender, EventArgs e)
{
    IMultiLayerMarkerSymbol multiLayerMarkerSymbol = new MultiLayerMarker SymbolClass();
    IPictureMarkerSymbol pictureMarkerSymbol = new PictureMarkerSymbolClass();
    ICharacterMarkerSymbol characterMarkerSymbol = new CharacterMarker Symbol();
    stdole.IFontDisp fontDisp = (stdole.IFontDisp)(new stdole.StdFontClass());
    IRgbColor rgbColor = new RgbColor();
    rgbColor = getRGB(0, 0, 0);
    fontDisp.Name = "arial";
    fontDisp.Size = 12;
    fontDisp.Italic = true;
    //创建字符符号
    characterMarkerSymbol.Angle = 0;
    characterMarkerSymbol.CharacterIndex = 97;
    characterMarkerSymbol.Color = rgbColor;
    characterMarkerSymbol.Font = fontDisp;
    characterMarkerSymbol.Size = 24;
    characterMarkerSymbol.XOffset = 0;
    characterMarkerSymbol.YOffset = 0;
    //创建图片符号
    string fileName = @"E: \qq.bmp";
    pictureMarkerSymbol.CreateMarkerSymbolFromFile(esriIPictureType.
    esriIPictureBitmap, fileName);
    pictureMarkerSymbol.Angle = 0;
    pictureMarkerSymbol.BitmapTransparencyColor = rgbColor;
    pictureMarkerSymbol.Size = 30;
    pictureMarkerSymbol.XOffset = 0;
    pictureMarkerSymbol.YOffset = 0;
    //添加图片、字符符号到组合符号中
    multiLayerMarkerSymbol.AddLayer(pictureMarkerSymbol);
    multiLayerMarkerSymbol.AddLayer(characterMarkerSymbol);
    multiLayerMarkerSymbol.Angle = 0;
    multiLayerMarkerSymbol.Size = 30;
    multiLayerMarkerSymbol.XOffset = 0;
    multiLayerMarkerSymbol.YOffset = 0;
    IPoint point1 = new PointClass();
    IPoint point2 = new PointClass();
    point1.PutCoords(5, 5);
```

```csharp
            point2.PutCoords(5, 10);
            this.axMapControl1.FlashShape(point1 as IGeometry, 3, 200, multiLayerMarkerSymbol);
            this.axMapControl1.FlashShape(point2 as IGeometry);
}
//添加实体对象到地图图层
private void addFeature(string layerName, IGeometry geometry)
{
    int i = 0;
    ILayer layer = null;
    for (i = 0; i < axMapControl1.LayerCount; i++)
    {
        layer = axMapControl1.Map.get_Layer(i);
        if (layer.Name.ToLower() == layerName)
        {
            break;
        }
    }
    IFeatureLayer featureLayer = layer as IFeatureLayer;
    IFeatureClass featureClass = featureLayer.FeatureClass;
    IDataset dataset = (IDataset)featureClass;
    IWorkspace workspace = dataset.Workspace;
    //开始空间编辑
    IWorkspaceEdit workspaceEdit = (IWorkspaceEdit)workspace;
    workspaceEdit.StartEditing(true);
    workspaceEdit.StartEditOperation();
    IFeatureBuffer featureBuffer = featureClass.CreateFeatureBuffer();
    IFeatureCursor featureCursor;
    //清除图层原有的实体对象
    featureCursor = featureClass.Search(null, true);
    IFeature feature;
    feature = featureCursor.NextFeature();
    while (feature != null)
    {
        feature.Delete();
        feature = featureCursor.NextFeature();
    }
    //开始插入新的实体对象
    featureCursor = featureClass.Insert(true);
    featureBuffer.Shape = geometry;
    object featureOID = featureCursor.InsertFeature(featureBuffer);
    //保存实体
    featureCursor.Flush();
    //结束空间编辑
    workspaceEdit.StopEditOperation();
    workspaceEdit.StopEditing(true);
    System.Runtime.InteropServices.Marshal.ReleaseComObject(featureCursor);
}
}
```

5.2.2.2 LineSymbol 对象

LineSymbol 对象是用于修饰线型几何对象的符号。ILineSymbol 作为每一种 LineSymbol 类都实现了接口，该接口定义了两个公共属性：Color 和 Width。LineSymbol 类是抽象类，该类泛化 8 个类：CartographicLineSymbol、HashLineSymbol、MarkerLineSymbol、MultiLayerLineSymbol、PictureLineSymbol、SimpleLine3DSymbol、SimpleLineSymbol 和 TextureLineSymbol。

SimpleLineSymbol 是简单线符号，该类继承了 IlineSymbol 接口，实现了 ISimpleLineSymbol 接口，提供了 Style 属性用于设置线的样式。

CartographicLineSymbol 是制图线符号，该类主要实现了两个接口：ICartographicLineSymbol 和 ILineProperties。ICartographicLineSymbol 接口主要用于设置线符号的节点属性，如 Cap 属性用于设置线的首尾点的形状，Join 属性用于设置线要素转折处的样式。ILineProperties 接口主要用于设置 dash-dot 类型的线要素符号属性。

MultiLayerLineSymbol 是多图层线符号，该对象可以使用重叠符号的方法生成新的线符号。

PictureLineSymbol 是图片线符号，该对象用位图来表现一条线对象，比如火车线路符号。

HashLineSymbol 是离散线符号，SimpleLine3Dsymbol 是 3D 线符号，MarkerLineSymbol 是点线符号，TextureLineSymbol 是纹理贴图线符号。

```csharp
//简单线符号
        private void button6_Click(object sender, EventArgs e)
        {
            ISimpleLineSymbol  simpleLineSymbol=new SimpleLineSymbolClass ();
            simpleLineSymbol.Style = esriSimpleLineStyle.esriSLSDashDotDot ;
            IPolyline polyline = new PolylineClass();
            IPoint point = new PointClass();
            point.PutCoords(1, 1);
            polyline.FromPoint = point;
            point.PutCoords(10, 10);
            polyline.ToPoint = point;
            simpleLineSymbol.Width =10;
            IRgbColor rgbColor = getRGB(255, 0, 0);
            simpleLineSymbol.Color = rgbColor;
            ISymbol symbol =simpleLineSymbol as ISymbol;
            symbol.ROP2 =esriRasterOpCode.esriROPNotXOrPen;
            IActiveView activeView = this.axMapControl1.ActiveView;

            activeView.ScreenDisplay.StartDrawing(activeView.ScreenDisplay.hDC,
 (short)esriScreenCache.esriNoScreenCache);
            activeView.ScreenDisplay.SetSymbol(symbol);
            activeView.ScreenDisplay.DrawPolyline(polyline as IGeometry);
            activeView.ScreenDisplay.FinishDrawing();
            activeView.ScreenDisplay.FinishDrawing();

        }
        //制图线符号
        private void button7_Click(object sender, EventArgs e)
        {
            ICartographicLineSymbol cartographicLineSymbol = new CartographicLine
            SymbolClass();
            cartographicLineSymbol.Cap = esriLineCapStyle.esriLCSButt;
            cartographicLineSymbol.Join = esriLineJoinStyle.esriLJSBevel;
            cartographicLineSymbol.Width = 10;
            cartographicLineSymbol.MiterLimit = 4;
            ILineProperties lineProperties;
            lineProperties = cartographicLineSymbol as ILineProperties;
            lineProperties.Offset = 0;
            double[] dob = new double[6];
            dob[0] = 0;
            dob[1] = 1;
            dob[2] = 2;
            dob[3] = 3;
            dob[4] = 4;
            dob[5] = 5;
            ITemplate template = new TemplateClass();
            template.Interval = 1;
            for (int i = 0; i < dob.Length; i+=2)
            {
```

```csharp
        template .AddPatternElement (dob[i],dob[i+1]);
    }
    lineProperties.Template =template ;

    IPolyline polyline = new PolylineClass();
    IPoint point = new PointClass();
    point.PutCoords(1, 1);
    polyline.FromPoint = point;
    point.PutCoords(10, 10);
    polyline.ToPoint = point;
    IRgbColor rgbColor = getRGB(0, 255, 0);
    cartographicLineSymbol.Color = rgbColor;
    IActiveView activeView = this.axMapControl1.ActiveView;
    activeView.ScreenDisplay.StartDrawing(activeView.ScreenDisplay.hDC,
    (short)esriScreenCache.esriNoScreenCache);
    activeView.ScreenDisplay.SetSymbol(cartographicLineSymbol as ISymbol );
    activeView.ScreenDisplay.DrawPolyline(polyline as IGeometry);
    activeView.ScreenDisplay.FinishDrawing();
    activeView.ScreenDisplay.FinishDrawing();

}
//多图层线符号
private void button8_Click(object sender, EventArgs e)
{
    IMultiLayerLineSymbol multiLayerLineSymbol = new MultiLayerLine SymbolClass();
    ISimpleLineSymbol simpleLineSymbol = new SimpleLineSymbolClass();
    simpleLineSymbol.Style = esriSimpleLineStyle.esriSLSDashDotDot;
    simpleLineSymbol.Width = 10;
    IRgbColor rgbColor = getRGB(255, 0, 0);
    simpleLineSymbol.Color = rgbColor;
    ISymbol symbol = simpleLineSymbol as ISymbol;
    symbol.ROP2 = esriRasterOpCode.esriROPNotXOrPen;

    ICartographicLineSymbol cartographicLineSymbol = new CartographicLine
    SymbolClass();
    cartographicLineSymbol.Cap = esriLineCapStyle.esriLCSButt;
    cartographicLineSymbol.Join = esriLineJoinStyle.esriLJSBevel;
    cartographicLineSymbol.Width = 10;
    cartographicLineSymbol.MiterLimit = 4;
    ILineProperties lineProperties;
    lineProperties = cartographicLineSymbol as ILineProperties;
    lineProperties.Offset = 0;
    double[] dob = new double[6];
    dob[0] = 0;
    dob[1] = 1;
    dob[2] = 2;
    dob[3] = 3;
    dob[4] = 4;
    dob[5] = 5;
    ITemplate template = new TemplateClass();
    template.Interval = 1;
    for (int i = 0; i < dob.Length; i += 2)
    {
        template.AddPatternElement(dob[i], dob[i + 1]);
    }
    lineProperties.Template = template;

    IPolyline polyline = new PolylineClass();
    IPoint point = new PointClass();
    point.PutCoords(1, 1);
    polyline.FromPoint = point;
    point.PutCoords(10, 10);
```

```csharp
        polyline.ToPoint = point;
        rgbColor = getRGB(0, 255, 0);
        cartographicLineSymbol.Color = rgbColor;
        multiLayerLineSymbol.AddLayer(simpleLineSymbol);
        multiLayerLineSymbol.AddLayer(cartographicLineSymbol);
        IActiveView activeView = this.axMapControl1.ActiveView;
        activeView.ScreenDisplay.StartDrawing(activeView.ScreenDisplay.hDC,
        (short)esriScreenCache.esriNoScreenCache);
        activeView.ScreenDisplay.SetSymbol(multiLayerLineSymbol as ISymbol);
        activeView.ScreenDisplay.DrawPolyline(polyline as IGeometry);
        activeView.ScreenDisplay.FinishDrawing();
        activeView.ScreenDisplay.FinishDrawing();
    }
    //离散线符号
    private void button9_Click(object sender, EventArgs e)
    {
        IHashLineSymbol hashLineSymbol = new HashLineSymbolClass();
        ILineProperties lineProperties = hashLineSymbol as ILineProperties;
        lineProperties.Offset = 0;
        double[] dob = new double[6];
        dob[0] = 0;
        dob[1] = 1;
        dob[2] = 2;
        dob[3] = 3;
        dob[4] = 4;
        dob[5] = 5;
        ITemplate template = new TemplateClass();
        template.Interval = 1;
        for (int i = 0; i < dob.Length; i += 2)
        {
            template.AddPatternElement(dob[i], dob[i + 1]);
        }
        lineProperties.Template = template;

        hashLineSymbol.Width = 2;
        hashLineSymbol.Angle = 45;
        IRgbColor hashColor = new RgbColor();
        hashColor = getRGB(0, 0, 255);
        hashLineSymbol.Color = hashColor;

        IPolyline polyline = new PolylineClass();
        IPoint point = new PointClass();
        point.PutCoords(1, 1);
        polyline.FromPoint = point;
        point.PutCoords(10, 10);
        polyline.ToPoint = point;
        IActiveView activeView = this.axMapControl1.ActiveView;
        activeView.ScreenDisplay.StartDrawing(activeView.ScreenDisplay.hDC,
        (short)esriScreenCache.esriNoScreenCache);
        activeView.ScreenDisplay.SetSymbol(hashLineSymbol as ISymbol);
        activeView.ScreenDisplay.DrawPolyline(polyline as IGeometry);
        activeView.ScreenDisplay.FinishDrawing();
        activeView.ScreenDisplay.FinishDrawing();
    }
    //点线符号
    private void button10_Click(object sender, EventArgs e)
    {
        IArrowMarkerSymbol arrowMarkerSymbol = new ArrowMarkerSymbolClass();
        IRgbColor rgbColor = getRGB(255, 0, 0);
        arrowMarkerSymbol.Color = rgbColor as IColor;
        arrowMarkerSymbol.Length = 10;
        arrowMarkerSymbol.Width = 10;
```

```
            arrowMarkerSymbol.Style = esriArrowMarkerStyle.esriAMSPlain;

            IMarkerLineSymbol markerLineSymbol = new MarkerLineSymbolClass();
            markerLineSymbol.MarkerSymbol = arrowMarkerSymbol;
            rgbColor = getRGB(0, 255, 0);
            markerLineSymbol.Color = rgbColor;
            IPolyline polyline = new PolylineClass();
            IPoint point = new PointClass();
            point.PutCoords(1, 1);
            polyline.FromPoint = point;
            point.PutCoords(10, 10);
            polyline.ToPoint = point;
            IActiveView activeView = this.axMapControl1.ActiveView;
            activeView.ScreenDisplay.StartDrawing(activeView.ScreenDisplay.hDC,
            (short)esriScreenCache.esriNoScreenCache);
            activeView.ScreenDisplay.SetSymbol(markerLineSymbol as ISymbol);
            activeView.ScreenDisplay.DrawPolyline(polyline as IGeometry);
            activeView.ScreenDisplay.FinishDrawing();
            activeView.ScreenDisplay.FinishDrawing();

        }
        //图片线符号
        private void button11_Click(object sender, EventArgs e)
        {
            IPictureLineSymbol pictureLineSymbol=new PictureLineSymbolClass ();
            //创建图片符号
            string fileName = @"E: \qq.bmp";
            pictureLineSymbol.CreateLineSymbolFromFile(esriIPictureType.
            esriIPictureBitmap, fileName);
            IRgbColor rgbColor = getRGB(0, 255, 0);
            pictureLineSymbol.Color = rgbColor;
            pictureLineSymbol.Offset = 0;
            pictureLineSymbol.Width = 10;
            pictureLineSymbol.Rotate = false;

            IPolyline polyline = new PolylineClass();
            IPoint point = new PointClass();
            point.PutCoords(1, 1);
            polyline.FromPoint = point;
            point.PutCoords(10, 10);
            polyline.ToPoint = point;
            IActiveView activeView = this.axMapControl1.ActiveView;
            activeView.ScreenDisplay.StartDrawing(activeView.ScreenDisplay.hDC,
            (short)esriScreenCache.esriNoScreenCache);
            activeView.ScreenDisplay.SetSymbol(pictureLineSymbol as ISymbol);
            activeView.ScreenDisplay.DrawPolyline(polyline as IGeometry);
            activeView.ScreenDisplay.FinishDrawing();
            activeView.ScreenDisplay.FinishDrawing();


        }
```

5.2.2.3 FillSymbol 对象

FillSymbol 是用来修饰多边形等具有面积的几何形体的符号对象。该对象实现了 IFillSymbol 接口，该接口只定义了两个属性 Color 和 OutLine，Color 用来设置填充符号的基本颜色，OutLine 用来设置符号的外边框，默认情况下是 Solid 类型的简单线符号。

（1）SimpleFillSymbol：简单填充符号。

SimpleFillSymbol 是最简单的填充符号，该类实现了 ISimpleFillSymbol 接口，使用它定义的属性和方法，可以设置一个简单填充符号不同的 Style 属性，SimpleFillSymbol 提供了 8 种类型。

（2）LineFillSymbol：线填充符号。

LineFillSymbol 中的填充符号是重复的线条，该类实现了 ILineFillSymbol 接口，定义了它的角度、偏移量和线之间的间隔距离。如 LineSymbol 属性用于设置线填充符号的线的样式；Angle 属性用于设置线与水平线的夹角，默认情况下这个角度值为 0；Offset 属性用于设置线的偏移量；Separation 属性用于设置单位距离内线出现的数量，如果这个值小于 LineDSymbol 对象的宽度，将会被覆盖。

（3）MarkerFillSymbol：点填充符号。

MarkerFillSymbol 是使用一个 Marker 符号作为背景填充符号，该类实现了两个接口：IMarkerFillSymbol 和 IfillProperties。IMarkerFillSymbol 接口提供了填充对象的 Marker 对象的属性，如 GridAngle 用于设置 Marker 点的角度，MarkerSymbol 属性用于设置这些 Marker 的类型。IFillProperties 接口提供了设置 Marker 在填充区域内的分布情况，默认情况下填充区的圆点处应该是一个 Marker 的圆心，不会发生偏移，但用户可以使用 XOffset 和 YOffset 属性来设置这个偏移量，XSeparation 和 Yseparation 用于设置 Marker 之间的水平和垂直距离，这 4 个属性都是使用像素作为单位，默认情况下是 12 个像素。

（4）GradientFillSymbol：渐变颜色填充符号。

GradientFillSymbol 使用渐变颜色带进行填充，该类实现了 IGradientFillSymbol 接口。ColorRamp 属性用于设置这个渐变填充符号的颜色带对象，在设置颜色条的时候，可使用 IntervalCount 属性来设置用户所要使用的颜色梯度。Style 属性用于设计渐变填充的样式，系统提供了 4 种样式：esriGFSLinear、esriGFSRectangular、esriGFSCircular 和 esriGFSBuffered，它们决定了这些渐变的填充是使用线、矩形、圆形还是缓冲的方法。

（5）PictureFillSymbol：图片填充符号。

PictureFillSymbol 对象使用图片来进行填充，该类实现了 IPictureFillSymbol 接口，如 CreateFillSymbolFromFile 通过指定图片的类型和路径获取系统外的图片，该对象只支持 EMP 和 BMP 两种图形；XScale、YScale 属性用于填充对象内部图片之间 X 方向和 Y 方向上的距离。

（6）MultilayerFillSymbol：多层填充符号。

MultilayerFillSymbol 使用多个填充符号进行叠加，以产生新的填充符号对象，该对象实现了 IMultilayerFillSymbol 接口，提供了移动、添加、删除和清空构成多层符号使用的填充符号，如 MoveLayer 方法用于移动这些符号，可以显示不同的效果。

（7）DotDensityFillSymbol：点密度填充符号。

DotDensityFillSymbol 是一种基于数据的填充符号，和 DotDensityRenderer 着色对象一起使用，该对象使用由 MarkerSymbol 组成的随机位置点来显示数据的属性，而单位面积内点的个数则由 DotDensityRenderer 对象计算出来。该对象实现了 IDotDensityFillSymbol 接口，通过该接口可以设置填充符号的外观，如 Marker 的数目、尺寸和填充符号的背景颜色等属性；BackgroundColor 用于设置填充区域的背景颜色，如果新建一个 MultilayerFillSymbol 对象，其中最上一层是 DotDensityFillSymbol，要使下面的图层可以看见，用户需要设置这个颜色为 NullColor 才行。Color 属性用于设置点的颜色，OutLine 属性则用于改变这些点的外框情况。

该类还实现了 ISymbolArray 接口，DotDensityFillSymbol 对象可以使用多个点对象作为填充符号，具体实现如下。

```
//简单填充符号
private void button12_Click(object sender, EventArgs e)
{
```

```csharp
    //简单填充符号
    ISimpleFillSymbol simpleFillSymbol = new SimpleFillSymbolClass();
    simpleFillSymbol.Style = esriSimpleFillStyle.esriSFSDiagonalCross;
    simpleFillSymbol.Color = getRGB(255, 0, 0);
    //创建边线符号
    ISimpleLineSymbol simpleLineSymbol = new SimpleLineSymbolClass();
    simpleLineSymbol.Style = esriSimpleLineStyle.esriSLSDashDotDot;
    simpleLineSymbol.Color =getRGB (0,255,0);
    ISymbol symbol = simpleLineSymbol as ISymbol;
    symbol.ROP2 = esriRasterOpCode.esriROPNotXOrPen;

    simpleFillSymbol.Outline = simpleLineSymbol;
    //创建面对象
    object Missing = Type.Missing;
    IPolygon polygon = new PolygonClass();
    IPointCollection pointCollection = polygon as IPointCollection;
    IPoint point = new PointClass();
    point.PutCoords(5, 5);
    pointCollection.AddPoint(point, ref Missing, ref Missing);
    point.PutCoords(5, 10);
    pointCollection.AddPoint(point, ref Missing, ref Missing);
    point.PutCoords(10, 10);
    pointCollection.AddPoint(point, ref Missing, ref Missing);
    point.PutCoords(10, 5);
    pointCollection.AddPoint(point, ref Missing, ref Missing);
    polygon.SimplifyPreserveFromTo();
    IActiveView activeView = this.axMapControl1.ActiveView;
    activeView.ScreenDisplay.StartDrawing(activeView.ScreenDisplay.hDC,
    (short)esriScreenCache.esriNoScreenCache);
    activeView.ScreenDisplay.SetSymbol(simpleFillSymbol as ISymbol);
    activeView.ScreenDisplay.DrawPolygon(polygon as IGeometry);
    activeView.ScreenDisplay.FinishDrawing();
}
//线填充符号
private void button13_Click(object sender, EventArgs e)
{
    ICartographicLineSymbol cartoLine = new CartographicLineSymbol();
    cartoLine.Cap = esriLineCapStyle.esriLCSButt;
    cartoLine.Join = esriLineJoinStyle.esriLJSMitre;
    cartoLine.Color = getRGB (255,0,0);
    cartoLine.Width = 2;
    //Create the LineFillSymbo
    ILineFillSymbol lineFill = new LineFillSymbol();
    lineFill.Angle = 45;
    lineFill.Separation = 10;
    lineFill.Offset = 5;
    lineFill.LineSymbol = cartoLine;
    object Missing = Type.Missing;
    IPolygon polygon = new PolygonClass();
    IPointCollection pointCollection = polygon as IPointCollection;
    IPoint point = new PointClass();
    point.PutCoords(5, 5);
    pointCollection.AddPoint(point, ref Missing, ref Missing);
    point.PutCoords(5, 10);
    pointCollection.AddPoint(point, ref Missing, ref Missing);
    point.PutCoords(10, 10);
    pointCollection.AddPoint(point, ref Missing, ref Missing);
    point.PutCoords(10, 5);
    pointCollection.AddPoint(point, ref Missing, ref Missing);
    polygon.SimplifyPreserveFromTo();
    IActiveView activeView = this.axMapControl1.ActiveView;
    activeView.ScreenDisplay.StartDrawing(activeView.ScreenDisplay.hDC,
```

```csharp
        (short)esriScreenCache.esriNoScreenCache);
    activeView.ScreenDisplay.SetSymbol(lineFill as ISymbol );
    activeView.ScreenDisplay.DrawPolygon(polygon as IGeometry);
    activeView.ScreenDisplay.FinishDrawing();
}
//点填充符号
private void button14_Click(object sender, EventArgs e)
{
    IArrowMarkerSymbol arrowMarkerSymbol = new ArrowMarkerSymbolClass();
    IRgbColor rgbColor = getRGB(255, 0, 0);
    arrowMarkerSymbol.Color = rgbColor as IColor;
    arrowMarkerSymbol.Length = 10;
    arrowMarkerSymbol.Width = 10;
    arrowMarkerSymbol.Style = esriArrowMarkerStyle.esriAMSPlain;
    IMarkerFillSymbol markerFillSymbol = new MarkerFillSymbolClass();
    markerFillSymbol.MarkerSymbol = arrowMarkerSymbol;
    rgbColor = getRGB(0, 255, 0);
    markerFillSymbol.Color = rgbColor;
    markerFillSymbol.Style = esriMarkerFillStyle.esriMFSGrid;
    IFillProperties fillProperties = markerFillSymbol as IFillProperties;
    fillProperties.XOffset = 2;
    fillProperties.YOffset = 2;
    fillProperties.XSeparation = 15;
    fillProperties.YSeparation = 20;
    object Missing = Type.Missing;
    IPolygon polygon = new PolygonClass();
    IPointCollection pointCollection = polygon as IPointCollection;
    IPoint point = new PointClass();
    point.PutCoords(5, 5);
    pointCollection.AddPoint(point, ref Missing, ref Missing);
    point.PutCoords(5, 10);
    pointCollection.AddPoint(point, ref Missing, ref Missing);
    point.PutCoords(10, 10);
    pointCollection.AddPoint(point, ref Missing, ref Missing);
    point.PutCoords(10, 5);
    pointCollection.AddPoint(point, ref Missing, ref Missing);
    polygon.SimplifyPreserveFromTo();
    IActiveView activeView = this.axMapControl1.ActiveView;
    activeView.ScreenDisplay.StartDrawing(activeView.ScreenDisplay.hDC,
        (short)esriScreenCache.esriNoScreenCache);
    activeView.ScreenDisplay.SetSymbol(markerFillSymbol as ISymbol);
    activeView.ScreenDisplay.DrawPolygon(polygon as IGeometry);
    activeView.ScreenDisplay.FinishDrawing();
}
//渐变色填充符号
private void button15_Click(object sender, EventArgs e)
{
    IGradientFillSymbol gradientFillSymbol = new GradientFillSymbolClass();
    IAlgorithmicColorRamp algorithcColorRamp = new AlgorithmicColorRamp Class();
    algorithcColorRamp.FromColor = getRGB(255, 0, 0);
    algorithcColorRamp.ToColor = getRGB(0, 255, 0);
    algorithcColorRamp.Algorithm = esriColorRampAlgorithm.esriHSVAlgorithm;
    gradientFillSymbol.ColorRamp = algorithcColorRamp;
    gradientFillSymbol.GradientAngle = 45;
    gradientFillSymbol.GradientPercentage = 0.9;
    gradientFillSymbol.Style = esriGradientFillStyle.esriGFSLinear;
    object Missing = Type.Missing;
    IPolygon polygon = new PolygonClass();
    IPointCollection pointCollection = polygon as IPointCollection;
    IPoint point = new PointClass();
    point.PutCoords(5, 5);
    pointCollection.AddPoint(point, ref Missing, ref Missing);
```

```csharp
            point.PutCoords(5, 10);
            pointCollection.AddPoint(point, ref Missing, ref Missing);
            point.PutCoords(10, 10);
            pointCollection.AddPoint(point, ref Missing, ref Missing);
            point.PutCoords(10, 5);
            pointCollection.AddPoint(point, ref Missing, ref Missing);
            polygon.SimplifyPreserveFromTo();
            IActiveView activeView = this.axMapControl1.ActiveView;
            activeView.ScreenDisplay.StartDrawing(activeView.ScreenDisplay.hDC,
                (short)esriScreenCache.esriNoScreenCache);
            activeView.ScreenDisplay.SetSymbol(gradientFillSymbol as ISymbol);
            activeView.ScreenDisplay.DrawPolygon(polygon as IGeometry);
            activeView.ScreenDisplay.FinishDrawing();
        }
        //图片填充符号
        private void button16_Click(object sender, EventArgs e)
        {
            IPictureFillSymbol pictureFillSymbol = new PictureFillSymbolClass();
            //创建图片符号
            string fileName = @"E:\qq.bmp";
            pictureFillSymbol.CreateFillSymbolFromFile(esriIPictureType.esriIPictureBitmap, fileName);
            pictureFillSymbol.Color = getRGB(0, 255, 0);
            ISimpleLineSymbol simpleLineSymbol = new SimpleLineSymbolClass();
            simpleLineSymbol.Style = esriSimpleLineStyle.esriSLSDashDotDot;
            simpleLineSymbol.Color = getRGB(255, 0, 0);
            ISymbol symbol = pictureFillSymbol as ISymbol;
            symbol.ROP2 = esriRasterOpCode.esriROPNotXOrPen;
            pictureFillSymbol.Outline = simpleLineSymbol;
            pictureFillSymbol.Angle = 45;
            object Missing = Type.Missing;
            IPolygon polygon = new PolygonClass();
            IPointCollection pointCollection = polygon as IPointCollection;
            IPoint point = new PointClass();
            point.PutCoords(5, 5);
            pointCollection.AddPoint(point, ref Missing, ref Missing);
            point.PutCoords(5, 10);
            pointCollection.AddPoint(point, ref Missing, ref Missing);
            point.PutCoords(10, 10);
            pointCollection.AddPoint(point, ref Missing, ref Missing);
            point.PutCoords(10, 5);
            pointCollection.AddPoint(point, ref Missing, ref Missing);
            polygon.SimplifyPreserveFromTo();
            IActiveView activeView = this.axMapControl1.ActiveView;
            activeView.ScreenDisplay.StartDrawing(activeView.ScreenDisplay.hDC,
                (short)esriScreenCache.esriNoScreenCache);
            activeView.ScreenDisplay.SetSymbol(pictureFillSymbol as ISymbol);
            activeView.ScreenDisplay.DrawPolygon(polygon as IGeometry);
            activeView.ScreenDisplay.FinishDrawing();

        }
        //多层填充符号
        private void button17_Click(object sender, EventArgs e)
        {
            IMultiLayerFillSymbol multiLayerFillSymbol = new MultiLayerFillSymbol Class();
            IGradientFillSymbol gradientFillSymbol = new GradientFillSymbolClass();
            IAlgorithmicColorRamp algorithcColorRamp = new AlgorithmicColorRamp Class();
            algorithcColorRamp.FromColor = getRGB(255, 0, 0);
            algorithcColorRamp.ToColor = getRGB(0, 255, 0);
            algorithcColorRamp.Algorithm = esriColorRampAlgorithm.esriHSVAlgorithm;
            gradientFillSymbol.ColorRamp = algorithcColorRamp;
            gradientFillSymbol.GradientAngle = 45;
            gradientFillSymbol.GradientPercentage = 0.9;
```

```csharp
gradientFillSymbol.Style = esriGradientFillStyle.esriGFSLinear;
ICartographicLineSymbol cartoLine = new CartographicLineSymbol();
cartoLine.Cap = esriLineCapStyle.esriLCSButt;
cartoLine.Join = esriLineJoinStyle.esriLJSMitre;
cartoLine.Color = getRGB(255, 0, 0);
cartoLine.Width = 2;
//Create the LineFillSymbo
ILineFillSymbol lineFill = new LineFillSymbol();
lineFill.Angle = 45;
lineFill.Separation = 10;
lineFill.Offset = 5;
lineFill.LineSymbol = cartoLine;
multiLayerFillSymbol.AddLayer(gradientFillSymbol);
multiLayerFillSymbol.AddLayer(lineFill);
object Missing = Type.Missing;
IPolygon polygon = new PolygonClass();
IPointCollection pointCollection = polygon as IPointCollection;
IPoint point = new PointClass();
point.PutCoords(5, 5);
pointCollection.AddPoint(point, ref Missing, ref Missing);
point.PutCoords(5, 10);
pointCollection.AddPoint(point, ref Missing, ref Missing);
point.PutCoords(10, 10);
pointCollection.AddPoint(point, ref Missing, ref Missing);
point.PutCoords(10, 5);
pointCollection.AddPoint(point, ref Missing, ref Missing);
polygon.SimplifyPreserveFromTo();
IActiveView activeView = this.axMapControl1.ActiveView;
activeView.ScreenDisplay.StartDrawing(activeView.ScreenDisplay.hDC,
(short)esriScreenCache.esriNoScreenCache);
activeView.ScreenDisplay.SetSymbol(multiLayerFillSymbol as ISymbol);
activeView.ScreenDisplay.DrawPolygon(polygon as IGeometry);
activeView.ScreenDisplay.FinishDrawing();
}
```

5.2.2.4 TextSymbol 对象

TextSymbol 对象用于修饰文字元素，该类主要实现了 3 个接口：ITextSymbol、IsimpleTextSymbol 和 IFormattedTextSymbol。

ITextSymbol 接口是定义文本字符对象的主要接口，如 Font 属性用来设置文字的字体，可以通过 IFontDisp 接口来设置字体的大小、粗体、斜体等。

ISimpleTextSymbol 接口提供了一些简单的设置属性，如 XOffset 和 YOffset 用来设置字符的偏移量。其中有一个重要的属性 TextPath，用于确定每一个字符的排列位置，它是一个抽象类，有 3 个子类：BezierTextPath、SimpleTextPath 和 OverposterTextPath。ITextPath 接口提供了方法来计算每一个字符沿文字路径的位置，如 SimpleTextPath 对象可以让文字路径是任何一种曲线轨迹，BezierTextPath 对象可以让文字的路径按照贝塞尔曲线设置。

IFormattedTextSymbol 接口用来设置背景对象、阴影等属性，如 ShowllowColor 用于设置阴影颜色，CharacterSpcing 和 Charterwidth 用于设置文本符号中单个字符之间的空隙和字符的宽度等，具体实现如下：

```csharp
//TextSymbol
private void button18_Click(object sender, EventArgs e)
{
    ITextSymbol textSymbol =new TextSymbolClass ();
    System.Drawing.Font drawFont =new System.Drawing.Font ("宋体",16,Font Style.Bold );
    stdole.IFontDisp fontDisp = (stdole.IFontDisp)(new stdole.StdFontClass ());
```

```csharp
textSymbol.Font =fontDisp ;
textSymbol.Color =getRGB (0,255,0);
textSymbol.Size = 20;
IPolyline polyline = new PolylineClass();
IPoint point = new PointClass();
point.PutCoords(1, 1);
polyline.FromPoint = point;
point.PutCoords(10, 10);
polyline.ToPoint = point;
ITextPath textPath=new BezierTextPathClass ();
//创建简单标注
ILineSymbol lineSymbol =new SimpleLineSymbolClass ();
lineSymbol.Color =getRGB (255,0,0);
lineSymbol.Width =5;
ISimpleTextSymbol simpleTextSymbol =textSymbol as ISimpleTextSymbol ;
simpleTextSymbol.TextPath =textPath ;
object oLineSymbol =lineSymbol ;
object oTextSymbol =textSymbol ;
IActiveView activeView = this.axMapControl1.ActiveView;
activeView.ScreenDisplay.StartDrawing(activeView.ScreenDisplay.hDC,
(short)esriScreenCache.esriNoScreenCache);
activeView.ScreenDisplay.SetSymbol(oLineSymbol as ISymbol);
activeView.ScreenDisplay.DrawPolyline (polyline as IGeometry);
activeView.ScreenDisplay.SetSymbol(oTextSymbol as ISymbol);
activeView.ScreenDisplay.DrawText(polyline as IGeometry,"简单标注");;
activeView.ScreenDisplay.FinishDrawing();

//创建气泡标注（两种风格，一种是有锚点，另一种是marker方式）
//锚点方式
ISimpleFillSymbol simpleFillSymbol = new SimpleFillSymbolClass();
simpleFillSymbol.Color = getRGB(0, 255, 0);
simpleFillSymbol.Style = esriSimpleFillStyle.esriSFSSolid;
IBalloonCallout balloonCallout = new BalloonCalloutClass();
balloonCallout.Style = esriBalloonCalloutStyle.esriBCSRectangle;
balloonCallout.Symbol = simpleFillSymbol;
balloonCallout.LeaderTolerance = 10;

point.PutCoords(5, 5);
balloonCallout.AnchorPoint = point;

IGraphicsContainer graphicsContainer = activeView as IGraphicsContainer;
IFormattedTextSymbol formattedTextSymbol = new TextSymbolClass();
formattedTextSymbol.Color = getRGB(0, 0, 255);
point.PutCoords(10, 5);
ITextBackground textBackground = balloonCallout as ITextBackground;
formattedTextSymbol.Background = textBackground;
activeView.ScreenDisplay.StartDrawing(activeView.ScreenDisplay.hDC,
(short)esriScreenCache.esriNoScreenCache);
activeView.ScreenDisplay.SetSymbol(formattedTextSymbol as ISymbol);
activeView.ScreenDisplay.DrawText(point as IGeometry, "气泡1");
activeView.ScreenDisplay.FinishDrawing();


//marker方式
textSymbol =new TextSymbolClass ();
textSymbol.Color=getRGB (255,0,0);
textSymbol.Angle =0;
textSymbol.RightToLeft =false ;
textSymbol.VerticalAlignment=esriTextVerticalAlignment.esriTVABaseline ;
textSymbol.HorizontalAlignment =esriTextHorizontalAlignment.esriTHAFull ;
```

```
        IMarkerTextBackground markerTextBackground = new MarkerTextBackground Class();
        markerTextBackground.ScaleToFit = true;
        markerTextBackground.TextSymbol = textSymbol;

        IRgbColor rgbColor = new RgbColorClass();
        IPictureMarkerSymbol pictureMarkerSymbol = new PictureMarkerSymbolClass();
        string fileName = @"E:\qq.bmp";
        pictureMarkerSymbol.CreateMarkerSymbolFromFile(esriIPictureType.
        esriIPictureBitmap, fileName);
        pictureMarkerSymbol.Angle = 0;
        pictureMarkerSymbol.BitmapTransparencyColor = rgbColor;
        pictureMarkerSymbol.Size = 20;
        pictureMarkerSymbol.XOffset = 0;
        pictureMarkerSymbol.YOffset = 0;

        markerTextBackground.Symbol = pictureMarkerSymbol as IMarkerSymbol;

         formattedTextSymbol = new TextSymbolClass();
        formattedTextSymbol.Color = getRGB(255, 0, 0);
        fontDisp.Size = 10;
        fontDisp.Bold  = true;
        formattedTextSymbol.Font = fontDisp;

        point.PutCoords(15, 5);

        formattedTextSymbol.Background = markerTextBackground;
        activeView.ScreenDisplay.StartDrawing(activeView.ScreenDisplay.hDC,
        (short)esriScreenCache.esriNoScreenCache);
        activeView.ScreenDisplay.SetSymbol(formattedTextSymbol as ISymbol);
        activeView.ScreenDisplay.DrawText(point as IGeometry, "气泡 2");
        activeView.ScreenDisplay.FinishDrawing();

}
```

5.2.2.5　3DChartSymbol 对象

3DChartSymbol 是一个抽象类，有 3 个子类：BarChart、PieChart 和 StaekedChart。一般用于 ChartSymbol 对象的着色，该对象主要是为 renderer 中的柱状和饼状着色。

3DChartSymbol 实现了多个接口，如 IChartSymbol、IBarChartSymbol、IpieChartSymbol 和 IStackedChartSymbol 等。

IChartSymbol 接口主要用于计算一个 ChartSymbol 对象中的柱状和饼状部分的尺寸，其中 Maximum 值是创建 3DChartSymbol 对象后必须设置的属性，着色一般是在一个要素图层上进行，因而可以从它的要素类中获得一系列的数值统计值。如果有 3 个数值字段参与着色，系统则必须对 3 个字段中的最大值进行比较，找出最大的那个值赋给 Maximum。Value 属性包含这一个值的数组，这个数组用来设置每个柱状高度或饼状宽度。使用 3D 符号着色的时候，符号可能不止一种，可以使用 ISymbolArray 接口来管理一个着色对象中的多个参与着色的符号对象。

BarChartSymbol 对象实现了 IBarChartSymbol 接口，该对象使用不同类型的柱子来代表一个要素中不同的属性，而柱子的高度则取决于属性值的大小。VerticalBars 属性用于确定使用的柱子是水平还是垂直排列；柱子之间的空隙可以通过 Width 和 Spacing 属性来调节；Axes 属性用来设置每个柱子的轴线，轴线是一个 ILineSymbol 对象，可以通过设置 ShowAxes 属性为 True 来显示轴线。

PieChartSymbol 对象实现了 IPieChartSymbol 接口，该对象是使用一个饼图来显示不同要素类中的不同属性，不同的属性按照它们的数值大小，占有一个饼图中的不同比例的扇形区域。该接口定义了设置饼图外观的属性，如 ClockWise 属性用于确定饼图中颜色的方向，当 ClockWise 为 True 时，饼图中的颜色块呈顺时针的方向分布；UseOutline 属性设置为 True 时，在饼图的外框可以设置外框线，该对象是一个 ILineSymbol 对象。

StackedChartSymbol 对象实现了 IStackedChartSymbol 接口，用于设置该对象的外观，如 Width 属性用于设置柱的宽度；Outline 和 UseOutline 用于设置符号的外框线；Fixed 设置为 True 时，每个柱子的长度是一样的，如果设置为 False，柱子的尺寸就会依据要素的属性来计算得出。

5.3 专题图制作

专题图是依据要素的一个或多个不同的属性而设置不同的符号，从而达到区分不同类型要素的目的。Arc Engine 提供了多个着色对象用于产生专题图，可以使用标准的着色方案，也可以自定义着色方案，Arc Engine 提供了 8 种标准的着色方案。这些方案都是对一个图层中所有的要素进行的。如果要对图层中的部分要素单独着色，可以在这些要素上绘制图形元素或者将需要着色的要素放入选择集中，然后为选择集中的对象创建一个新的图层，然后对这个新图层着色。

5.3.1 SimpleRenderer 专题图

该专题图是使用单一符号进行着色分类，不涉及对要素图层的数据进行处理。如在 MapControl 中打开一幅地图时，同一图层内的所有元素都是一种符号。通过 SimpleRenderer 对象对 Symbol 进行设置后，赋予 IGeoFeatureLayer 接口的 Renderer 属性，对象实现了 ITransparencyRenderer 接口，通过该接口的属性，可以根据要素的某个数值字段值来设置要素的显示透明度。该对象还实现了 ISimpleRenderer 接口，提供了两个重要的属性：Description 和 Label，这两个属性用于设置图例。

5.3.2 ClassBreakRenderer 专题图

该专题图也称分级专题图，通过要素图层中要素的某个数值字段的值，根据用户的要求，将这些值分为多个级别，每个级别用不同的 Symbol 显示。该对象实现了 IClassBreakRenderer 接口，提供了实现分级显示的属性和方法，如 Field 属性用于设置分级着色的字段，BreakCount 属性用于设置分级的数目。

5.3.3 UniqueValueRenderer 专题图

该专题图可以采用多种着色颜色，依据要素图层的要素中某个数值字段的不同值，给每个要素一个单独的颜色，因此可以区分存在的每一个要素。UniqueValueRenderer 实现了 IUniqueValueRenderer 接口，提供了各种方法和属性，如 AddValue 方法用于将单个要素的某个字段值和与之相匹配的着色符号加入到 UniqueValueRenderer 对象，如：AddValue 方法，该方法将单个要素的某个字段值和与之相匹配的着色符号加入 UniqueValueRenderer 对象。

5.3.4 ProportionalSymbolRenderer 专题图

该专题图也称为梯度着色法，该对象实现了 IProportionalSymbolRenderer 接口。这种着色方式，用户需要知道最大和最小的圆点各自代表的字段值、着色基于的字段和着色点使用的符号，以及它在 Legend 要出现的级别数目。

5.3.5 ChartRenderer 专题图

ChartRenderer 对象使用饼图和柱状图来比较一个要素的多个属性。该专题图有两种表现形式，一种是水平排列,另一种是累积排列。该对象实现了 IChartRenderer 接口,定义了一系列的方法,如 ChartSymbol 方法用于设置着色对象的着色符号,Label 属性用于设置 Legend 的标签。

该着色方法是比较一个要素中的不同属性,因此需要获取着色图层的单个或多个字段,通过 RendererField 对象来操作。该类实现了 IRendererField 接口,可通过 AddField 方法来添加字段。

该着色法是使用饼图来表现要素的多个属性之间的比率关系。该对象实现了 IPicChartRenderer 接口,使用 PieChartSymbol 符号来修饰要素。

5.3.6 DotDensityRenderer 专题图

DotDensityRenderer 对象使用 DotDensityFillSymbol 符号对 Polygon 类型的要素进行着色。该专题图使用随机分布的点的密度来表现要素某个属性值的大小。也可以对图层的多个属性着色,通过指定不同的点符号来区分。

DotDensityRenderer 类实现了 IDotDensityRenderer 接口,定义了使用点密度着色的方法和属性,如 DotDensitySymbol 用于确定着色用的点符号,CreateLegend 方法用于产生图例,具体实现如下。

```csharp
//获取颜色对象
private IRgbColor getRGB(int r, int g, int b)
{
    IRgbColor pColor;
    pColor = new RgbColorClass();
    pColor.Red = r;
    pColor.Green = g;
    pColor.Blue = b;
    return pColor;
}
//获取 GeoFeaturelayer 图层
private IGeoFeatureLayer getGeoLayer(string layerName)
{
    ILayer layer;
    IGeoFeatureLayer geoFeatureLayer ;
    for (int i = 0; i < this.axMapControl1.LayerCount; i++)
    {
        layer = this.axMapControl1.get_Layer(i);
        if (layer != null && layer.Name == layerName)
        {
            geoFeatureLayer = layer as IGeoFeatureLayer ;
            return geoFeatureLayer;
        }
    }
    return null;
}
//简单渲染专题图
private void button1_Click(object sender, EventArgs e)
{
    //简单填充符号
    ISimpleFillSymbol simpleFillSymbol = new SimpleFillSymbolClass();
    simpleFillSymbol.Style = esriSimpleFillStyle.esriSFSDiagonalCross;
    simpleFillSymbol.Color = getRGB(255, 0, 0);
    //创建边线符号
    ISimpleLineSymbol simpleLineSymbol = new SimpleLineSymbolClass();
    simpleLineSymbol.Style = esriSimpleLineStyle.esriSLSDashDotDot;
    simpleLineSymbol.Color = getRGB(0, 255, 0);
    ISymbol symbol = simpleLineSymbol as ISymbol;
    symbol.ROP2 = esriRasterOpCode.esriROPNotXOrPen;
    simpleFillSymbol.Outline = simpleLineSymbol;
```

```csharp
    ISimpleRenderer simpleRender =new SimpleRendererClass ();
    simpleRender .Symbol =simpleFillSymbol as ISymbol ;
    simpleRender.Label ="continent";
    simpleRender.Description ="简单渲染";

    IGeoFeatureLayer geoFeatureLayer;
    geoFeatureLayer = getGeoLayer("Continents");
    if (geoFeatureLayer != null)
    {
        geoFeatureLayer.Renderer = simpleRender as IFeatureRenderer ;
    }
    this.axMapControl1.Refresh();
}
//创建颜色带
private IColorRamp CreateAlgorithmicColorRamp(int count )
{
    //创建一个新AlgorithmicColorRampClass 对象
    IAlgorithmicColorRamp algColorRamp = new AlgorithmicColorRampClass();
    IRgbColor fromColor = new RgbColorClass();
    IRgbColor toColor = new RgbColorClass();
    //创建起始颜色对象
    fromColor.Red = 255;
    fromColor.Green = 0;
    fromColor.Blue = 0;
    //创建终止颜色对象
    toColor.Red = 0;
    toColor.Green = 0;
    toColor.Blue = 255;
    //设置AlgorithmicColorRampClass 的起止颜色属性
    algColorRamp.ToColor = fromColor;
    algColorRamp.FromColor = toColor;
    //设置梯度类型
    algColorRamp.Algorithm = esriColorRampAlgorithm.esriCIELabAlgorithm;
    //设置颜色带颜色数量
    algColorRamp.Size = count ;
    //创建颜色带
    bool bture = true;
    algColorRamp.CreateRamp(out bture);
    return algColorRamp;

}
//分级专题图
private void button2_Click(object sender, EventArgs e)
{
    int classCount = 6;
    ITableHistogram tableHistogram;
    IBasicHistogram basicHistogram;
    ITable table;
    IGeoFeatureLayer geoFeatureLayer;
    geoFeatureLayer = getGeoLayer("Continents");
    ILayer layer = geoFeatureLayer as ILayer;
    table = layer as ITable;
    tableHistogram =new BasicTableHistogramClass ();
     //按照数值字段分级
    tableHistogram.Table =table ;
    tableHistogram.Field ="sqmi";
    basicHistogram =tableHistogram as IBasicHistogram ;
    object    values;
    object    frequencys;
    //先统计每个值和各个值出现的次数
```

```csharp
        basicHistogram.GetHistogram (out values ,out frequencys );
        //创建平均分级对象
        IClassifyGEN classifyGEN =new QuantileClass ();
        //用统计结果进行分级，级别数目为 classCount
        classifyGEN.Classify (values ,frequencys , ref classCount );
        //获得分级结果,是个双精度类型数组
        double [] classes;
        classes=classifyGEN.ClassBreaks as double [];

        IEnumColors enumColors = CreateAlgorithmicColorRamp(classes.Length ). Colors;
        IColor color;

        IClassBreaksRenderer classBreaksRenderer = new ClassBreaksRendererClass();
        classBreaksRenderer.Field = "sqmi";
        classBreaksRenderer.BreakCount = classCount;
        classBreaksRenderer.SortClassesAscending = true;

        ISimpleFillSymbol simpleFillSymbol;
        for (int i = 0; i < classes.Length-1 ; i++)
        {
            color = enumColors.Next();
            simpleFillSymbol = new SimpleFillSymbolClass ();
            simpleFillSymbol.Color = color;
            simpleFillSymbol.Style = esriSimpleFillStyle.esriSFSSolid;

            classBreaksRenderer.set_Symbol(i, simpleFillSymbol as ISymbol);
            classBreaksRenderer.set_Break(i, classes[i]);

        }

        if (geoFeatureLayer != null)
        {
            geoFeatureLayer.Renderer = classBreaksRenderer as IFeatureRenderer;
        }

        this.axMapControl1.ActiveView.Refresh();
}
//单一值专题图
private void button3_Click(object sender, EventArgs e)
{
        IGeoFeatureLayer geoFeatureLayer;
        geoFeatureLayer = getGeoLayer("Continents");
        IUniqueValueRenderer uniqueValueRenderer =new UniqueValueRendererClass ();
        uniqueValueRenderer.FieldCount =1;
        uniqueValueRenderer.set_Field (0,"continent");

        //简单填充符号
        ISimpleFillSymbol simpleFillSymbol = new SimpleFillSymbolClass();
        simpleFillSymbol.Style = esriSimpleFillStyle.esriSFSSolid ;

        IFeatureCursor featureCursor = geoFeatureLayer.FeatureClass.Search(null, false);
        IFeature feature;

        if (featureCursor != null)
        {
            IEnumColors enumColors = CreateAlgorithmicColorRamp(8).Colors;
            int fieldIndex =geoFeatureLayer.FeatureClass.Fields.FindField ("continent");
            for (int i = 0; i <8; i++)
            {
                feature = featureCursor.NextFeature();
                string nameValue = feature.get_Value(fieldIndex).ToString ();
                simpleFillSymbol = new SimpleFillSymbolClass();
```

```csharp
            simpleFillSymbol.Color = enumColors.Next();
            uniqueValueRenderer.AddValue(nameValue, "continent", simpleFill
            Symbol as ISymbol);
        }
    }

    geoFeatureLayer.Renderer = uniqueValueRenderer as IFeatureRenderer;
    this.axMapControl1.Refresh();
}
//梯度着色法专题图
private void button4_Click(object sender, EventArgs e)
{

    IGeoFeatureLayer geoFeatureLayer;
    IFeatureLayer featureLayer ;
    IProportionalSymbolRenderer proportionalSymbolRenderer ;
    ITable table ;
    ICursor cursor;
    IDataStatistics dataStatistics ;
    IStatisticsResults statisticsResult;
    stdole.IFontDisp fontDisp;

    geoFeatureLayer = getGeoLayer("Continents");
    featureLayer =geoFeatureLayer as IFeatureLayer ;
    table =geoFeatureLayer as ITable ;
    cursor =table.Search (null,true );
    dataStatistics =new DataStatisticsClass ();
    dataStatistics.Cursor =cursor ;
    dataStatistics.Field ="sqmi";
    statisticsResult =dataStatistics.Statistics ;
    if (statisticsResult !=null)
    {
        IFillSymbol  fillSymbol =new SimpleFillSymbolClass ();
        fillSymbol.Color =getRGB (0,255,0);
        ICharacterMarkerSymbol characterMarkerSymbol=new CharacterMarker
        SymbolClass ();
        fontDisp =new stdole.StdFontClass () as stdole.IFontDisp ;
        fontDisp.Name ="arial";
        fontDisp.Size =20;
        characterMarkerSymbol.Font =fontDisp ;
        characterMarkerSymbol.CharacterIndex =90;
        characterMarkerSymbol.Color =getRGB (255,0,0);
        characterMarkerSymbol.Size =8;
        proportionalSymbolRenderer=new ProportionalSymbolRendererClass ();
        proportionalSymbolRenderer.ValueUnit =esriUnits.esriUnknownUnits;
        proportionalSymbolRenderer .Field ="sqmi";
        proportionalSymbolRenderer.FlanneryCompensation =false ;
        proportionalSymbolRenderer.MinDataValue =statisticsResult.Minimum ;
        proportionalSymbolRenderer.MaxDataValue =statisticsResult.Maximum ;
        proportionalSymbolRenderer.BackgroundSymbol =fillSymbol ;
        proportionalSymbolRenderer.MinSymbol =characterMarkerSymbol as ISymbol ;
        proportionalSymbolRenderer.LegendSymbolCount =10;
        proportionalSymbolRenderer.CreateLegendSymbols ();
        geoFeatureLayer.Renderer =proportionalSymbolRenderer as Ifeature Renderer ;
    }
    this.axMapControl1.Refresh();
}
//BarChartSymbol
private void button5_Click(object sender, EventArgs e)
{
    IGeoFeatureLayer geoFeatureLayer ;
    IFeatureLayer featureLayer ;
```

```csharp
            ITable table ;
            ICursor cursor;
            IRowBuffer rowBuffer;
            //设置渲染要素
            string field1="sqmi";
            string field2="sqkm";
            //获取渲染图层
            geoFeatureLayer = getGeoLayer("Continents");
            featureLayer =geoFeatureLayer as IFeatureLayer ;
            table =featureLayer as ITable ;
            geoFeatureLayer.ScaleSymbols =true;
            IChartRenderer chartRenderer =new ChartRendererClass ();
            IRendererFields rendererFields=chartRenderer as IRendererFields ;
            rendererFields.AddField (field1 ,field1 );
            rendererFields.AddField (field2,field2);
            int [] fieldIndexs=new int [2];
            fieldIndexs [0]=table.FindField (field1);
            fieldIndexs [1]=table.FindField (field2);
            //获取要素最大值
            double fieldValue =0.0,maxValue=0.0;
            cursor = table.Search(null, true);
            rowBuffer =cursor .NextRow ();
            while (rowBuffer!=null  )
            {
                for (int i =0;i<2;i++)
                {
                    fieldValue =double .Parse (rowBuffer.get_Value (fieldIndexs[i]).ToString ()) ;
                    if (fieldValue >maxValue )
                    {
                        maxValue =fieldValue ;
                    }
                }
                rowBuffer =cursor.NextRow ();
            }
            //创建水平排列符号
            IBarChartSymbol barChartSymbol =new BarChartSymbolClass ();
            barChartSymbol.Width =10;
            IMarkerSymbol markerSymbol =barChartSymbol as IMarkerSymbol ;
            markerSymbol.Size =50;
            IChartSymbol chartSymbol=barChartSymbol as IChartSymbol ;
            chartSymbol.MaxValue =maxValue ;
            //添加渲染符号
            ISymbolArray symbolArray =barChartSymbol as ISymbolArray ;
            IFillSymbol fillSymbol =new SimpleFillSymbolClass ();
            fillSymbol.Color =getRGB (255,0,0);
            symbolArray.AddSymbol (fillSymbol as ISymbol );
            fillSymbol =new SimpleFillSymbolClass ();
            fillSymbol .Color =getRGB (0,255,0);
            symbolArray .AddSymbol (fillSymbol as ISymbol );
            //设置柱状图符号
            chartRenderer.ChartSymbol =barChartSymbol as IChartSymbol ;
            fillSymbol =new SimpleFillSymbolClass ();
            fillSymbol.Color =getRGB (0,0,255);
            chartRenderer.BaseSymbol =fillSymbol as ISymbol ;
            chartRenderer.UseOverposter =false ;
            //创建图例
            chartRenderer.CreateLegend ();
            geoFeatureLayer.Renderer =chartRenderer as IFeatureRenderer ;
            this.axMapControl1.Refresh();
        }
        //StackedChartSymbol
```

```csharp
private void button6_Click(object sender, EventArgs e)
{
    IGeoFeatureLayer geoFeatureLayer;
    IFeatureLayer featureLayer;
    ITable table;
    ICursor cursor;
    IRowBuffer rowBuffer;
    //设置渲染要素
    string field1 = "sqmi";
    string field2 = "sqkm";
    //获取渲染图层
    geoFeatureLayer = getGeoLayer("Continents");
    featureLayer = geoFeatureLayer as IFeatureLayer;
    table = featureLayer as ITable;
    geoFeatureLayer.ScaleSymbols = true;
    IChartRenderer chartRenderer = new ChartRendererClass();
    IRendererFields rendererFields = chartRenderer as IRendererFields;
    rendererFields.AddField(field1, field1);
    rendererFields.AddField(field2, field2);
    int[] fieldIndexs = new int[2];
    fieldIndexs[0] = table.FindField(field1);
    fieldIndexs[1] = table.FindField(field2);
    //获取要素最大值
    double fieldValue = 0.0, maxValue = 0.0;
    cursor = table.Search(null, true);
    rowBuffer = cursor.NextRow();
    while (rowBuffer != null)
    {
        for (int i = 0; i < 2; i++)
        {
            fieldValue = double.Parse(rowBuffer.get_Value(fieldIndexs[i]).ToString());
            if (fieldValue > maxValue)
            {
                maxValue = fieldValue;
            }
        }
        rowBuffer = cursor.NextRow();
    }
    //创建累积排列符号
    IStackedChartSymbol stackedChartSymbol = new StackedChartSymbolClass();

    stackedChartSymbol.Width = 10;
    IMarkerSymbol markerSymbol = stackedChartSymbol as IMarkerSymbol;
    markerSymbol.Size = 50;
    IChartSymbol chartSymbol = stackedChartSymbol as IChartSymbol;
    chartSymbol.MaxValue = maxValue;
    //添加渲染符号
    ISymbolArray symbolArray = stackedChartSymbol as ISymbolArray;
    IFillSymbol fillSymbol = new SimpleFillSymbolClass();
    fillSymbol.Color = getRGB(255, 0, 0);
    symbolArray.AddSymbol(fillSymbol as ISymbol);
    fillSymbol = new SimpleFillSymbolClass();
    fillSymbol.Color = getRGB(0, 255, 0);
    symbolArray.AddSymbol(fillSymbol as ISymbol);
    //设置柱状图符号
    chartRenderer.ChartSymbol = stackedChartSymbol as IChartSymbol;
    fillSymbol = new SimpleFillSymbolClass();
    fillSymbol.Color = getRGB(0, 0, 255);
    chartRenderer.BaseSymbol = fillSymbol as ISymbol;
    chartRenderer.UseOverposter = false;
    //创建图例
```

```csharp
            chartRenderer.CreateLegend();
            geoFeatureLayer.Renderer = chartRenderer as IFeatureRenderer;
            this.axMapControl1.Refresh();
}
//PieChartRenderer
private void button7_Click(object sender, EventArgs e)
{
    IGeoFeatureLayer geoFeatureLayer;
    IFeatureLayer featureLayer;
    ITable table;
    ICursor cursor;
    IRowBuffer rowBuffer;
    //设置饼图的要素
    string field1 = "sqmi";
    string field2 = "sqkm";
    //获取渲染图层
    geoFeatureLayer = getGeoLayer("Continents");
    featureLayer = geoFeatureLayer as IFeatureLayer;
    table = featureLayer as ITable;
    geoFeatureLayer.ScaleSymbols = true;
    IChartRenderer chartRenderer = new ChartRendererClass();
    IPieChartRenderer pieChartRenderer =chartRenderer as IPieChartRenderer ;
    IRendererFields rendererFields = chartRenderer as IRendererFields;
    rendererFields.AddField(field1, field1);
    rendererFields.AddField(field2, field2);
    int[] fieldIndexs = new int[2];
    fieldIndexs[0] = table.FindField(field1);
    fieldIndexs[1] = table.FindField(field2);
    //获取渲染要素的最大值
    double fieldValue = 0.0, maxValue = 0.0;
    cursor = table.Search(null, true);
    rowBuffer = cursor.NextRow();
    while (rowBuffer != null)
    {
        for (int i = 0; i < 2; i++)
        {
            fieldValue = double.Parse(rowBuffer.get_Value(fieldIndexs[i]).ToString());
            if (fieldValue > maxValue)
            {
                maxValue = fieldValue;
            }
        }
        rowBuffer = cursor.NextRow();
    }
    //设置饼图符号
    IPieChartSymbol pieChartSymbol =new PieChartSymbolClass ();
    pieChartSymbol.Clockwise =true ;
    pieChartSymbol.UseOutline =true ;
    IChartSymbol chartSymbol =pieChartSymbol as IChartSymbol ;
    chartSymbol.MaxValue =maxValue ;
    ILineSymbol lineSymbol =new SimpleLineSymbolClass ();
    lineSymbol.Color =getRGB (255,0,0);
    lineSymbol.Width =2;
    pieChartSymbol.Outline =lineSymbol ;
    IMarkerSymbol markerSymbol=pieChartSymbol as IMarkerSymbol ;
    markerSymbol.Size =30;
    //添加渲染符号
    ISymbolArray symbolArray=pieChartSymbol as ISymbolArray ;
    IFillSymbol fillSymbol =new SimpleFillSymbolClass ();
    fillSymbol.Color =getRGB (0,255,0);
    symbolArray.AddSymbol (fillSymbol as ISymbol );
    fillSymbol =new SimpleFillSymbolClass ();
```

```csharp
        fillSymbol .Color=getRGB  (0,0,255);
        symbolArray.AddSymbol (fillSymbol as ISymbol );
        chartRenderer.ChartSymbol =pieChartSymbol as IChartSymbol  ;
        fillSymbol=new SimpleFillSymbolClass ();
        fillSymbol.Color =getRGB (100,100,100);
        chartRenderer.BaseSymbol =fillSymbol as ISymbol ;
        chartRenderer.UseOverposter =false ;
        //创建图例
        chartRenderer.CreateLegend ();
        geoFeatureLayer.Renderer =chartRenderer as IFeatureRenderer ;
        this.axMapControl1.Refresh ();
    }
    //点密度专题图
    private void button8_Click(object sender, EventArgs e)
    {
        IGeoFeatureLayer geoFeatureLayer ;
        IDotDensityFillSymbol dotDensityFillSymbol;
        IDotDensityRenderer dotDensityRenderer ;
        //获取渲染图层
        geoFeatureLayer = getGeoLayer("Continents");
        dotDensityRenderer=new DotDensityRendererClass ();
        IRendererFields rendererFields =dotDensityRenderer as IRendererFields ;
        //设置渲染字段
        string field1="sqmi";
        rendererFields.AddField (field1 ,field1 );
        //设置填充颜色和背景色
        dotDensityFillSymbol =new DotDensityFillSymbolClass ();
        dotDensityFillSymbol.DotSize =3;
        dotDensityFillSymbol.Color =getRGB (255,0,0);
        dotDensityFillSymbol.BackgroundColor =getRGB (0,255,0);
        //设置渲染符号
        ISymbolArray symbolArray =dotDensityFillSymbol as ISymbolArray ;
        ISimpleMarkerSymbol simpleMarkerSymbol =new SimpleMarkerSymbolClass ();
        simpleMarkerSymbol.Style =esriSimpleMarkerStyle.esriSMSCircle ;
        simpleMarkerSymbol.Size =2;
        simpleMarkerSymbol.Color =getRGB (0,0,255);
        symbolArray.AddSymbol (simpleMarkerSymbol as ISymbol );
        dotDensityRenderer.DotDensitySymbol =dotDensityFillSymbol ;
        //设置渲染密度
        dotDensityRenderer.DotValue =50000;
        //创建图例
        dotDensityRenderer.CreateLegend ();
        geoFeatureLayer .Renderer =dotDensityRenderer as IFeatureRenderer ;
        this.axMapControl1.Refresh ();
    }
}
```

5.4 地图打印输出

地图输出分为两种情况,一种是地图的打印输出,另一种是地图的转换输出。Arc Engine 提供了不同的对象来支持这两种输出,打印输出使用 Printer 对象,转换输出使用 Exporter 对象。

5.4.1 Printer 对象

在 Arc Engine 中,打印地图使用 Printer 类。该类是抽象类,有 3 个子类:EmfPrinter、ArcPressPrinter 和 PsPrinter 。EmfPrinter 对象是通过使用 EMF 数据格式来作为打印驱动,PsPrinter

对象是使用 PostScript 作为驱动来打印输出地图，ArcPressPrinter 则是使用 ArcPressPrinterDriver 来输出地图。

IPrinter 接口定义了所有打印对象的一般方法和属性，如 Paper 属性用于初始化与系统关联的默认打印机，是一个 Paper 对象，可以产生一个独立的 Paper 对象去使用另一台打印机；PrintToFile 属性用于设置是否输出一个文件；StartPrinting 方法用于返回一个打印设备的 hDC；FinishPrinting 用于清除打印后的缓存对象。

（1）EmfPrinter 对象。EMF 即 Enhanced Windows Metalfile。EmfPrinter 是使用 Windows Metalfile 格式作为打印驱动的打印对象，支持 IEmfPrinter 接口，该接口仅指明用户是否使用这个对象进行打印。

（2）ArcPressPrinter 对象。是一种使用 ESEI ArcPress 打印驱动的打印对象。ArcPress 将标准的图形交换格式和内置的打印机语言表达成打印文件传递到工业标准的宽幅格式打印机上。

（3）PsPrinter 对象。是使用 PostScript 作为驱动的打印对象。PostScript 是一种与设备无关的打印机语言，在定义图像时不需要考虑输出设备的特性，而是通过打印机描述文件实现各种打印机的不同特性。

5.4.2 Paper 对象

Paper 对象是 Printer 对象要求的一个关键属性对象，它的属性和方法用于维持 Printer 对象使用的打印机和打印纸张之间的联系。该对象实现了 IPaper 接口，Orientation 属性用于获取和设置打印的方向，1 表示纵向打印，2 表示横向打印；QueryPapersize 方法用于获得页面的尺寸。

5.4.3 在控件中打印输出

IPageLayoutControl 的 PrintPageLayout 方法用于打印控件中的布局视图。在打印前需要对控件的 Page 对象进行设置，如果控件中的页面 Page 无法与打印对象的 Paper 相匹配，则可使用 IPage 接口的 PageToPrinterMapping 来决定是否让页面和打印文本相匹配。

5.4.4 地图的转换输出

当地图不需要被打印出来，或者希望保存为图片格式文件时，可以使用 Export 对象来实现。该对象实现了 IExport 接口。地图转换输出的文件格式分为两大类，一类是栅格格式的 ExportImage，另一类是基于矢量的 ExportVector。

5.4.4.1 基于影像格式的输出

ExportImage 类及其子类的对象都是用于将地图输出为栅格格式文件的对象。该类是抽象类，实现了 IExportImage 接口，定义了所有操作基于栅格格式文件的一般方法和属性，IExportImage 定义了一般属性和方法，如 BackgroundColor 用于确定文件的背景颜色，Width 和 Height 用于确定文件的尺寸，ImageType 用于设置和获取图片的类型。

ExportImage 类还实现了 IWorldFiesettings 接口，通过接口可以设置或查询 MapExtent 和 OutputworldFile 属性。该类有 5 个子类：ExportBMP、ExportJPEG、ExportPNG、ExportTIFF 和 ExportGIF，不同子类的操作方法基本一致，只是生成的图片格式分别为 BMP、JPEG、PNG、TIFF 和 GIF。

5.4.4.2 基于矢量格式的输出

ExportVector 类及其子类对象都是将地图输出为矢量文件。该对象是一个抽象类，实现了多个接口用于设置这些矢量转换输出对象的一般方法和属性，如 IExportVectorColorSpaceSettings 接口的 ColorSpace 用于设置输出文件的颜色空间。

ExprotVector 的 5 个子类分别是 ExportEMF、ExportAI、ExportPDF、ExportPS 和 ExportSVG。

5.4.5　ExportFileDialog 对象

ExportFileDialog 对象提供了可视化的对话框，将地图保存为任何 ArcEngine 支持的格式。该对象实现了 IExportFileDialog 接口，如 DoModal 用于弹出对话框，Export 用于返回一个转换对象。

5.5　本章小结

本章用较大篇幅介绍地图的两种标注，以及地图制图使用的颜色对象和符号库，给出了专题图制作。地理信息系统的优点是将空间信息以图形的方式直观明了的展示出来，然而地图的显示空间有限而信息量却很多，如何将信息以更好的方式鲜明的体现信息的内容，是地图制图的目的，通过对信息的着色、符号化、专题化等多种方式，将信息在有限的地图显示空间内展示出来。

第 6 章　空间数据管理

6.1 SDE 及空间数据

6.1.1 SDE 介绍

ArcSDE 是数据库系统中管理地理数据库的接口,通过该接口可以往关系数据中加入空间数据,提供地理要素的空间位置及形状等信息,是 ArcGIS 与关系数据库之间的 GIS 通道。它允许用户在多种数据管理系统中管理地理信息,并使所有的 ArcGIS 应用程序都能够使用这些数据。

ArcSDE 是多用户 ArcGIS 系统的一个关键部件,它为 DBMS 提供了一个开放的接口,允许 ArcGIS 在多种数据库平台上管理地理信息,这些平台包括 Oracle、Oracle with Spatial/Locator、Microsoft SQL Server、IBM DB2 和 Informix。通过 ArcSDE,ArcGIS 可以在 DBMS 中轻而易举地管理一个共享的、多用户的空间数据库。ArcSDE 的具体功能如下。

（1）高性能的 DBMS 通道。

ArcSDE 是多种 DBMS 的通道。它本身并非一个关系数据库或数据存储模型。它是一个能在多种 DBMS 平台上提供高级的、高性能的 GIS 数据管理的接口。

（2）开放的 DBMS 支持。

ArcSDE 允许在多种 DBMS 中管理地理信息：Oracle、Oracle with Spatial or Locator、Microsoft SQL Server、Informix 以及 IBM DB2。

多用户 ArcSDE 为用户提供了大型空间数据库支持,并且支持多用户编辑。

（3）连续、可伸缩的数据库。

ArcSDE 可以支持海量的空间数据库和任意数量的用户,直至 DBMS 的上限。GIS 工作流和长事务处理 GIS 中的数据管理工作流,例如多用户编辑、历史数据管理、check-out/check-in,以及松散耦合的数据复制等,都依赖于长事务处理和版本管理。ArcSDE 为 DBMS 提供了这种支持。

（4）丰富的地理信息数据模型。

ArcSDE 保证了存储于 DBMS 中的矢量和栅格几何数据的高度完整性。这些数据包括：矢量和栅格几何图形,支持 x、y、z 和 x、y、z、m 的坐标,以及曲线、立体、多行栅格、拓扑、网络、注记、元数据、空间处理模型、地图、图层等。

（5）灵活的配置。

ArcSDE 通道可以让用户在客户端应用程序内,或跨网络、跨计算机地对应用服务器进行多种多层结构的方案配置。ArcSDE 支持 Windows、UNIX、Linux 等多种操作系统。

ArcSDE 能够让同样的功能在所有的 DBMS 上得到实现。尽管所有的关系数据库都支持 SQL,并能使用相似的方法处理简单的 SQL,但是不同数据库的数据库服务器在实现细节上却有着显著的差别,这些差别包括性能和索引、支持的数据类型、集成管理工具和复杂查询的执行,还包括在

DBMS 中对空间数据类型的支持。

标准的 SQL 并不支持空间数据。 ISO SQL/MM Spatial 和 OGC 的简单要素 SQL 规范扩展了 SQL，并且为不同的矢量数据定义了标准的 SQL 支持。 DB2 和 Informix 直接支持这些 SQL 类型。 Oracle 使用的是自己的标准，其空间类型系统是核心数据库系统上的一个独立的可选扩展。而微软的 SQL Server 不提供空间类型的支持。ArcSDE 不但灵活地支持了每个 DBMS 提供的独特功能，而且能为底层 DBMS 提供它们所不具备的功能的支持。

ArcSDE 支持高性能的空间数据的管理，它支持的数据库如下。

① Oracle（带压缩二进制）。
② Oracle（带 Locator 和 Spatial）。
③ 微软 SQL Server（带压缩二进制）。
④ IBM DB2（带 Spatial Extender）。
⑤ IBM Informix（带 Spatial Database）。

ArcSDE 是为了解决 DBMS 的多样性和复杂性而存在的。ArcSDE 的体系结构给用户提供了巨大的灵活性，它允许用户自由地选择 DBMS 来存储空间数据。ArcSDE 分摊了 DBMS 和 GIS 之间对管理空间数据的职责，对空间数据的管理职责是由 GIS 软件和常规 DBMS 软件所共同承担的。某些空间数据的管理功能，例如磁盘存储、属性类型定义、查询处理，以及多用户事务处理等，是由 DBMS 来完成的。当然一些 DBMS 引擎本身也扩展了对空间数据的支持，它们具备索引和搜索的功能。GIS 软件负责为特定的 DBMS 提供各种地理数据的表达。从实际效用上看，DBMS 是被作为一个空间数据的实现机制而存在的。

ArcSDE 是基于多层体系结构的（应用和存储）。数据的存储和提取由存储层（DBMS）实现，而高端的数据整合和数据处理功能则由应用层（ArcGIS）提供。ArcSDE 支持 ArcGIS 应用层，并提供 DBMS 通道技术，使得空间数据可以存储于多种 DBMS 中。ArcSDE 用于高效地存储、索引、访问和维护 DBMS 中的矢量、栅格、元数据及其他空间数据。ArcSDE 同时能保证所有的 GIS 功能可用，而无需考虑底层的 DBMS。使用 ArcSDE，用户在 DBMS 中即可有效地管理自己的地理数据资源。ArcSDE 使用 DBMS 支持的数据类型，以表格的形式管理底层存储的空间数据，并可使用 SQL 在 DBMS 中访问这些数据。ArcSDE 同时也提供了开放的客户端开发接口（C API 和 Java API），通过这些接口，用户定制的应用程序也可以完全访问底层的空间数据表。

6.1.2 空间数据

空间数据（Spatial Data）是指用来表示空间实体的位置、形状、大小及其分布特征诸多方面信息的数据。它可以用来描述来自现实世界的目标，它具有定位、定性、时间和空间关系等特性。定位是指在一个已知的坐标系里，空间目标都具有唯一的空间位置；定性是指有关空间目标的自然属性，它伴随着目标的地理位置；时间是指空间目标是随时间的变化而变化；空间关系通常又称拓扑关系。

空间数据适用于描述所有呈二维、三维甚至多维分布的关于区域的现象。空间数据不仅能够表示实体本身的空间位置及形态信息，而且还有表示实体属性和空间关系(如拓扑关系)的信息。在空间数据中不可再分最小单元现象称为空间实体，空间实体是对存在于这个自然世界中地理实体的抽象，主要包括点、线、面以及实体等基本类型。如把一根电线杆抽象成为一个点，该点可以包含电线杆所处的位置信息、电线杆的高度信息和其他一些相关信息；可以把一条道路抽象为一条线，该线可以包含这条道路的长度、宽度、起点、终点，以及道路等级等相关信息；可以把一个湖泊抽象为一个面，该面可以包含湖泊的周长、面积和湖水的质量信息等。在空间对象建立后，还可以进一步定义其相互之间的关系，这种相互关系被称为"空间关系"，又称为"拓扑关系"，如可以定义点

—线关系、线—线关系、点—面关系等。因此可以说空间数据是一种可以用点、线、面以及实体等基本空间数据结构来表示人们赖以生存的自然世界的数据。

归纳起来它具有以下 5 个基本特征。

（1）空间特征。

每个空间对象都具有空间坐标，即空间对象隐含了空间分布特征。这意味着在空间数据组织方面，要考虑它的空间分布特征。除了通用性数据库管理系统或文件系统关键字的索引和辅关键字索引以外，一般需要建立空间索引。

（2）非结构化特征。

在当前通用的关系数据库管理系统中，数据记录一般是结构化的。即它满足关系数据模型的第一范式要求，每一条记录是定长的，数据项表达的只能是原子数据，不允许嵌套记录。而空间数据则不能满足这种结构化要求。若将一条记录表达一个空间对象，它的数据项可能是变长的。例如，其一，1 条弧段的坐标，其长度是不可限定的，它可能是 2 对坐标，也可能是 10 万对坐标。其二，1 个对象可能包含另外的 1 个或多个对象，例如 1 个多边形，它可能含有多条弧段。若 1 条记录表示 1 条弧段，在这种情况下，1 条多边形的记录就可能嵌套多条弧段的记录，所以它不满足关系数据模型的范式要求，这也就是为什么空间图形数据难以直接采用通用的关系数据管理系统的主要原因。

（3）空间关系特征。

空间数据除了前面所说的空间坐标隐含了空间分布关系外，空间数据中记录的拓扑信息则表达了多种空间关系。这种拓扑数据结构一方面方便了空间数据的查询和空间分析，另一方面也给空间数据的一致性和完整性维护增加了复杂性。特别是有些几何对象，没有直接记录空间坐标的信息。如拓扑的面状目标，仅记录了组成它的弧段的标识，因而进行查找、显示和分析操作时都要操纵和检索多个数据文件方能得以实现。

（4）分类编码特征。

一般而言，每一个空间对象都有一个分类编码，而这种分类编码往往属于国家标准、行业标准或地区标准，每一种地物的类型在某个 GIS 中的属性项个数是相同的。因而在许多情况下，一种地物类型对应于一个属性数据表文件。当然，如果几种地物类型的属性项相同，多种地物类型也可以共用一个属性数据表文件。

（5）海量数据特征。

空间数据量是巨大的，通常称为海量数据。之所以称为海量数据，是指它的数据量比一般的通用数据库要大得多。一个城市地理信息系统的数据量可能达几十 GB，如果考虑影像数据的存储，可能达几百个 GB。这样的数据量在城市管理的其他数据库中是很少见的。正因为空间数据量大，所以需要在二维空间上划分块或者图幅，在垂直方向上划分层来进行组织。

空间数据库系统是一个存储空间和非空间数据的数据库系统，其数据模型和查询语言能支持空间数据类型和空间索引，并且提供进行空间查询和其他空间分析的方法。空间数据库中存储的信息包含两部分：一部分是和空间有关的信息，如点、线、矩形、多边形、多面体等占有空间的对象，称之为空间数据；另一部分是与空间数据有关的各个属性，如点所表示的城市的面积、人口等，称之为非空间数据（属性数据）。

6.2 空间数据库及组织

空间数据库系统是描述、存储和处理空间数据及其属性数据的数据库系统。空间数据库是随着地理信息系统的开发和应用而发展起来的数据库新技术。目前，空间数据库系统尚不是独立存在的系统，它与应用紧密结合，大多数是作为地理信息系统的基础和核心的形式出现。由于空间数据的复杂性和特殊性，一般的商用数据库管理系统难以满足要求。因而，围绕空间数据管理方法，出现了几种不同的组织方式。当前空间数据库的组织方式主要有两种：混合型空间数据库和集成型空间数据库，它们的区别在于是否对空间数据和属性数据进行一体化组织。

6.2.1 混合型空间数据库

目前大多数的地理信息开发平台就是采用混合结构型结构，即非空间数据存储在关系数据库中，空间数据存放在系统文件中。在空间数据库技术领域，比较有名的是 MapInfo 公司和 ESRI 公司，其中的许多产品都是采用混合型的空间数据库，如商用系统 MapInfo 系列，ESRI 的 ArcView、ArcInfo 系列等。这些产品分别使用不同的模块存储空间数据和属性数据。其空间数据在垂直方向上是分层进行组织的，空间数据在水平方向上是按图幅组织的。不同的是：MapInfo 公司设计的产品中每一个图层由 5 个文件组成，而 ESRI 公司设计的产品中每一个图层只由 3 个文件组成。

这种混合管理的模式（即用文件系统管理几何图形数据，用商用关系数据库管理系统管理属性数据），它们之间是通过目标标识或者内部连接码进行连接。在这种管理模式中，几何图形数据与属性数据除了它们的 ObjectID 作为连接关键字段以外，两者几乎是独立地组织、管理与检索。就几何图形而言，由于 GIS 系统采用高级语言编程，可以直接操纵数据文件，所以图形用户界面与图形文件处理是一体的，中间没有裂缝。但对属性数据来说，则因系统和历史发展而异。早期系统中的属性数据必须通过关系数据库管理系统，图形处理的用户界面和属性的用户界面是分开的，它们只是通过一个内部码连接，导致这种连接方式的主要原因是早期的数据库管理系统不提供编程的高级语言，如 Fortran 或 C 的接口，而只能采用数据库操纵语言。这样通常要同时启动两个系统（GIS 图形系统和关系数据库管理系统），甚至两个系统来回切换，使用起来很不方便。

最近几年，随着数据库技术的发展，越来越多的数据库管理系统提供有高级编程语言 C 和 Fortran 等接口，使得地理信息系统可以在 C 语言的环境下直接操纵属性数据，并通过 C 语言的对话框和列表框显示属性数据，或通过对话框输入 SQL 语句，并将该语句通过 C 语言与数据库的接口查询属性数据库，在 GIS 的用户界面下显示查询结果。对于这种工作模式，并不需要启动一个完整的数据库管理系统，用户甚至不知道何时调用了关系数据库管理系统，图形数据和属性数据的查询与维护完全在一个界面之下。在 ODBC（开放性数据库连接协议）推出之前，每个数据库厂商提供一套自己的与高级语言的接口程序，这样 GIS 软件商就要针对每个数据库开发一套与 GIS 的接口程序，所以往往在数据库的使用上会受到限制。

在推出了 ODBC 之后，GIS 软件商只要开发 GIS 与 ODBC 的接口软件，就可以将属性数据与任何一个支持 ODBC 协议的关系数据库管理系统连接。无论是通过 C，还是通过 ODBC 与关系数据库连接，GIS 用户都是在一个界面下处理图形和属性数据，它比前面分开的界面要方便得多。但这种采用文件与关系数据库管理系统的混合管理模式，还不能说建立了真正意义上的空间数据库管理系统，因为文件管理系统的功能较弱，特别是在数据的安全性、一致性、完整性、并发控制以及数据损坏后的恢复等方面缺少基本的功能。多用户操作的并发控制比起商用数据库管理系统来要逊色得多，因而 GIS 软件商一直在寻找采用商用数据库管理系统来同时管理图形和属性数据。

文件系统结构简单，在数据存取过程中几乎没有额外开销，并且可以按照用户的需求任意定制数据存储格式或存储复杂数据结构。但其缺点也很明显，即数据冗余度大、难以共享数据、容易造成数据的不一致性、程序与数据缺乏独立性、系统不易扩充。

这些产品虽然在功能上达到了空间信息处理的要求，但由于没有从 DBMS 的核心部分，即数据结构部分来考虑空间信息的特殊性，所以处理的效率受到了很大的限制。

6.2.2 集成型空间数据库

近来国外提出的一种集成型结构,是将所有的数据都存储于一个数据库中。由于其采用的数据库原型不同,所以集成性空间数据库的类型也大不相同。

(1) 全关系型数据库模型。

全关系型空间数据库管理系统是指图形和属性数据都用现有的关系数据库管理系统管理。关系数据库管理系统的软件厂商不作任何扩展,由 GIS 软件商在此基础上进行开发,使之不仅能管理结构化的属性数据,而且能管理非结构化的图形数据。其设计思想是把构成空间对象的每个基本元素(点、线段)对作为一行记录存入表中,通过关系模型来维护其空间属性的内部结构以及空间对象之间的拓扑关系。

用关系数据库管理系统管理图形数据的模式有两种,一种是基于关系模型的方式,图形数据按照关系数据模型组织。这种组织方式由于涉及一系列关系连接运算,相当费时,因此在处理空间目标方面效率不高。另一种方式是将图形数据的变长部分处理成 Binary 二进制块 Block 字段。目前大部分关系数据库管理系统都提供了二进制块的字段域,以便管理多媒体数据或可变长文本字符。GIS 利用这种功能,通常把图形的坐标数据,当做一个二进制块,交由关系数据库管理系统存储和管理。这种存储方式虽然省去了前面所述的大量关系连接操作,但是二进制块的读写效率要比定长的属性字段慢得多,特别是涉及对象的嵌套时,速度则更慢。

关系型数据库(RDB)有强大的关系理论支持,无疑是数据库产品中最成熟的。由于它是基于几种简单的基本类型和简单的二维表,因此它能完成非常复杂的关系操作,并能得到几乎任何期望的信息组合。它的优点也是很明显的,第一,物理数据存储与逻辑数据库结构的独立性;第二,多样、简易的数据访问能力,从而提供了高效数据访问的可能性;第三,相当灵活的数据库设计;第四,数据存储和冗余的减少。

采用关系数据库来管理地理信息是 GIS 与数据库结合的初步尝试,但其中存在两个难题:首先,难以在关系数据库中存储空间数据,为了存储和表示结构复杂多变的空间数据,需要设计复杂的 E−R 模型,并在数据库中存放大量的表,这样会增大系统的复杂性,而且还无法利用数据库提供的索引机制;其次,难以保持地理信息的数据一致性,由于地理信息的空间数据和属性数据通常不能存放在一张表里,因此这样会割裂逻辑上为一个整体的地理信息,而且拓扑关系也难以维护。

由于存在上述难题,虽然关系数据库在与 GIS 的结合中做出了有益的尝试,但最终还是没能成为主流。为此,随着面向对象型数据库(OODB)的出现与发展,人们提出了面向对象型数据库与 GIS 相结合的方案。

(2) 面向对象数据库。

面向对象模型最适于空间数据的表达和管理,它不仅支持变长记录,而且支持对象的嵌套、信息的继承与聚集。面向对象的空间数据库管理系统允许用户定义对象和对象的数据结构以及它的操作。这样,我们就可以将空间对象根据 GIS 的需要,定义出合适的数据结构和一组操作。这种空间数据结构可以是不带拓扑关系的面条数据结构,也可以是拓扑数据结构,当采用拓扑数据结构时,往往会涉及对象的嵌套、对象的连接和对象与信息的聚集。

虽然面向对象数据库屏蔽了原有的关系理论,具有完全面向对象的特征,具有很强的理论优势,但其商业化产品则侧重于程序设计语言对复杂数据的访问,同时还存在查询表达能力不强、查询效率比较低的问题。而且由于其中的许多技术细节还不够成熟,因此目前 OODB 还难以用于与地理信息系统应用相关的领域。

(3) 对象—关系模型数据库。

对象—关系模型数据库(ORDB)是面向对象思想与数据库管理系统相结合的一个折中产物,

它与 OODB 的区别在于，它没有抛弃关系理论，而是在关系模型的基础上，进一步支持了面向对象思想。这样不但继承了关系模型的优良特性，并兼容了原有的使用关系数据库的软件系统；而且提供了支持面向对象的类型与接口，也满足了面向对象软件系统的要求。不论在科学研究中，还是在实际应用中，它都占有重要的地位，是当前数据管理技术的中流砥柱。ORDB 在与 GIS 结合的过程中，具有如下明显的优势：① 支持基本类型扩充；② 支持复杂对象；③ 支持继承；④ 支持规则。这样，通过对对象存储访问的支持，不仅解决了关系型数据库所面临的难题，同时通过兼容关系数据库系统，保证了对现有运行系统升级的可能性。而最为重要的是，ORDB 已经拥有比较成熟的商业化产品，这为它与 GIS 的结合打下了良好的基础。

对象—关系模型把关系理论与面向对象思想有机结合，不但继承了关系系统优良的特性，还提供使用面向对象方法建模的能力，满足了用户对数据库表达能力的要求。

① 空间扩展模型。

此方案是基于对象—关系模型数据库而建立的。它通过建立一个空间对象表达模型，把空间数据和属性数据存储统一在 DBMS 之下，从而把 GIS 的数据管理部分的功能统一到 DBMS 之中，由 DBMS 完成数据层功能。其模型的重点在数据库上，希望通过增强数据库的空间功能而从根本上解决问题。这种扩展的空间对象管理模块，主要解决了空间数据变长记录的管理。由于由数据库软件商进行扩展，因此效率要比前面所述的二进制块的管理高得多。空间扩展模型体现出了两个优势：一是由数据库厂商支持的数据分析操作可以得到数据库系统内部优化机制的支持，因而效率较高。二是该模型在 DBMS 层为访问空间对象提供了统一的接口，也方便了系统从 RDB 到 ORDB 的移植。

但是，它还存在着技术不成熟的问题，如它仍然没有解决对象的嵌套问题，空间数据结构也不能由用户任意定义，不能提供对象—关系系统应具有的所有功能。同时空间扩展模型是数据库厂商提供的解决方案，它希望通过增强数据库的空间功能来从根本上解决问题。由于该方案是由数据库厂商开发的，因此能够比较充分地利用底层数据库，从而达到比较好的性能。但是，数据库的空间扩展毕竟是数据库系统的附属产品，它不但继承了底层数据库的优势，也受到底层数据库的制约。例如，只有几个大的数据库厂商提供对空间扩展模型的支持，而这些数据库产品又只能应用于比较大的应用系统中，这样空间扩展模型也就难以投入到中小应用中去。因此这种模型的使用仍然受到了一定程度的限制。

② 基于中间件的数据库模型。

此方案也是基于对象—关系模型数据库而建立的，其本质就是在数据库和最终用户之间增加中间件，通过中间件层实现从空间对象模型到数据库存储的映射，从而避免了对 DBMS 内核的直接修改，以消除数据库和最终用户间接口的差异，这样一方面能充分利用数据库提供的功能，另一方面也能满足用户最终的需求。这种模型的典型例子是 ESRI 的 ArcSDE。

中间件的主要任务就是分析并执行空间对象访问命令，为 GIS 应用提供一个一致且稳定的接口。基于中间件的数据库模型一般包括用户、中间件、数据存储层等 3 个层次。其中，用户使用地理信息，数据存储层负责存储地理信息，而中间件则负责用户与数据库之间的数据访问管理。对上层（GIS 应用及 SDSS 应用）的接口，提供完整的空间对象管理功能，统一管理空间属性和用户属性。对下层（数据库）的接口，则将空间属性与用户属性分开管理，空间属性的管理功能由中间件和 DBMS 共同实现，用户属性的管理功能则由 DBMS 直接实现。由于支持面向对象方法，因此中间件为空间对象提供了统一的访问接口，同时中间件可以实现 DBMS 不能完成但效率不高的部分。

基于中间件的数据库模型则是 GIS 厂商提供的解决方案，它希望通过提供灵活方便的接口来适应多变的应用需求，即通过设计中间件来封装数据管理部分，以提供对多种类型数据源的支持。这虽然降低了 GIS 应用开发的复杂度，但是，增加中间件也势必在空间数据存取中增加了额外开销，从而影响系统的整体效率，另外，构造并维护中间件的人力物力投入也不容忽视。与空间扩展模型相比，中间件模型是一种以性能换功能的方法，但它在当前的应用中却又是相当有效的。其特点也是很明显的，主要表现在以下几个方面。

- GIS 应用接口稳定，使得基于它的 GIS 应用也具有较强的稳定性。
- 易于裁剪，采用构件方式实现可以进行多层次的裁剪，以取得良好的性能/功能比。
- 数据库接口统一，不依赖于特定的数据库产品。
- 易于扩展，可以针对效率问题在中间件层提供一定的优化，如智能 CACHE 技术、空间索引过滤和查询优化，可以针对领域应用扩展功能，比如为了满足空间数据库在 SDSS 中的特殊要求，还可以在中间件层添加辅助空间数据挖掘的调用模块和支持空间知识库系统的调用模块等。

从本质上看，基于中间件的数据库模型和空间扩展模型的目的都是一样的，即都要为 GIS 应用提供方便高效的服务，都要提供一个空间对象的统一访问接口。而它们的区别在于，基于中间件的数据库模型是对应用需求的直接响应，也是空间扩展模型的基础，而空间扩展模型则是基于中间件的数据库模型的总结与改进。相比之下，由于基于中间件的数据库模型具有灵活的结构，所以它不但在研究方面很有潜力，而且可以方便地实验各种数据管理技术。

6.3 空间数据模型

空间数据模型是地理信息系统的基础，它不仅决定了系统数据管理的有效性，而且是系统灵活性的关键。空间数据模型是在实体概念的基础上发展起来的，它包含两个基本内容，即实体组和它们之间的相关关系。实体和相关关系可以通过性质和属性来说明。空间数据模型可以被定义为一组由相关关系联系在一起的实体集。

结合空间数据的具体特点进行空间数据模型的设计，是地理信息系统的关键。空间数据模型的设计，主要是构建一个能够用真实世界的抽象提取来代表该真实世界的模型。由于空间数据模型的设计与计算机硬件、系统软件和工具软件的发展现状密切相关，所以，就目前的发展现状而言，很难用一个统一的数据模型来表达复杂多变的地理空间实体。例如，某些空间数据模型可能很适合于绘图，但它们对于空间分析来说效率却十分低；有些数据模型有利于空间分析，但对图形的处理则不理想。

目前，与 GIS 设计有关的空间数据模型主要有矢量模型、栅格模型、数字高程模型、面向对象模型、矢量和栅格的混合数据模型等。前 4 种模型属于定向性模型，在模型设计时只包括与应用目标有关的实体及其相互关系，而混合模型的设计则包括所有能够指出的实体及其相互关系。就目前的应用现状而言，矢量模型、栅格模型、数字高程模型已相当成熟（目前成熟的商业化 GIS 主要是采用这 3 类模型），而其他模型，特别是混合模型则处于大力发展之中。

6.3.1 矢量模型（vector model）

矢量模型是利用边界或表面来表达空间目标对象的面或体要素，通过记录目标的边界，同时采用标识符（Identifier）表达它的属性来描述空间对象实体。矢量模型能够方便地用于比例尺变换、投影变换以及图形的输入和输出。矢量模型处理的空间图形实体是点（point）、线（line）和面（area）。矢量模型的基本类型起源于"Spaghetti"模型。在 Spaghetti 模型中，点用空间坐标对表示，线由一串坐标对表示，面则是由线形成的闭合多边形。CAD 等绘图系统大多采用 Spaghetti 模型。

GIS 的矢量数据模型与 Spaghetti 模型的主要区别是：前者通过拓扑结构数据来描述空间目标之间的空间关系，而后者则没有。在矢量模型中，拓扑关系是进行空间分析的关键。在 GIS 的拓扑数据模型中，与点、线、面相对应的空间图形实体主要有结点（node）、弧段（arc）和多边形（polygon），多边形的边界被分割成一系列的弧和结点，结点、弧、多边形间的空间关系在数据结构或属性表中加以定义。GIS 的矢量数据模型具有以下几个特点。

① 通过对结点、弧、多边形拓扑关系的描述，相邻弧段的公用结点和相邻多边形的公用弧段，在计算机中只需记录一次。而在 Spaghetti 模型中，记录的次数则大于 1。

② 空间图形实体的拓扑关系，如拓扑邻接、拓扑关联、拓扑包含等，不会随着诸如移动、缩放、旋转等变换而变化，而空间坐标及一些几何属性（如面积、周长、方向等）则会受到影响。

③ 一般情况下，通过矢量模型所表达的空间图形实体数据文件占用的存储空间要比栅格模型小。

④ 能够精确地表达图形目标，精确地计算空间目标的参数（如周长、面积）。

6.3.2 栅格模型（raster model）

栅格模型直接采用面域或空域枚举来直接描述空间目标对象。在栅格模型中，点（点状符号）是由一个或多个像元，线是由一串彼此相连的像元构成。在栅格模型中，每一像元的大小是一致的（一般是正方形），而且每一个栅格像元层记录着不同的属性（如植被类型等）。像元的位置由纵横坐标（行列）决定。所以，每个像元的空间坐标不一定要直接记录，因为像元记录的顺序已经隐含了空间坐标。

栅格模型具有以下几个特点。

① 栅格的空间分辨率是指一个像元在地面所代表的实际面积大小（一个正方形的面积）。

② 对于同一幅图形或图像来说，随着分辨率的增大，存储空间也随之增大。例如每一像元占用一个字节，而且分辨率为 100m，那么，一个面积为 10km×10km=100km 的区域就有 1000×1000=1000000 个像元，所占，存储空间为 1000000 个字节；如果分辨率为 10m，那么，同样面积的区域就有 10000×10000=1 亿个像元，所占存储空间近 100MB。

③ 表达空间目标、计算空间实体相关参数的精度与分辨率密切相关，分辨率越高，精度越高。

④ 常适合进行空间分析。例如，同一地区多幅遥感图像的叠加操作等。

⑤ 不适合进行比例尺变化和投影变换等。

6.3.3 数字高程模型（DEM，Digital Elevation Model）

数字高程模型是采用规则或不规则多边形拟合面状空间对象的表面，主要是对数字高程表面的描述。根据多边形的形状，可以把数字高程模型分为两种，即网格模型和不规则三角网模型。

（1）网格模型(Grid model)。

与栅格模型相似，同样是直接采用面域或空域枚举来描述空间目标对象。一般情况下，栅格模型的每一像元或像元的中心点代表一定面积范围内空间对象或实体的各种空间几何特征和属性几何特征，而网格模型通常以行列的交点特征值代表交点附近空间对象或实体的各种空间几何特征和属性几何特征。栅格模型主要用于图像的分析和处理，而网格模型则主要进行等值线的自动生成、坡度、坡向的分析等。栅格模型处理的数据主要来源于航空、航天摄影以及视频图像等，而网格模型则主要来源于原始空间数据的插值。

（2）TIN 模型。

TIN 模型是利用不规则三角形来描述数字高程表面。在 TIN 模型中，同样可以建立三角形顶点（数据点）、三角形边、三角形个体间的拓扑关系。如果建立了 TIN 模型图形实体（三角形顶点、三角形边、二角形）的拓扑关系，将大大加快处理三角形的速度。

归纳起来，数字高程模型的主要优点是能够方便地进行空间分析和计算，如对地表坡度、坡向的计算等。

6.3.4 面向对象的数据模型（Object-Oriented Data Model）

面向对象表示方法的最大优点是便于表达复杂的目标。面向对象的方法为数据模型的建立提供了分类、概括、联合和聚集等 4 种数据处理技术，这些技术对复杂空间数据的表达较为理想。在概括、联合、聚集等技术的运用中，都要涉及对象或类型的属性值或属性结构在不同级或者层之间的传递或继承。为此，面向对象的方法提供了继承和传播两种工具。

在类型的层次结构中，子类的属性结构或操作方法可以部分地从超类中获取，这就是继承。继承可以减少数据冗余，并且有助于保持数据的完整性。

无论是联合还是聚集，它们都有一个共同之处，那就是将一组对象合并成为一个更加复杂的对象，这一类复杂对象的属性值来源于两个方面。

① 一部分属性值由该复杂对象本身定义，与构成它的成员对象（组件对象）无关。

② 另一部分属性值依赖于这些成员对象。

为此，复杂对象必须具备获取成员对象属性值的能力。传播正是一种用来描述复杂对象的依赖性并获取成员对象属性值的工具，这种工具所基于的最基本的原理就是成员对象的相关属性只能存储一次，然后再将这些属性值传给复杂对象，这样当成员对象的属性值被改变后，复杂对象的属性值则无需修改。显然，这种工具的使用大大地减少了数据冗余量，并保证了数据的一致性。

通过采用继承和传播两种工具，将使得原本十分复杂的描述变得自然和简单。例如，一个城市由若干个区组成，因此，城市的人口应为各区人口的总和。对于城市这样一个复杂对象，它的人口数由它的成员对象，即各区的人口数求和派生而成。

6.3.5 混合数据模型（Hybrid Model）

由于矢量模型、栅格模型、数字高程模型在处理空间对象时都有各自的优缺点，所以，能否在一个统一模型的基础上充分利用相关模型的优点就是目前 GIS 界研究的方向之一。到目前为止，有代表性的研究成果主要有 TIN 与矢量的一体化模型，栅格与矢量一体化的多级网格模型。下面简单介绍一些这方面的内容。

（1）矢量与 TIN 的一体化的数据模型。

发展矢量与 TIN 一体化数据模型的主要原因是：它既能发挥 TIN 的空间分析和计算功能，又能方便地查询属性信息。因为 TIN 的三角形顶点和边（如测量点、河流走向线、陆上径流路径、封闭洼地边界线等）可能为矢量图形的特征点和线的一部分，所以它们的系统代码或内部码是完全一致的。模型将空间数据类型抽象为点状目标、线状目标、面状目标，这 3 类图形目标又通过特殊点、结点、一般线段、弧以及多边形等加以表达。

TIN 与矢量之间存在着部分（part of）或包含的关系，实际上也是通过这种关系实现一体化的。由于 TIN 是以矢量系统的矢量图形为基础生成的，所以，单个三角形顶点、边、三角形本身包含的空间区域可能是矢量系统点、线、面状图形目标的一部分。如果在 TIN 和矢量数据结构中保持点和顶点、线和边、面和三角形有关标识号的统一，那么，就能实现 TIN 和矢量的一体化。

（2）多级网格模型。

多级网格模型的本质是在基本网格的基础上细分为 256×256 或 16×16 个网格，然后基本网格和细分网格都采用四叉树的编码方法去表达点、线、面的有关参数。具体来讲，就是要遵循如下 3 点约定：

① 点状地物是仅有空间位置，没有形状和面积，在计算机内仅有一个位置数据；

② 线性地物是有形状，没有面积，在计算机内由一组原子填满路径；

③ 面状地物是有形状和面积，在计算机内是由一组填满路径的原子表达的边界线。

6.4 Geodatabase 体系结构

6.4.1 Geodatabase 介绍

随着关系数据库技术的完善,变长信息的存储成为可能,人们开始采用关系数据库技术管理空间数据,并逐渐成为主导。利用关系数据库管理空间几何信息、属性信息、拓扑信息、三维信息以及多媒体信息,使得 GIS 软件能够充分地利用商用数据库中已经成熟的众多特征,如快速索引、数据完整性和一致性、并发控制等机制。在此情况下,ESRI 公司推出了 Geodatabase,该数据模型采用全关系数据库管理空间数据,与各种商用关系数据库兼容。Geodatabase 充分利用面向对象技术,将空间要素的属性和行为有机地结合在一起,采用现代化的、标准的方式实现对象,对象都定义为一个 COM 组件,并允许用户在这些基本模型的基础上扩展自己的面向对象数据模型。

Geodatabase 是面向对象的空间数据存储模型,ArcSDE 是 ESRI 公司针对空间数据的存储问题推出的一套空间数据库管理软件。通过 SDE,用户可以将多种数据产品按照 Geodatabase 模型存储于商用数据库系统中,并获得管理和检索服务。

Geodatabase 提供了可伸缩的空间数据存储方案,提供了两种存储方式:personal database 和 Enterprise Geodatabase。Personal database 适用于单机环境下的 GIS 应用,实际上就是一个 Access 数据库,Personal database 通过 Microsoft Jet Engine 将空间数据存放在 Access 数据库中。Enterprise Geodatabase 主要用于在网络环境下工作的 GIS,适合大型数据库在网络环境下进行多用户并发操作和版本控制。GIS 用户可以通过建立版本来并发访问 Geodatabase 中的空间数据。

Geodatabase 提供了 4 种空间数据的表述形式:描述要素的矢量数据、描述影像和网格的栅格数据、描述表面不规则三角网络、用于地址定位的 Addresses 和 Locator。

6.4.2 Geodatabase 的体系结构

Geodatabase 的基本体系结构包括要素数据集、栅格数据集、TIN 数据集、独立的对象类、独立的关系类和属性域等。其中要素数据集又由对象类、要素类、关系类等构成。

(1) 对象类(Object Class)。

对象类表示的是非空间实体,不在地图上直接表示,但与地图上的直接表示的图形要素关联。对象类可以组织到一个要素数据集中,也可以单独存在。

(2) 要素类(Feature Class)。

要素类表示的是具有相同几何形状的空间实体。要素类可以独立存在,也可以具有拓扑关系。当不同的要素类之间存在拓扑关系时,应组织到一个要素数据集中。要素类可细分为点状要素类、线状要素类、面状要素类等。

要素类与对象类的主要区别在于:要素类中存储了空间信息,而对象类中却没有。

(3) 要素数据集(Feature Datasets)。

要素数据集由一组具有相同空间参考(spatial reference)的要素类组成,用于存放矢量数据。

(4) 栅格数据集(Raster Datasets)。

栅格数据集用于存放栅格数据,可以支持海量栅格数据,支持影像镶嵌。可以通过建立"金字塔"索引,从而提高海量栅格数据的检索速度和显示效率。

(5) TIN 数据集(Tin Datasets)。

TIN 数据集由一系列不规则的三角形构成,这些不规则的三角形表示了地表的起伏。

(6) 关系类(Relationship Class)。

关系类用于定义两个不同的要素类或对象类之间的关联关系。

（7）属性域（Domain）。

属性域用于定义属性的有效取值范围，可以是连续的变化区间，也可以是离散的。

（8）几何网络（Geometric Network）。

几何网络是在若干要素类的基础上建立的一个新的类。几何网络包括网络的边要素和点要素，例如供水系统中的几何网络，水管是边要素，点要素就是阀门。

6.5 Geodatabase 对象模型

6.5.1 Geodatabase 中的主要类

Geodatabase 中的主要类可以分为工作区部分和数据集部分，这两部分包含了 Geodatabase 中最基础、最主要的类。工作区部分主要负责对 Geodatabase 中的各种数据源进行宏观管理，而数据集部分主要是用于对数据库中各种具体的数据进行描述和管理。

（1）WorkspaceFactory 类。

该类用于生成 Workspace，该类允许用户通过事先指定的连接属性连接一个 Workspace，这些连接属性一般使用 PropertySet 对象定义，并且可以存放在一个连接文件中。WorkspaceFactory 是一个可创建的对象，具有一个连接池，该连接池存放了与当前应用连接、处于激活状态的 Workspace。该类还提供了浏览、管理基于文件系统的 Workspace 的方法，同时还提供了管理远程数据库中的数据的功能。

该类实现了 IWorkspaceFactory 和 IWorkspaceFactory2 两个接口，提供了用于创建、打开 Workspace 和获取一个 Workspacefactory 的属性的方法。该类还派生 ArcInfoWorkspaceFactory、CadWorkspaceFactory、IMsWorkspaceFactory、ShapefileworkspaceFactory、SdeWorkspaceFactory 和 AccessWorkspaceFactory 等。

IworkspaceFactory 接口提供了创建和打开 Workspace 的方法，同时提供了获取 Workspace 的相关属性信息的方法。该接口提供了 10 个方法和两个属性。如 Copy 方法，用于将一个 Workspace 复制到指定的目录；GetWorkspaceName 方法，用于获得一个具有指定文件名列表对应的 Workspace 的名称；Move 方法，用于将一个 Workspace 移到指定的文件夹下；Open 方法，用于打开一个具有指定连接属性的 Workspace，这个方法用于打开 ArcSDE 数据库；OpenFromFile 方法，用于根据指定的文件名或者目录打开一个 Workspace，这个方法通常用于打开文件系统工作区或本地数据库工作区，也可以根据连接属性文件打开 SDE 数据库工作区。

IWorkspaceFactory2 接口的作用和 IWorkspaceFactory 接口基本一致，只是这个接口另外提供了直接根据一个描述连接属性的字符串来打开一个 Workspace 的方法 OpenFromString。

（2）Workspace 类。

Workspace 是一个用于存放空间数据和非空间数据的容器，可以存放 FeatureDatasets、Raster datasets 和 Tables 等各种数据。一个 Workspace 就是一个数据集集合，该类提供了列举、添加、删除、复制和更改它所包含的数据集的方法，还提供了创建数据集的方法。根据枚举 esriWorkspaceType，Workspace 可以分为 esriFileSystemWorkspace（文件系统工作区）、esriLocalDatabaseworkspace（本地数据库工作区）、esriRemoteDatabaseWorkspace（分布式远程数据库工作区）等。Shapefiles 和 ArcInfo workspaces 就是文件系统工作区类型的 Workspace。该类实现了 IWorkspace、IWorkspace2、IWorkspaceDomains2 等众多接口。

一个 Workspace 可以被看做是文件系统中的目录，也可以被看做一个关系数据库。在访问一个

Workspace 中的数据库之前,用户首先必须打开这个 Workspace。用户不能直接创建一个 Workspace,一般是由 WorkspaceFactory 生成。

IWorkspace 接口提供了获取 Workspace 各种信息的方法,该接口提供了 3 个方法和 6 个属性。如 Exists 用于判断当前 Workspace 是否存在,IsDirectoy 用于判断当前 Workspace 是否是文件系统下的一个目录,WorkspaceFactory 用于获取当前 Workspace 对应的 WorkspaceFactory。

(3)FeatureDataset 组件类。

FeatureDataset 组件类是一个用来存储具有与该 FeatureDataset 同空间参考的多个 Feature Class 的数据集。FeatureDataset 还可以存储 Geometric Network 和 Relationship Class。该类实现了 IFeatureDataset 接口,该接口提供继承 IDataset 接口。IFeatureDataset 提供了在一个 Dataset 中创建一个新的 Feature Class 的功能,共有 7 个方法和 9 个属性,如 CreateFeatureClass 方法用于创建一个新的 Feature Class。

6.5.2 Geodatabase 中的其他常用类

Geodatabase 的 OMD 中还提供了一些常用类,如与空间数据入库相关的类、与查询相关的类。空间入库相关类还同时实现了各种数据格式的转换,如 FeatureDataConverter、FieldChecker、EnumFieldError 等。其中 FeatureDataConverter 类是最核心的类,所有的数据导入功能都是通过这个类实现的。FeatureDataConverter 类主要用于实现类似 ArcCatalog 的功能;FieldChecker 类用于对属性字段进行有效性判定;EnumFieldError 类是一个枚举,用来记录所有数据转换失败的无效字段的名称。

FeatureDataConverter 类实现了 IfeatureDataConverter 和 IFeatureDataConverter2 两个接口。IFeatureDataConverter 接口用于对不同格式的数据进行相互转换,提供了 3 个方法,如 ConvertFeatureClass 用于将 FeatureClass 导入到数据库中,ConvertFeatureDataset 用于将要素数据集导入到数据库中,ConvertTable 用于将一个表格数据导入到数据库中。IFeatureDataConverter2 接口在功能上基本与 IFeatureDataConverter 接口一致。

FieldChecker 类实现了 IFieldChecker 接口,用于对属性字段进行有效性判定,如 Validate 方法用于检查一系列属性字段的有效性。

查询相关组件类有 QueryFilter 和 SpatialFilter,这两个类将在第 10 章介绍。

6.6 Geodatabase 的使用与开发

6.6.1 空间数据库连接

空间数据库企业级 Geodatabase 的连接主要是使用 IworkspaceFactory、Iworkspace 和 IpropertySet 等 3 个接口,通过 IPropertySet 设置数据库的连接属性,由 IworkspaceFactory 创建一个 IWorkspace 的接口指针并返回该指针。6.6.3 小节的代码演示了如何进行空间数据库的连接。

6.6.2 创建新的数据集

数据集是由一组具有相同空间参考的要素类组成,用于存放数据。创建新的数据集主要使用 IWorkspace、IFeatureWorkspace、IfeatureDataset 和 IspatialReference 等 4 个接口,其中,IWorkspace 用于生成 IFeatureWorkspace,ISpatialReference 用于设置要素数据集的空间参考属性(该属性为可选属性),整个过程返回 IFeatureDataset 接口指针,即新生成的要素数据集。

6.6.3 空间数据的入库

下面代码使用 SDE 空间数据库,因此 ArcEngine 的 Lisence 应使用企业数据库类型,在 Form1_Load 事件中初始化 lisence,不能使用 lisence 控件设置,否则提示没有许可 lisence。

```csharp
private void Form1_Load(object sender, EventArgs e)
{
    IAoInitialize pao = new AoInitializeClass();
    pao.Initialize(esriLicenseProductCode.esriLicenseProductCodeEngineGeoDB);

}
// 工作空间
IWorkspace workspace;
//矢量数据工作空间
IFeatureWorkspace featureWorkspace;
//影像数据工作空间
IRasterWorkspaceEx rasterWorkspace;
//矢量数据集
IFeatureDataset featureDataset;
//影像数据集
IRasterDataset rasterDataset;
private void button1_Click(object sender, EventArgs e)
{
    // SDE 空间连接属性
    IPropertySet propertySet = new PropertySetClass();
    propertySet.SetProperty("server", this.textBox1.Text );
    propertySet.SetProperty("instance",this.textBox2.Text );
    propertySet.SetProperty ("database",this.textBox3.Text );
    propertySet.SetProperty("user", this.textBox4.Text );
    propertySet.SetProperty ("password",this.textBox5.Text );
    propertySet.SetProperty("version", "SDE.DEFAULT");
    IWorkspaceFactory workspaceFactory = new SdeWorkspaceFactory();
    //打开 SDE 工作空间

    workspace = workspaceFactory.Open(propertySet, 0);
    MessageBox.Show("连接 SDE 空间数据库成功");

}
//创建数据集(矢量数据集和影像数据集)
private void button2_Click(object sender, EventArgs e)
{
    featureWorkspace =workspace as IFeatureWorkspace ;
    rasterWorkspace = workspace as IRasterWorkspaceEx;
    //定义空间参考
    ISpatialReferenceFactory spatialReferenceFactory = new SpatialReferenceEnvironmentClass();
    ISpatialReference spatialReference = spatialReferenceFactory.CreateGeographicCoordinateSystem((int)esriSRGeoCSType.esriSRGeoCS_Beijing1954);
    spatialReference.SetDomain(-1000, -1000, 1000, 1000);

    IEnumDatasetName enumDatasetName;
    IDatasetName    datasetName;
    string dsName = "";
    enumDatasetName = workspace.get_DatasetNames(esriDatasetType.esriDTFeatureDataset);
    datasetName = enumDatasetName.Next();
    bool isExist = false;
    //创建矢量数据集
    dsName = "SDE."+this.textBox6.Text;
    while (datasetName != null)
```

```csharp
        {
            if (dataSetName.Name == dsName)
            {
                isExist = true;
            }
            datasetName = enumDatasetName.Next();
        }
        if (isExist ==false )
        {
            featureDataset = featureWorkspace.CreateFeatureDataset(this.textBox6.Text,
            spatialReference);
        }
        //创建影像数据集
        isExist = false;
        enumDatasetName = workspace.get_DatasetNames(esriDatasetType.esriDTRaster
        Dataset );
        datasetName = enumDatasetName.Next();

        dsName = "SDE." + this.textBox6.Text;
        while (datasetName != null)
        {
            if (dataSetName.Name == dsName)
            {
                isExist = true;
            }
            datasetName = enumDatasetName.Next();
        }
        if (isExist == false)
        {
            //设置存储参数
            IRasterStorageDef rasterStorageDef = new RasterStorageDefClass();
            rasterStorageDef.CompressionType = esriRasterCompressionType.
            esriRasterCompressionUncompressed;
            rasterStorageDef.PyramidLevel = 1;
            rasterStorageDef.PyramidResampleType = rstResamplingTypes.RSP_
            BilinearInterpolation;
            rasterStorageDef.TileHeight = 128;
            rasterStorageDef.TileWidth = 128;
            //设置坐标系统
            IRasterDef rasterDef = new RasterDefClass();
            ISpatialReference rasterDpatialRefrence = new UnknownCoordinateSystem Class();
            rasterDef.SpatialReference = rasterDpatialRefrence;

            IGeometryDef geometryDef = new GeometryDefClass();
            IGeometryDefEdit geometryDefedit = (IGeometryDefEdit)geometryDef;
            geometryDefedit.AvgNumPoints_2 = 5;
            geometryDefedit.GridCount_2 = 1;
            geometryDefedit.set_GridSize(0, 1000);
            geometryDefedit.GeometryType_2 = esriGeometryType.esriGeometry Polygon;
            ISpatialReference spatialReference2 = new UnknownCoordinateSystem Class();
            geometryDefedit.SpatialReference_2 = spatialReference2;
            rasterDataset = rasterWorkspace.CreateRasterDataset(this.textBox7.Text, 1,
            rstPixelType.PT_LONG, rasterStorageDef, "DEFAULTS", rasterDef, geometry Def);

        }

}
//加载矢量数据到 SDE 数据库
private void button3_Click(object sender, EventArgs e)
{
    featureWorkspace = workspace as IFeatureWorkspace;
    this.openFileDialog1.Filter = "shp file (*.shp)|*.shp";
```

```csharp
this.openFileDialog1.Title = "打开矢量数据";
this.openFileDialog1.Multiselect = false;
string fileName = "";
if (this.openFileDialog1.ShowDialog() == DialogResult.OK)
{
    fileName = this.openFileDialog1.FileName;
    string filepath;
    string file;
    int lastIndex;
    lastIndex = fileName.LastIndexOf(@"\");
    filepath = fileName.Substring(0, lastIndex );
    file = fileName.Substring(lastIndex+1);
    //读取SHP数据
    IWorkspaceFactory shpwpf = new ShapefileWorkspaceFactoryClass();
    IWorkspace shpwp = shpwpf.OpenFromFile(filepath, 0);
    IFeatureWorkspace shpfwp = shpwp as IFeatureWorkspace;
    IFeatureClass shpfc = shpfwp.OpenFeatureClass(file);

    //导入SDE数据库
    IFeatureClass sdeFeatureClass=null;
    IFeatureClassDescription featureClassDescription = new FeatureClass
    DescriptionClass();
    IObjectClassDescription objectClassDescription = featureClassDescription
    as IObjectClassDescription;
    IFields fields = shpfc.Fields;
    IFieldChecker fieldChecker = new FieldCheckerClass();
    IEnumFieldError enumFieldError = null;
    IFields validateFields = null;
    fieldChecker.ValidateWorkspace = featureWorkspace as IWorkspace;
    fieldChecker.Validate(fields, out enumFieldError, out validateFields);
    featureDataset = featureWorkspace.OpenFeatureDataset(this.textBox6.Text);
    try
    {
        sdeFeatureClass = featureWorkspace.OpenFeatureClass(shpfc.Alias Name);
    }
    catch (Exception ex)
    {
    }
    //在SDE数据库中创建矢量数据集
    if (sdeFeatureClass == null)
    {
        sdeFeatureClass = featureDataset.CreateFeatureClass(shpfc. AliasName,
        validateFields,objectClassDescription.InstanceCLSID,objectClassDescription.
        ClassExtensionCLSID, shpfc.FeatureType, shpfc.ShapeFieldName, "");
    }
    IFeatureCursor featureCursor = shpfc.Search(null, true);
    IFeature feature = featureCursor.NextFeature();
    IFeatureCursor sdeFeatureCursor=  sdeFeatureClass.Insert(true);
    IFeatureBuffer sdeFeatureBuffer;
    //添加实体对象
    while (feature != null)
    {
        sdeFeatureBuffer = sdeFeatureClass.CreateFeatureBuffer();
        IField shpField = new FieldClass();
        IFields shpFields = feature.Fields;
        for (int i = 0; i < shpFields.FieldCount; i++)
        {
            shpField = shpFields.get_Field(i);
            int index = sdeFeatureBuffer.Fields.FindField(shpField.Name);
            if (index != -1)
            {
                sdeFeatureBuffer.set_Value(index, feature.get_Value(i));
```

```csharp
            }
        }
        sdeFeatureCursor.InsertFeature(sdeFeatureBuffer);
        sdeFeatureCursor.Flush();
        feature = featureCursor.NextFeature();
    }
    //加载数据到 Mapcontrol
    IFeatureLayer sdeFeatureLayer = new FeatureLayerClass();
    sdeFeatureLayer.FeatureClass = sdeFeatureClass;
    this.axMapControl1.Map.AddLayer(sdeFeatureLayer as ILayer);
    this.axMapControl1.Extent = this.axMapControl1.FullExtent;
    this.axMapControl1.Refresh();
  }

}
//加载影像数据到 SDE 数据库
private void button4_Click(object sender, EventArgs e)
{
    this.openFileDialog1.Filter = "TIFF file (*.tif)|*.tif";
    this.openFileDialog1.Title = "打开影像数据";
    this.openFileDialog1.Multiselect = false;
    string fileName = "";
    if (this.openFileDialog1.ShowDialog() == DialogResult.OK)
    {
        fileName = this.openFileDialog1.FileName;
        string filepath;
        string file;
        int lastIndex;
        lastIndex = fileName.LastIndexOf(@"\");
        filepath = fileName.Substring(0, lastIndex);
        file = fileName.Substring(lastIndex + 1);

        //导入 SDE 数据库
        rasterWorkspace = workspace as IRasterWorkspaceEx;
        IWorkspaceFactory tifwpf = new RasterWorkspaceFactoryClass();
        IWorkspace tifwp = tifwpf.OpenFromFile(filepath, 0);
        IRasterWorkspace tifrwp = tifwp as IRasterWorkspace;
        IRasterDataset rasterDataset = tifrwp.OpenRasterDataset(file);
        IRasterDataset sdeRasterDataset = null;
        lastIndex = file.LastIndexOf(@".");
        file = file.Substring(0, lastIndex);
        try
        {
            sdeRasterDataset = rasterWorkspace.OpenRasterDataset(file);
        }
        catch (Exception Ex)
        {
        }
        if (sdeRasterDataset == null)
        {
            IGeoDataset geoDataset =rasterDataset as IGeoDataset ;
            IRasterSdeServerOperation rasterSdeServeroperation ;

            IBasicRasterSdeConnection sdeCon = new BasicRasterSdeLoader();
            IPropertySet propertySet = new PropertySetClass();
            propertySet = workspace.ConnectionProperties;
            //建立与 SDE 数据库的连接
            sdeCon.ServerName = propertySet.GetProperty("server").ToString();
            sdeCon.Instance = propertySet.GetProperty("instance").ToString();
            sdeCon.UserName = propertySet.GetProperty("user").ToString();
            sdeCon.Password = "sde";
            sdeCon.Database = propertySet.GetProperty("database").ToString();
```

```csharp
                    sdeCon.SdeRasterName = file;
                    sdeCon.InputRasterName = fileName;
                    rasterSdeServeroperation = sdeCon as IRasterSdeServerOperation;
                    //保存影像数据到 SDE 数据库中
                    rasterSdeServeroperation.Create ();
                    rasterSdeServeroperation.Update ();
                    rasterSdeServeroperation.ComputeStatistics ();
                    IRasterLayer rasterLayer=new RasterLayerClass ();
                    sdeRasterDataset = rasterWorkspace.OpenRasterDataset(file);
                    rasterLayer.CreateFromDataset(sdeRasterDataset);
                    this.axMapControl1.Map.AddLayer(rasterLayer as ILayer);
                    this.axMapControl1.Extent = this.axMapControl1.FullExtent;
                    this.axMapControl1.Refresh();
                }
            }
        }
    }
}
```

6.7 本章小结

　　空间数据库是地理信息系统的基础，空间数据库性能优劣直接关系到系统的性能。通过几十年的发展，空间数据库也逐步发展成熟，本章介绍空间数据库特征、并着重介绍了空间数据库的组织：混合型数据库和集成型空间数据库。空间数据库只管理和装载空间数据的容器，空间数据具有多种模型：矢量模型、栅格模型、数字高程模型、面向对象模型、混合数据模型。这些数据模型都有着各自的优点和缺点，在空间数据库的构建时应综合考虑这些模型的优劣，并进行权衡考虑。

　　ESRI 的 Geodatabase 是面向对象的空间数据存储模型，该模型提供了 4 种空间数据的表述形式：描述要素的矢量数据、描述影像和格网的栅格数据、描述表面不规则三角网络、用于地址定位的 Addresses 和 Locator。也是初学者应重点掌握的内容。

第 7 章　数据编辑

7.1 简介

空间数据编辑是 GIS 的基本功能之一，ArcEngine 提供了丰富的编辑功能。其中，IWorkspaceEdit 接口是空间数据编辑功能最重要的接口，通过它可以启动或停止一个编辑流程。GIS 的编辑操作往往是长事务，因此需要确保用户在这段时间内所做的编辑能够恢复。该接口提供了 13 种方法，如 StartEditing 方法用于启动一个编辑流程，WithUndoRedo 参数用来确定工作空间是否支持"恢复/取消恢复"操作；AbortEditoperation 方法用于在编辑过程中如果出现异常，则取消所有的编辑操作；StopEditoperation 方法用来确保编辑操作的完成；UndoEditoperation 方法用于编辑操作的回滚；StopEditing 方法用来完成编辑。本章通过两个例子程序介绍要素的编辑和高级编辑。

7.2 捕捉功能设计与实现

捕捉是新建地理空间对象与已有地理空间对象的一种关联关系。例如，要在自来水管线上增加一个阀门，来控制管线中水的流动，因此阀门应该位于管线之上，并在此处打断管线，如何将阀门准确地定位到管线上，有没有偏离？这需要用捕捉来实现。当用户在创建阀门的时候，应当尽可能将输入点置于管线附近，如果阀门与管线的距离小于一定的范围（即容差），就认为用户捕捉到管线，同时返回阀门向管线引垂线与管线交点的坐标。

Engine 提供了 3 个接口 ImovePointFeedback、ImoveLineFeedback、ImovePolygonFeedback，分别用于点、线、面的捕捉，捕捉功能的实现原理是，鼠标光标移动时系统自动根据设置的容差、间隔时间、捕捉图层，在捕捉图层上自动找到最近的点：

```
bool bCreateElement = true;
int internalTime = 5;                          //时间间隔
int snapTime = 10;                             //捕捉初始值
IElement m_element = null;                     //界面绘制点元素
IPoint currentPoint = new PointClass();        //当前鼠标光标位置
IPoint snapPoint = null;                       //捕捉到的点
IMovePointFeedback movePointFeedback = new MovePointFeedbackClass();
string snapLayer ="";
public Form1()
{
    InitializeComponent();
}
private void Form1_Load(object sender, EventArgs e)
{
    //加载地图文档
    loadMapDocument();
```

```csharp
    for (int i = 0; i < this.axMapControl1.LayerCount; i++)
    {
        cbLayerName.Items.Add(this.axMapControl1.get_Layer(i).Name);
    }
    cbLayerName.Text = cbLayerName.Items[0].ToString();
    snapLayer =cbLayerName.Text ;
}

//鼠标光标移动事件
private void axMapControl1_OnMouseMove(object sender, ESRI.ArcGIS.Controls.ImapControlEvents2_OnMouseMoveEvent e)
{
    currentPoint.PutCoords(e.mapX, e.mapY);

    snapTime++;
    snapTime = snapTime % internalTime;
    ILayer layer = GetLayerByName(snapLayer, axMapControl1);
    IFeatureLayer featureLayer = layer as IFeatureLayer;
    if (bCreateElement)
    {
        CreateMarkerElement(currentPoint);
        bCreateElement = false;
    }
    if (snapPoint == null)
        ElementMoveTo(currentPoint);
    //鼠标光标自动捕获顶点
    if (snapTime == 0)
        snapPoint = Snapping(e.mapX, e.mapY, featureLayer );
    if (snapPoint != null && snapTime == 0)
        ElementMoveTo(snapPoint);

}
//捕捉
public IPoint Snapping(double x, double y, IFeatureLayer featureLayer)
{
    IMap map = this.axMapControl1.Map;
    IActiveView activeView = this.axMapControl1.ActiveView;
    IFeatureClass featureClass = featureLayer.FeatureClass;
    IPoint point = new PointClass();
    point.PutCoords(x, y);

    IFeature  feature = featureClass.GetFeature(0);
    IPoint hitPoint1 = new PointClass();
    IPoint hitPoint2 = new PointClass();
    IHitTest hitTest = feature.Shape as IHitTest;
    double hitDist = 0;
    int partIndex = 0;
    int vertexIndex = 0;
    bool bVertexHit = false;

    double tol = ConvertPixelsToMapUnits(activeView , 8);
    if (hitTest.HitTest(point, tol, esriGeometryHitPartType.esriGeometryPartBoundary,
        hitPoint2, ref hitDist, ref partIndex, ref vertexIndex, ref bVertexHit))
    {
        hitPoint1 = hitPoint2;
    }
    axMapControl1.ActiveView.Refresh();
    return hitPoint1;
}
public void CreateMarkerElement(IPoint point)
{
    IActiveView activeView = this.axMapControl1.ActiveView ;
```

```csharp
    IGraphicsContainer graphicsContainer = axMapControl1.Map as Igraphics Container;
    //建立一个marker元素
    IMarkerElement markerElement = new MarkerElement() as IMarkerElement ;
    ISimpleMarkerSymbol simpleMarkerSymbol = new SimpleMarkerSymbol();
    //符号化元素
    IRgbColor rgbColor1 = new RgbColor();
    rgbColor1.Red = 255;
    rgbColor1.Blue = 0;
    rgbColor1.Green = 0;
    simpleMarkerSymbol.Color = rgbColor1;
    IRgbColor rgbColor2 = new RgbColor();
    rgbColor2.Red = 0;
    rgbColor2.Blue = 255;
    rgbColor2.Green = 0;
    simpleMarkerSymbol.Outline = true;
    simpleMarkerSymbol.OutlineColor = rgbColor2 as IColor;
    simpleMarkerSymbol.OutlineSize = 1;
    simpleMarkerSymbol.Size = 5;
    simpleMarkerSymbol.Style = esriSimpleMarkerStyle.esriSMSCircle;
    ISymbol symbol = simpleMarkerSymbol as ISymbol;
    symbol.ROP2 = esriRasterOpCode.esriROPNotXOrPen;
    markerElement.Symbol = simpleMarkerSymbol;
    m_element = markerElement as IElement;
    m_element.Geometry = point as IGeometry;
    graphicsContainer.AddElement(m_element, 0);
    activeView.PartialRefresh(esriViewDrawPhase.esriViewGraphics, m_element, null);
    IGeometry geometry = m_element.Geometry;
    movePointFeedback.Display = activeView.ScreenDisplay;
    movePointFeedback.Symbol = simpleMarkerSymbol as ISymbol;
    movePointFeedback.Start(geometry as IPoint, point);
}
//移动元素到新的位置
public void ElementMoveTo(IPoint point )
{
    //移动元素
    movePointFeedback.MoveTo(point);
    IGeometry geometry1 = null;
    IGeometry geometry2 = null;
    if (m_element != null)
    {
        geometry1 = m_element.Geometry;
        geometry2 = movePointFeedback.Stop();
        m_element.Geometry = geometry2;
        //更新该元素的位置
        this.axMapControl1.ActiveView.GraphicsContainer.UpdateElement(m_element);
        //重新移动元素
        movePointFeedback.Stop();//(geometry1 as IPoint, point);
        this.axMapControl1.ActiveView.PartialRefresh(esriViewDrawPhase.
            esriViewGeography, null, null);
    }
}
//通过名称获取图层
public ILayer GetLayerByName(string layerName, AxMapControl axMapControl1)
{
    for (int i = 0; i < axMapControl1.LayerCount; i++)
    {
        if (axMapControl1.get_Layer(i).Name.Equals(layerName))
            return axMapControl1.get_Layer(i);
    }
    return null;
}
```

```csharp
//转换像素到地图单位
public double ConvertPixelsToMapUnits(IActiveView activeView, double pixelUnits)
{
    double realDisplayExtent;
    int pixelExtent;
    double sizeOfOnePixel;
    pixelExtent = activeView.ScreenDisplay.DisplayTransformation.get_Device
    Frame().right - activeView.ScreenDisplay.DisplayTransformation.get_
    DeviceFrame().left;
    realDisplayExtent = activeView.ScreenDisplay.DisplayTransformation.Visible
    Bounds.Width;
    sizeOfOnePixel = realDisplayExtent / pixelExtent;
    return pixelUnits * sizeOfOnePixel;
}

//加载地图文档
private void loadMapDocument()
{
    System.Windows.Forms.OpenFileDialog openFileDialog;
    openFileDialog = new OpenFileDialog();
    openFileDialog.Title = "打开地图文档";
    openFileDialog.Filter = "map documents(*.mxd)|*.mxd";
    openFileDialog.ShowDialog();
    string filePath = openFileDialog.FileName;
    if (axMapControl1.CheckMxFile(filePath))
    {
        axMapControl1.MousePointer = esriControlsMousePointer.esriPointerHourglass;
        axMapControl1.LoadMxFile(filePath, 0, Type.Missing);
        axMapControl1.MousePointer = esriControlsMousePointer.esriPointerDefault;
    }
    else
    {
        MessageBox.Show(filePath + "不是有效的地图文档");
    }
}

//更改当前捕捉图层名
private void cbLayerName_TextChanged(object sender, EventArgs e)
{
    snapLayer = cbLayerName.Text;
}
```

7.3 要素编辑

空间数据编辑是 GIS 的重要功能之一，ArcEngine 提供了丰富的编辑功能，本节分类介绍编辑功能。

7.3.1 开始编辑

IWorkspaceEdit 接口是空间数据编辑功能的重要接口，可以通过它来启动或停止一个编辑流程，在这个编辑流程内，可以对地理数据进行更新操作。由于 GIS 的编辑操作往往都是长事务过程，所以依次在流程内的编辑操作可以恢复。

开始编辑使用该接口的 StartEditing 方法，该方法的 withUndoRedo 参数用来确定工作空间是否支持"恢复/取消恢复"操作。如果需要用到此功能，则应将所有对象相关的更改放在一个流程内。

启动编辑后，可以使用 StartEditOperation 方法开始编辑操作，这个操作不是一个流程，可以理解为只是数据更改的一个动作。但在编辑的过程中出现异常时，可以使用 AbortEditOperation 方法取消所有的编辑操作。

7.3.2 结束编辑

用户完成数据的编辑后，可以使用 StopEditOperation 方法来确保编辑操作的完成。UndoEditOperation 方法可以用于处于编辑状态的回滚操作，如果发现编辑过程有误，通过执行这个方法可以恢复到最近变化前的状态。StopEditing 方法用来完成编辑，执行该方法后，先前所做的编辑操作都不能再回滚，类似数据库中的 Commit 命令。

前面章节中的代码示例都用到了这些接口，这里完整地介绍这些接口的使用，编辑操作开始时，首先启动一个工作空间工厂，通过该工厂打开个人数据库、SDE 数据库、SHP 文件的路径目录，工作空间工厂在打开数据库后，返回一个工作空间，通过该工作空间，可以打开 SHP 文件、图层。根据工作空间工厂开发的数据库类型不同，返回的工作空间（IWorkspace）相应地继承接口的一个实例，例如，打开的是一个 SHP 文件，则返回的是工作空间（IWorkspace）的继承接口（IFeatureWorkspace）的一个实例，该接口（IFeatureWorkspace）也继承了 IWorkspaceEdit，因此可以通过 IWorkspaceEdit 接口定义的启动编辑、启动操作、撤销操作、结束操作、结束编辑，进行数据编辑控制，编辑的过程必须先启动编辑然后启动操作，再以结束操作然后结束编辑这样的顺序执行。

```
private void WorkspaceEdit()
{
    IWorkspaceFactory pWorkspaceFactory;
    pWorkspaceFactory = new AccessWorkspaceFactoryClass();
    IFeatureWorkspace pFeatureWorkspace;
    pFeatureWorkspace = pWorkspaceFactory.OpenFromFile(@"D:\Usa.mdb", 0) as IFeatureWorkspace;
    IFeatureClass pFeatureClass;
    pFeatureClass = pFeatureWorkspace.OpenFeatureClass("States");
    IWorkspaceEdit pWorkspaceEdit;
    pWorkspaceEdit = pFeatureWorkspace as IWorkspaceEdit;
    IFeature pFeature ;
    bool bHasEdits = true;
    pWorkspaceEdit.StartEditing(true);
    pWorkspaceEdit.RedoEditOperation();
    pFeature = pFeatureClass.GetFeature(1);
    pFeature.Delete();
    pWorkspaceEdit.StopEditOperation();
    DialogResult iResponse;
    iResponse = MessageBox.Show("Edit Operation", "Undo operation?", MessageBoxButtons.YesNo);
    if (iResponse == DialogResult.Yes)
    {
        pWorkspaceEdit.UndoEditOperation();
    }
    pWorkspaceEdit.HasEdits(ref bHasEdits);
    if (bHasEdits)
    {
        iResponse = MessageBox.Show("Edit Operation","Save edits?", MessageBoxButtons.YesNo);
        if (iResponse == DialogResult.Yes)
        {
            pWorkspaceEdit.StopEditing(true);
        }
        else
```

```
            {
                pWorkspaceEdit.StopEditing(false);
            }
        }
    }
```

7.3.3 图形编辑

图形编辑是数据编辑的主要实现，即对空间要素几何形状所做的修改，如移动位置、旋转、删除节点、增加节点。主要用到以下几个接口：IFeature 、IFeatureEdit、IWorkspaceEdit、IFeatureClass、IFeatureCursor。

IFeatureEdit 中的 MoveSet、RotateSet、DeleteSet 分别是移动、旋转、删除由一个或者多个要素组成的要素集。Split 主要用来分割几何形体，通过点分割线，通过多义线分割多边形；分割后旧的被删除，新的要素自动产生。因为每个要素都有对应的属性，因此要素改变了，其对应的属性也会相应地改变。Splitattributes 用来分割要素属性字段中的值，该方法可以在 Split 方法执行后自动执行。

Feature 又派生出很多类，如 networkfeature、rastercatalogfeature、coverageannotationfeature、annotationfeature、dimensionfeature。feature 的主要接口是 Ifeature，其中的 Extent 用来返回要素对象的包络线，显示要素的空间范围；Featuretype 用来返回要素的类型；Shape 用来返回要素的几何形体对象；ShapeCopy 用来得到几何形体对象的一个复制。

图形编辑涉及图形的增、删、改、查，在 GIS 系统中，改操作涉及图形的移动，Engine 提供了 IDisplayFeedback 接口，该接口只定义了 4 个方法和属性。

Display：用于显示当前操作的 Feedback 对象。
MoveTo：用于将对象移动到一个新位置。
Refresh：用于刷新对象。
Symbol：用于移动对象的显示符号。

Engine 中所有 Feedback 类均派生于 IdisplayFeedback，这些 Feedback 类大致可以分成两类，一类是创建新对象的 Feedback，如 NewArcFeedback、NewEllipseFeedback，另一类是基与显示目的的 Feedback，如 MovePointFeedback、MovePolygonFeedback、MoveLineFeedback。创建新对象的 Feedback 作用在于创建新对象时，图形的创建过程在地图上可见，基于显示目的的 Feedback 大多用于移动到新位置操作，可以看到对象跟随鼠标的光标移动而移动。

本示例演示线、面对象节点增加、删除，点、线、面要素的移动，用到基于显示目的的 Feedback。

```
//操作类型
string strOperator="";
//当前地图视图
IActiveView m_activeView = null;
//当前操作图层
IFeatureLayer m_FeatureLayer = null;
//当前操作实体
IFeature m_Feature = null;
 //当前点移动反馈对象
IMovePointFeedback m_MovePointFeedback = new MovePointFeedbackClass();
//当前线移动反馈对象
IMoveLineFeedback m_MoveLineFeedback = new MoveLineFeedbackClass();
//当前面移动反馈对象
IMovePolygonFeedback m_MovePolygonFeedback = new MovePolygonFeedbackClass();

//初始化反馈对象参数
```

```csharp
private void button3_Click(object sender, EventArgs e)
{
    strOperator ="move";
    m_MovePointFeedback = new MovePointFeedbackClass();
    m_MoveLineFeedback = new MoveLineFeedbackClass();
    m_MovePolygonFeedback = new MovePolygonFeedbackClass();
}
//删除节点
private void button4_Click(object sender, EventArgs e)
{
    IPointCollection pointCollection;
    if (m_Feature.Shape.GeometryType == esriGeometryType.esriGeometryPolyline)
    {
        pointCollection = new PolylineClass();
        IPolyline polyline = m_Feature.Shape as IPolyline;

        pointCollection = polyline as IPointCollection;
        //如果点个数少于两个就无法构成线
        if (pointCollection.PointCount > 2)
        {
            //移除指定的节点
            pointCollection.RemovePoints(pointCollection.PointCount - 1, 1);
        }
    }
    else if (m_Feature.Shape.GeometryType == esriGeometryType.esriGeometryPolygon)
    {
        pointCollection = new PolygonClass ();
        IPolygon polygon = m_Feature.Shape as IPolygon ;
        pointCollection = polygon as IPointCollection;
        //如果点个数少于3个就无法构成面
        if (pointCollection.PointCount > 3)
        {
            //移除指定的节点
            pointCollection.RemovePoints(pointCollection.PointCount - 1, 1);
        }
    }
    IWorkspaceEdit workspaceEdit;
    IWorkspace workspace;
    IDataset dataset = m_FeatureLayer.FeatureClass as IDataset;
    workspace = dataset.Workspace;
    workspaceEdit = workspace as IWorkspaceEdit;
    //开始编辑
    workspaceEdit.StartEditing(true);
    workspaceEdit.StartEditOperation();
    //保存数据
    m_Feature.Store();
    //结束编辑
    workspaceEdit.StopEditOperation();
    workspaceEdit.StopEditing(true);
    m_activeView.Refresh();
}

private void button5_Click(object sender, EventArgs e)
{
    IPointCollection pointCollection;
    IPoint point = new PointClass();
    IPoint fromPoint = new PointClass();
    IPoint toPoint = new PointClass();
    if (m_Feature.Shape.GeometryType == esriGeometryType.esriGeometryPolyline)
    {
        pointCollection = new PolylineClass();
```

```csharp
            IPolyline polyline = m_Feature.Shape as IPolyline;

            object missing1 = Type.Missing;
            object missing2 = Type.Missing;
            pointCollection = polyline as IPointCollection;
            //获取线对象的最后两个点
            fromPoint = pointCollection.get_Point(pointCollection.PointCount - 2);
            toPoint = pointCollection.get_Point(pointCollection.PointCount - 1);
            //根据线最后两个点创建一个新点
            point.PutCoords((fromPoint.X + toPoint.X) / 2, (fromPoint.Y + toPoint.Y) / 2 + 50);
            //将新点添加到线对象的点集合中
            pointCollection.AddPoint(point, ref missing1, ref missing2);

        }
        else if (m_Feature.Shape.GeometryType == esriGeometryType.esriGeometryPolygon)
        {
            pointCollection = new PolygonClass();
            IPolygon polygon = m_Feature.Shape as IPolygon;

            object missing1 = Type.Missing;
            object missing2 = Type.Missing;
            pointCollection = polygon as IPointCollection;
            //获取面对象点集最后两个点
            fromPoint = pointCollection.get_Point(pointCollection.PointCount - 2);
            toPoint = pointCollection.get_Point(pointCollection.PointCount - 1);
            //根据线最后两个点创建一个新点
            point.PutCoords((fromPoint.X + toPoint.X) / 2, (fromPoint.Y + toPoint.Y) / 2 + 50);
            //将新点添加到线对象的点集合中
            pointCollection.AddPoint(point, ref missing1, ref missing2);
        }
        IWorkspaceEdit workspaceEdit;
        IWorkspace workspace;
        IDataset dataset = m_FeatureLayer.FeatureClass as IDataset;
        workspace = dataset.Workspace;
        workspaceEdit = workspace as IWorkspaceEdit;
        //开始编辑
        workspaceEdit.StartEditing(true);
        workspaceEdit.StartEditOperation();
        //保存数据
        m_Feature.Store();
        //结束编辑
        workspaceEdit.StopEditOperation();
        workspaceEdit.StopEditing(true);
        m_activeView.Refresh();
    }

    private void axMapControl1_OnMouseDown(object sender, IMapControlEvents2_OnMouseDownEvent e)
    {
        IPoint point = new PointClass();
        IFeatureClass featureClass = null;
        if (m_Feature == null) return;
        switch (strOperator)
        {
            case "move":
                //将当前鼠标光标位置的点转换为地图上的坐标
                point = m_activeView.ScreenDisplay.DisplayTransformation.ToMapPoint(e.x, e.y);
                if (m_Feature.Shape.GeometryType == esriGeometryType.esriGeometryPoint)
```

```csharp
            {
                //设置显示对象，并启动移动
                m_MovePointFeedback.Display = m_activeView.ScreenDisplay ;
                m_MovePointFeedback.Start(m_Feature.Shape as IPoint , point);
            }
            else if (m_Feature.Shape.GeometryType == esriGeometryType.esriGeometryPolyline)
            {
                //设置显示对象，并启动移动
                m_MoveLineFeedback.Display = m_activeView.ScreenDisplay;
                m_MoveLineFeedback.Start(m_Feature.Shape as IPolyline, point);
            }
            else if (m_Feature.Shape.GeometryType == esriGeometryType.esriGeometryPolygon)
            {
                //设置显示对象，并启动移动
                m_MovePolygonFeedback.Display = m_activeView.ScreenDisplay;
                m_MovePolygonFeedback.Start(m_Feature.Shape as IPolygon, point);
            }
            break;
    }
}

private void axMapControl1_OnMouseMove(object sender, IMapControlEvents2_OnMouseMoveEvent e)
{
    IPoint point = new PointClass();
    switch (strOperator)
    {
        case "move":
            if (m_Feature.Shape.GeometryType == esriGeometryType.esriGeometryPoint)
            {
                if (m_MovePointFeedback != null)
                {
                    //将当前鼠标光标位置的点转换为地图上的坐标
                    point = m_activeView.ScreenDisplay.DisplayTransformation.ToMapPoint(e.x, e.y);
                    //移动对象到当前鼠标光标位置
                    m_MovePointFeedback.MoveTo(point);
                }
            }
            else if (m_Feature.Shape.GeometryType == esriGeometryType.esriGeometryPolyline)
            {
                if (m_MoveLineFeedback != null)
                {
                    //将当前鼠标光标位置的点转换为地图上的坐标
                    point = m_activeView.ScreenDisplay.DisplayTransformation.ToMapPoint(e.x, e.y);
                    //移动对象到当前鼠标光标位置
                    m_MoveLineFeedback.MoveTo(point);
                }
            }
            break;
    }
}

private void axMapControl1_OnMouseUp(object sender, IMapControlEvents2_OnMouseUpEvent e)
{
    if (m_Feature == null) return;
```

```csharp
            IGeometry resultGeometry = null;
            switch (strOperator)
            {
                case "move":
                    if (m_Feature.Shape.GeometryType == esriGeometryType.esriGeometryPoint)
                    {
                        //停止移动
                        resultGeometry = m_MovePointFeedback.Stop() as IGeometry;
                        m_Feature.Shape = resultGeometry;
                    }
                    else if (m_Feature.Shape.GeometryType == esriGeometryType.esriGeometryPolyline)
                    {
                        //停止移动
                        resultGeometry = m_MoveLineFeedback.Stop() as IGeometry;
                        m_Feature.Shape = resultGeometry;
                    }
                    else if (m_Feature.Shape.GeometryType == esriGeometryType.esriGeometryPolygon)
                    {
                        //停止移动
                        resultGeometry = m_MovePolygonFeedback.Stop() as IGeometry;
                        m_Feature.Shape = resultGeometry;
                    }
                    IWorkspaceEdit workspaceEdit;
                    IWorkspace workspace;
                    IDataset dataset = m_FeatureLayer.FeatureClass as IDataset;
                    workspace = dataset.Workspace;
                    workspaceEdit = workspace as IWorkspaceEdit;
                    //开始编辑
                    workspaceEdit.StartEditing(true);
                    workspaceEdit.StartEditOperation();
                    //保存实体
                    m_Feature.Store();
                    //结束编辑
                    workspaceEdit.StopEditOperation();
                    workspaceEdit.StopEditing(true);
                    m_MovePointFeedback = null;
                    m_MoveLineFeedback = null;
                    m_MovePolygonFeedback = null;
                    break;
            }
            m_activeView.Refresh();
            this.axMapControl1.Map.ClearSelection();
        }
    }
}
```

7.4 高级编辑

ArcEngine 高级编辑操作主要会用到 ITopologicalOperator 接口，通过该接口可以实现要素分解、要素合并、生成平行线、打断线、延长线、裁剪等操作，ItopologicalOperator 将在第 10 章介绍。

7.5 本章小结

GIS 的显示内容主要是由各种要素构成，在实际应用中，经常需要捕获地图上的要素，这些捕获这些要素可以通过空间查询或捕捉来实现，本章分析了地图捕捉的使用场景，并给出了线、面捕捉的实例。GIS 的二次开发中，地图编辑是一个重要的功能，地图是在分布式环境下应用，因此编辑的长事务处理是地图编辑应该着重关心和考虑的问题，通过 ArcGISEngine 提供的工作空间编辑接口可以实现地图要素编辑的长事务管理，确保地图编辑的完整性和一致性。

　　本章最后讨论了地图交互对象 DisplayFeedback 对象，这些对象可以实现移动、旋转、节点编辑等重要高级功能，ArcEngine 平台提供了丰富的 DisplayFeedback 对象，本文将这些对象划分为两大类：产生新对象 DisplayFeedback 和基于显示目的的 DisplayFeedback，并将无法划分到这两大类的 DisplayFeedback 对象进行了简单介绍。

第 8 章　栅格数据

8.1 简介

栅格数据是由一系列的规则格网单元组成的,用于表达专题、光谱以及图像等数据。它由多个等间距矩形栅格组成,用于模拟在一定的空间范围内连续变化的地理现象或者图片数据。这种格式常见,如卫星图片、扫描图和航片等,能表达各种地表类型,如高程、植被等。ArcEngine 中的栅格子模块主要包含访问和处理栅格的对象,如 Rasters、Rasterdatasets 和 Raster catalogs 等。这些对象不仅可以访问基于文件的栅格数据,也能访问存储在 GeoDatabase 中的栅格数据。

Raster 数据分为两种类型,一种是专题数据,可用于进行地理分析;另一种是影像数据,一般用于地图的背景显示。ArcGIS 所支持的栅格数据有:GRID、TIFF、ERDAS IMAGE、JPEG 等。在 ArcEngine 中,提供了一种统一的方式来管理任何类型的栅格数据。

栅格数据模型用连续空间的像元来代表要素,像元表现为一个矩形的区域,宽度和高度相等,其大小取决于原始地图的比例尺和绘制地图时最小的单位。栅格是一个二维的动态像元矩阵,每个栅格单元都有一个值,用于表示某个位置的某种属性,如高程、反射率、颜色等。

栅格数据集是存储在文件或数据库中的以某种栅格格式存在的数据对象,通过使用栅格图层将数据加载到地图控件中。

8.2 访问和创建栅格数据

对于栅格数据而言,基于文件的工作空间、基于 ACCESS 的个人数据、基于企业数据库的工作空间等都是其工作空间,在访问栅格数据时,必须打开一个工作空间。工作空间必须通过工作空间工厂来创建,可以通过 RasterWorkspaceFactory 对象来创建。

RasterWorkspaceFactory 对象通常用于生成一个 RasterWorkspace 对象,RasterWorkspace 对象就是一个栅格数据集容器,RasterWorkspaceFactory 类实现了 IWorkspaceFactory 和 IWorkspaceFactory2 接口。RasterWorkspace 对象实现了 IrasterWorkspace 和 IRasterWorkspace2 接口,这两个接口定义了用于打开和设置一个数据集对象,如:

• Copy 方法用于复制一个工作空间中所有的栅格数据集到一个新的或已经存在的工作空间中去;

• OpenRasterDataset 方法用于打开一个栅格数据集;

• CreateRasterDataset 方法用于产生一个新的基于文件的栅格数据集对象,这个数据格式可以为 GRID、TIFF 和 ERDAS Image 之一。

下面的代码片段演示了如何使用这些接口加载栅格数据。

```
IWorkspaceFactory workspaceFactory = new RasterWorkspaceFactoryClass();
IWorkspace workspace;
string filePath = @"E: \data";
IRasterWorkspace2 rasterWorkspace;
IRasterDataset rasterDataset;
rasterWorkspace = workspaceFactory.OpenFromFile(filePath, 0) as IRasterWorkspace2 ;
rasterDataset = rasterWorkspace.OpenRasterDataset( @"\NewTif");
IRasterLayer rasterLayer = new RasterLayerClass();
rasterLayer.CreateFromDataset(rasterDataset);
this.axMapControl1.Map.AddLayer(rasterLayer as ILayer);
this.axMapControl1.Extent = this.axMapControl1.FullExtent;
```

8.3 栅格数据配准

地理信息系统与遥感技术在应用中多少要涉及栅格数字图像的校正配准处理。例如 TM 影像图与数字高程模型 DEM 图的配准，配准后的 TM 图再用于地面上各类资源的空间分析研究与实际的生产应用。再如，经扫描输入微机的 Tiff 栅格图，由于存在人工操作及仪器误差，其图像总会有变形，因此必要时就得经过几何校正才能投入到应用中去。由此，数字图像的配准与校正技术在 GIS 及 RS 的应用研究中就显得尤为必要。

栅格数据的配准主要是使用 IGeoReference 接口来实现。通过 ActiveView 进行坐标转换，将 MapControl 中点的位置转换为地图和影像图上的坐标。该接口被 RasterLayer 类实现，该接口提供有栅格数据配准 11 个方法和 1 个属性，如：

- CanGeoRef 属性用于监测该图层是否可以做配准；
- PointTransform 方法用于将鼠标的位置转换为栅格文件上的相对坐标；
- Rectify 方法用于将纠正的结果保存为一个新的栅格文件；
- Register 方法用于将纠正的结果生成 Word 文件，和栅格文件保存在同一个目录下；
- Reset 方法用于取消纠正，但是对 Register 之前的操作不能恢复；
- Shift 方法可提供一点纠正，即平移；
- TwoPointsAdjust 方法可提供两点配准，有一定的缩放；
- Warp 方法可提供 3 点或以上的配准。

8.4 栅格数据处理

栅格数据是 GIS 最基本的数据之一。早期的计算机系统由于存储容量小、运算速度慢，因此栅格数据的实用性受到了极大的限制。在现在的计算机环境下，计算机的运算能力、寻址能力、外部设备的性能大为改善，栅格数据的优势也逐渐显现出来。随着 GIS 技术应用的普及与深入，出现了大量以遥感信息为主的数据源系统，如环境监测、火灾跟踪等，这些遥感数据大多为栅格数据形式的影像数据和数据流。本节介绍栅格数据处理的一些功能。

8.4.1 栅格数据转换

栅格数据转换是指将用矢量形式表示的空间数据表示为栅格格式，即将 GeoDatabase 中的 FeatureClass 或是基于文件的矢量数据栅格化（ShapeFile、Coverage 等）。在 ArcEngine 的 GeoAnalyst 类库中提供有 IConversionOp 接口，该接口提供有 5 个方法：RasterDataToLineFeatureData、RasterDataToPointFeatureData、RasterDataToPolygonFeatureData、ToFeatureData、ToRasterDataset。其中前 4 个方法用于将栅格数据转换为矢量数据，实现从栅格数据图层中提取点、线、面矢量要素，

可用于提取矢量要素的栅格对象有 Raster、RasterDataset、RasterBand 和 RasterDescriptor 等。ToRasterDataset 方法用于实现矢量要素栅格化，可用于栅格化的矢量要素有 FeatureClass、FeatureClassDescriptior 和 Featurelayer 等。该接口为 RasterConversionOpClass 类所实现。

下面的代码片段演示了如何将栅格数据导出为点状矢量数据。

```
string filePath=@"E:\data";
IConversionOp  conversionOP =new RasterConversionOpClass ();
IRasterWorkspace2 rasterWorkspace;
RasterDataset rasterDataset ;
rasterWorkspace = workspaceFactory.OpenFromFile(filePath, 0) as IRasterWorkspace2 ;
rasterDataset = rasterWorkspace.OpenRasterDataset( @"\NewTif");
IRasterLayer rasterLayer = new RasterLayerClass();
rasterLayer.CreateFromDataset(rasterDataset);
IRaster raster =rasterLayer.Raster ;

IGeoDataset geoDataset ;
IWorkspaceFactory workspaceFactory ;
IWorkspace workspace ;
workspace =workspaceFactory.OpenFromFile (@"e:\temp",0);
string outFeatureClassName ="newShapeFile.shp";
geoDataset    =conversionOP.RasterDataToPointFeatureData   (raster   ,workspace   ,
outFeatureClassName );
```

ToRasterDataset 方法用于将矢量数据转换为栅格数据，该方法生成的是 IRasterDataset 对象，方法的输入参数要求矢量数据为 FeatureClass、FeatureClassDescriptor、FeatureLayer 类型的对象，并支持输出 3 种栅格类型：ESRI GRID、ERDAS Imagine 和 TIFF。在方法的参数中用 "GRID" 代表 ESRI GRID 类型，生成没有扩展名的栅格文件；"TIFF"、"TIF" 代表 TIFF 格式，生成扩展名为 ".tiff"、".tif" 的栅格文件；"ERDAS Imagine" 生成 ".img" 格式的文件。下面的代码片段演示了如何把矢量数据转换为栅格数据。

```
ILayer layer = this.axMapControl1.get_Layer(0);
IFeatureLayer featureLayer = layer as IFeatureLayer;
//获取 FeatureClass
IFeatureClass featureClass = featureLayer.FeatureClass;
//获取转换的字段
IFeatureClassDescriptor featureClassDescriptor = new FeatureClassDescriptorClass ();
//转换字段
featureClassDescriptor.Create(featureClass, null, "code");
//获取第 1 个参数
IGeoDataset geoDataset = featureClassDescriptor as IGeoDataset;
string outPutPath = @"e:\temp";
IWorkspaceFactory workspaceFactory = new RasterWorkspaceFactoryClass();
IWorkspace workspace = workspaceFactory.OpenFromFile(outPutPath, 0);
IConversionOp conversionOp = new RasterConversionOpClass();
IRasterAnalysisEnvironment rasterAnalysisEnvironment = conversionOp as IrasterAnalysis
Environment;
//栅格的大小，通过宽度或者高度较小的值与 250 的比值
double dCellSize = 0.024668;
object oCellSize = dCellSize as object;
//先设置栅格大小，再进行转换
rasterAnalysisEnvironment.SetCellSize(esriRasterEnvSettingEnum.esriRasterEnvValue, ref
oCellSize);
//设置存储位置、转换字段以及格式
IRasterDataset    rasterDataset    =    conversionOp.ToRasterDataset(geoDataset,    "GRID",
workspace, "NewRaster");
```

8.4.2 栅格数据变换

栅格数据变换是指将一个栅格数据层，通过某种算法进行变换，得到新的栅格数据图层。例如，将用栅格数据表示的植被、土壤分类数据简单地转换为上一级的分类数据，将用栅格数据表示的高程数据转换为坡度、坡向数据，将用栅格数据表示的农作物分布数据转换为某一类农作物的分布数据等。

8.4.3 栅格数据叠置分析

栅格数据叠置分析，是指将多个栅格数据层，以相同空间位置上的栅格单元为处理单元，通过某种算法进行变换，得到新的栅格数据层。例如，将表示地貌、高程、土壤、水分等一系列地表性质的数据，转换为土地等级或土地评价数据等。

栅格图层叠置分析用到的主要接口有 IMathOp、ILogicalOp、ITrigOp 等。

IMathOp 接口提供有进行栅格数学运算的方法，组件类 RasterMathOps 封装了该接口，接口提供有 21 个方法。如：

- Abs 方法用于计算栅格像元的绝对值，并写入新数据集；
- Divide 方法用于把两个输入栅格图层像元值相除，并写入新数据集；
- Exp 方法用于计算栅格像元以 e 为底的次方值，并写入新数据集；
- Exp10 方法用于计算栅格像元以 10 为底的次方值，并写入新数据集；
- Exp2 方法用于计算栅格像元以 2 为底的次方值，并写入新数据集；
- Times 方法用于将两个输入栅格图层像元值相乘，并写入数据集。

ILogicalOp 接口提供有进行栅格逻辑运算的功能，提供有 15 个方法。如：

- BooleanAnd 方法用于两个 Rasters 对象之间进行布尔"and"操作；
- BooleanOr 方法用于两个 Rasters 对象之间进行布尔"Or"操作；
- CombinatorialAnd 方法用于两个 Rasters 对象之间进行组合"and"操作。

ITrigOp 接口提供有进行三角函数运算的功能，提供有 13 个方法。如：

- ACos 方法用于计算反余弦；
- ASin 方法用于计算反正弦。

下面的代码片段演示了两个栅格图层的相乘运算。

```
IMathOp  mathOp =new RasterMathOpsClass ();
IGeoDataset resultDataset= new RasterDataset ();
ILayer layer ;
IRasterLayer rasterLayer1,rasterLayer2;
layer =this.axMapControl1.get_Layer (0);
rasterLayer1 =layer as IRasterLayer ;
layer =this.axMapControl1.get_Layer (1);
rasterLayer2 =layer as IRasterLayer ;
resultDataset =mathOp.Times (rasterLayer1.Raster as IGeoDataset ,rasterLayer2. Raster as
IGeoDataset );
```

8.4.4 栅格数据与矢量数据叠加分析

栅格数据与矢量数据叠加分析，是指将一个栅格数据层通过与用矢量数据表示的面要素叠加，得到矢量数据层中地理对象的某一属性信息。例如，将一个表示农作物种植分布的栅格数据层与用矢量表示的行政数据层叠置，即可获得该区域范围内各行政单位的各类农作物种植面积。

这种叠置分析，往往是针对矢量数据的点数据和线数据进行，一般是通过对缓冲区分析等空间分析方法，生成点对象或线对象一定范围内的缓冲区多边形，然后用此多边形与栅格数据层进行叠置。

8.5 栅格图层渲染

专题图是依据要素的一个或多个属性而设置不同的符号，从而达到区分不同类型要素的目的。Arc Engine 提供有 RasterRender 抽象类，用于实现栅格数据的渲染。该类有 5 个子类，负责进行不同类型的着色运算，这些类都实现了 IRasterRender 接口，该接口定义了栅格图层符号化的公共属性和方法。通过 IRasterLayer.Renderer 属性获取的一个栅格图层的符号化对象模型如图 8-1 所示。

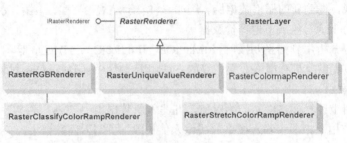

▲图 8-1

8.5.1 RasterRGBRenderer（栅格 RGB 符号化）

RasterRGBRenderer 通过绘制红、绿、蓝等 3 个波段的栅格数据集来显示，该组件类主要实现 IRasterRenderer、IRasterRGBRenderer2。通过 IRasterRenderer 提供的属性 Raster 设置渲染的栅格数据集对象，通过 IRasterRGBRenderer2 提供的 3 个波段索引属性：RedBandIndex、GreenBandIndex、BlueBandIndex 进行渲染。下面的代码片段演示了栅格 RGB 如何符号化。

```
private void ChangeRGBRenderer (IRasterLayer rasterLayer)
{
    IRaster raster =rasterLayer.Raster ;
    IRasterBandCollection rasterBandCollection ;
    rasterBandCollection=raster as IRasterBandCollection ;
        //如果波段小于 3 个，则不进行渲染
    if (rasterBandCollection.Count <3)
        return ;
    IRasterRGBRenderer2 rasterRGBRender=new RasterRGBRendererClass ();
    IRasterRenderer rasterRender =rasterRGBRender  as IRasterRenderer ;
    rasterRender.Raster =raster ;
    rasterRender.Update ();
    rasterRGBRender.RedBandIndex =2;
    rasterRGBRender.GreenBandIndex =1;
    rasterRGBRender.BlueBandIndex =0;
    rasterRender.Update ();
    rasterLayer.Renderer =rasterRGBRender ;
}
```

8.5.2 RasterUniqueValueRenderer（唯一值符号化）

唯一值符号化采用多种着色颜色，依据图层中的某个字段值给每个值一个单独的颜色，因此可以区分存在的每一个要素。该组件类实现 IRasterUniqueValueRenderer 接口，提供有 4 个方法和 11 个属性。如：

• AddValue 方法，用于将单个要素的某个字段值和与之相匹配的着色符号加入 RasterUniqueValueRenderer 对象；

• Symbol 属性用于设置每个值显示的符号；

- Field 属性用于设置取值字段；
- HeadingCount 属性用于设置渲染的标题数目；
- set_ClassCount 属性用于设置渲染的类别数目，即唯一值的个数。

下面的代码演示了如何使用该组件类实现唯一值符号化。

```
private void ChangeRender2UniqueValueRender(IRasterLayer rasterLayer)
{
    IRaster raster =rasterLayer.Raster ;
    ITable table;
    IRasterBand rasterBand;
    IRasterBandCollection rasterBandCollection;
    rasterBandCollection = raster as IRasterBandCollection;
    rasterBand = rasterBandCollection.Item(0);
    bool tablExist;
    rasterBand.HasTable(tablExist);
    if (!tablExist)
        return;
    table = rasterBand.AttributeTable;
    int numOfValues = table.RowCount(null);
    int fieldIndex;
    string fieldName="value";
    fieldIndex = table.FindField(fieldName);
    IRandomColorRamp randomColorRamp=new RandomColorRampClass ();
    randomColorRamp.Size = numOfValues;
    randomColorRamp.Seed = 100;
    randomColorRamp.CreateRamp(true);
    ISimpleFillSymbol simpleFillSymbol;
    IRasterUniqueValueRenderer rasterUniqueValueRender = new RasterUniqueValueRenderer
    Class();
    IRasterRenderer rasterRender = rasterUniqueValueRender as IRasterRenderer;
    rasterRender.Raster = raster;
    rasterRender.Update();
    rasterUniqueValueRender.HeadingCount = 1;
    rasterUniqueValueRender.set_Heading(0, "all data values");
    rasterUniqueValueRender.set_ClassCount(0, numOfValues);
    rasterUniqueValueRender.Field = fieldName;
    int i;
    IRow row;
    object labelValue;
    for (i = 0; i < numOfValues - 1;i++ )
    {
        row = table.GetRow(i);
        labelValue = row.get_Value(fieldIndex);
        rasterUniqueValueRender.AddValue(0, i, labelValue);
        rasterUniqueValueRender.set_Label(0, i, labelValue.ToString());
        simpleFillSymbol = new SimpleFillSymbolClass();
        simpleFillSymbol.Color = randomColorRamp.Colors.Next();
        rasterUniqueValueRender.set_Symbol(0, i, simpleFillSymbol);
    }
    rasterRender.Update();
    rasterLayer.Renderer = rasterUniqueValueRender;
}
```

8.5.3 RasterClassfyColorRampRenderer（分类符号化）

分类符号化是将栅格图层划分为指定个数的类，每个类用一种颜色渲染。该类实现 IRasterClassfyColorRampRenderer 接口，提供有 8 个属性。如：

- ClassCount 属性用于设置或获取分类的数目；
- ClassField 属性用于获取或设置用于渲染的字段；

- Symbol 属性用于设置或获取用于渲染的填充符号；
- Description 属性用于获取或设置指定索引类的锚；
- Label 属性用于获取或设置指定类的标签。

下面的代码片段演示了如何使用该组件类对象。

```
private void ChangeRender2ClassfyColorRampRenderer(IRasterLayer rasterLayer)
{
    IRaster raster = rasterLayer.Raster;
    //创建分类
    IRasterClassifyColorRampRenderer rasterClassifyColorRampRender = new RasterClassify
    ColorRampRendererClass();
    IRasterRenderer rasterRender = rasterClassifyColorRampRender as IRasterRenderer;
    rasterClassifyColorRampRender.ClassCount = 20;
    IRgbColor fromColor = new RgbColorClass();
    IRgbColor toColor = new RgbColorClass();
    fromColor = getRGB(255, 0, 0);
    toColor = getRGB(0, 255, 0);
    IAlgorithmicColorRamp algorithmicColorRamp = new AlgorithmicColorRampClass();
    algorithmicColorRamp.Size = 20;
    algorithmicColorRamp.FromColor = fromColor;
    algorithmicColorRamp.ToColor = toColor;
    algorithmicColorRamp.CreateRamp(true);
    IFillSymbol fillSymbol = new SimpleFillSymbolClass();
    for (int i =0;i<rasterClassifyColorRampRender.ClassCount -1;i++)
    {
        fillSymbol.Color =algorithmicColorRamp.Colors.Next ();
        rasterClassifyColorRampRender.set_Symbol (i,fillSymbol as ISymbol );
        rasterClassifyColorRampRender.set_Label (i,"Class"+i.ToString (0));
    }
    rasterRender.Update ();
    rasterLayer.Renderer =rasterClassifyColorRampRender ;
}
```

8.5.4　RasterStretchColorRampRenderer

RasterStretchColorRampRenderer 组件类使用一个 AlgorithmicColorRamp 颜色带，来渲染栅格图层。该颜色带是通过指定起始颜色、终止颜色来确定一个颜色带，起始、终止颜色使用 HSV 模型。该组件类主要实现 IRasterStretchColorRampRenderer 接口，提供有 1 个方法和 6 个属性。如：

- BandIndex 属性用于设置栅格数据集的波段索引；
- ColorRamp 属性用于设置渲染的颜色带。

下面的代码片段演示了如何使用该组件类对象。

```
//获取颜色对象
private IRgbColor getRGB(int r, int g, int b)
{
    IRgbColor pColor;
    pColor = new RgbColorClass();
    pColor.Red = r;
    pColor.Green = g;
    pColor.Blue = b;
    return pColor;
}
private void ChangeRender2RaseterStretchColorRampRender(IRasterLayer rasterLayer)
{
    IRaster raster =rasterLayer.Raster ;
    IRasterStretchColorRampRenderer rasterStretchColorRampRender = new RasterStretchColor
    RampRendererClass();
    IRasterRenderer rasterRender = rasterStretchColorRampRender as IRasterRenderer;
```

```
rasterRender.Raster = raster;
rasterRender.Update();
IColor fromColor = new RgbColorClass();
fromColor = getRGB(255, 0, 0);
IColor toColor = new RgbColorClass();
toColor = getRGB(0, 255, 0);
//创建起止颜色带
IAlgorithmicColorRamp algorithmicColorRamp = new AlgorithmicColorRampClass();
algorithmicColorRamp.Size = 255;
algorithmicColorRamp.FromColor = fromColor;
algorithmicColorRamp.ToColor = toColor;
algorithmicColorRamp.CreateRamp(true);
//选择拉伸颜色带符号化的波段
rasterStretchColorRampRender.BandIndex = 0;
//设置拉伸颜色带符号化所采用的颜色带
rasterStretchColorRampRender.ColorRamp = algorithmicColorRamp;
rasterRender.Update();
rasterLayer.Renderer = rasterStretchColorRampRender ;
}
```

8.5.5 RasterDiscreteColorRenderer（点密度符号化）

RasterDiscreteColorRenderer 组件类非常简单，主要实现 IRasterDiscreteColorRenderer 接口，只有两个属性，其中 Colormap 属性用于设置颜色带对象，NumColors 属性用于设置颜色的数目。

8.6 本章小结

栅格数据是在 GIS 中经常需要处理的另一大类地理数据，用于模拟连续分布特征的对象。ArcEngine 在数据处理中使用了一个统一的数据模型，因此栅格数据的操作方法与矢量数据有很大的相似性。本章简单介绍了栅格数据的创建与访问、栅格数据处理、栅格图层渲染等，并给出一些代码片段演示了各个组件类的使用。其中的 RasterColormapRenderer 渲染方法在"帮助"中介绍的不够详细，需要使用该渲染方法的读者，可以参考其他书籍。

第 9 章 三维可视化

9.1 简介

三维可视化即三维显示,是地理信息系统软件具备的一项基本功能。三维可视化能够为我们提供更多的知识,而这些知识是很难从二维平面地图上得到的。例如,通过三维可视化可以实际地看到山谷的存在,而不需要从等高线数据推断山谷的存在。

ArcScene 用于创建多图层的场景,并且可以控制每个图层场景以及在三维空间的定位和渲染等功能。可以控制整个场景的某些属性,也可以在场景中通过特征的属性或该特征相对于其他特征的位置来选取它们,同时可以从任意一个视角浏览整个三维场景。

9.2 数据的三维显示

在 ArcEngine 中提供了与地图控件类似的三维显示控件 SceneViewer Control,通过该控件,可以方便地显示三维数据。可以在三维场景中进行缩放、拖动、旋转等操作。在该控件中,鼠标光标默认是漫游状态,该控件直接提供了基本的三维操作。

9.2.1 DEM 数据的加载

DEM 数据加载需要用到两个组件类:Scene 和 SceneGraph。Scene 是一个矢量、栅格和图形数据显示与处理的容器,该类实现了 IScene 接口,提供了控制 Scene 的方法和属性,如 AddLayer 方法用于向场景中增加一个图层,SelectionCount 属性用于获取选择的实体数目。SceneGraph 是一个记录在 Scene 中出现的数据和事件的容器,该类实现了 ISceneGraph 接口,提供了控制和处理 Scene 中图形的方法和属性,如 Locate 方法用于通过单击场景中的任意点定位一个对象,RefreshViewers 用于重绘所有的视图,Remove 用于删除角色的一个对象。

9.2.2 叠加纹理数据

加载完 DEM 数据后,需要在三维物体上描绘细节,即纹理;建立 DEM 表面的点与纹理空间的点之间的关系,即纹理映射。

9.2.2.1 叠加影像数据

叠加影像数据就是把影像看做纹理,将其贴在地形表面,让其具有地形起伏的三维效果。叠加影像数据需要用到 4 个组件类:TinLayer、RasterLayer、Tin3DProperties 和 Raster3DProperties。TinLayer 组件类用于创建 Tin 图层对象,RasterLayer 组件类用于创建 Raster 图层对象,Tin3DProperties 组件类用于创建 Tin 数据的三维场景属性对象,Raster3DProperties 组件类用于创建

Raster 数据的三维场景属性对象。

TinLayer 和 RasterLayer 类都实现了 ILayerExtensions 接口，提供了两个方法和两个属性，如 ExtensionCount 用于获取当前图层对象扩展项的个数，RemoveExtension 用于移动指定的扩展项。

Tin3DProperties 和 Raster3DProperties 组件类实现了 I3DProperties 接口，提供了一个方法和一些属性，如 Apply3DProperties 用于应用三维属性到某个对象，BaseOption 用于设置赋予对象基本属性的方式。

9.2.2.2 叠加矢量数据

叠加矢量数据就是将具有高程属性矢量图层的数据加载到三维场景中，这样加载后的矢量数据会作为纹理自动贴在地形表面，具有起伏的三维显示效果。加载矢量数据使用的接口在前面已作过介绍，这里不再重复。

9.2.3 分层设色

根据地面高度划分的高程层，逐层设置不同的颜色，称为地貌分层设色法。通过分层设色，可使地貌高程分布及其相互对照鲜明。实现分层设色需要使用以下几个组件类：AlgorithmicColorRamp、RgbColor、RasterClassifyColorRampRenderer、SimpleFillSymbol 和 TinElevationRenderer。

AlgorithmicColorRamp 类实现了 IAlgorithmicColorRamp 接口，提供了一个方法和一些属性，如 CreateRamp 方法用于创建一定长度的颜色坡面，Size 属性用于设置产生的颜色数目。

RasterClassifyColorRampRenderer 类实现了 IRasterClassifyColorRampRenderer 接口，提供了按类别渲染 Raster 数据的相关属性，如 ClassCount 用于设置类别数目，Symbol 就用于设置类别使用的符号。该类还实现了 IRasterRenderer 接口，提供了渲染的方法和属性，如 Update 用于更新修改后的渲染效果，Raster 就用于设置需要渲染的 Raster 对象。

TinElevationRenderer 类实现了 ITinColorRampRenderer 接口，提供了基于连续属性值，如高程、坡度等。该接口提供了多个属性，如 BreakCount 用于设置分类数目，Break 用于设置分类的类别值。该类还实现了 IClassBreaksUIProperties 和 ITinRenderer 接口。

下面的代码示例演示"加载 DEM 数据"、"加载 TIN 数据"、"叠加 TIN 数据"、"分层设色"功能，"加载 DEM 数"功能用于将 RASTER 数据加载到 SceneControl 控件；"加载 TIN 数"功能用于将 TIN 数据加载到 SceneControl 控件；"叠加 TIN 数"功能用于将 TIN 数据应用到 Raster 数据，使之具有三维起伏效果；"分层设色"功能可以用一个颜色带来显示 TIN 数据。

```
//加载 DEM 数
private void button1_Click(object sender, EventArgs e)
{
    ISceneGraph pSceneGraph = this.axSceneControl1.SceneGraph;
    IScene pScene = pSceneGraph.Scene;
    IRasterLayer   pRasterLayer =new RasterLayerClass ();
    ILayer pLayer ;

    this.openFileDialog1.Title ="Raster layer";
    this.openFileDialog1.DefaultExt =".TIF";
    this.openFileDialog1.Filter ="(*.tif)|*.tif";
    if (this.openFileDialog1.ShowDialog() == DialogResult.OK)
    {
        string pPathName = this.openFileDialog1.FileName;
        string pPath = pPathName.Substring(0, pPathName.LastIndexOf('\\'));
        string fileName = pPathName.Substring(pPath.Length+1, pPathName. Length - pPath.Length-1);
        IWorkspaceFactory pwsf = new RasterWorkspaceFactoryClass();
```

```csharp
            IRasterWorkspace pRasterWorkspace;
            if (pwsf.IsWorkspace(pPath))
            {
                pRasterWorkspace = pwsf.OpenFromFile(pPath, 0) as IRasterWorkspace;
                IRasterDataset pRasterDataset = pRasterWorkspace.OpenRasterDataset
                (fileName);
                //pRasterDataset.OpenFromFile(pPath);
                pRasterLayer.CreateFromDataset(pRasterDataset);
                pLayer = pRasterLayer as ILayer;
                pScene.AddLayer(pLayer, true);
                //pScene.ExaggerationFactor = 6;
                pSceneGraph.RefreshViewers();
            }
        }
    }
    //加载TIN数
    private void button2_Click(object sender, EventArgs e)
    {
        ISceneGraph pSceneGraph = this.axSceneControl1.SceneGraph;
        IScene pScene = pSceneGraph.Scene;
        ITinLayer tinLayer = new TinLayerClass();
        ILayer pLayer;
        FileInfo fileInfo;
        string tinPath;
        IWorkspaceFactory tinWorkspaceFactory = new TinWorkspaceFactoryClass();
        ITinWorkspace tinWorkspace;
        ITin tin;

        if (this.folderBrowserDialog1.ShowDialog() == DialogResult.OK)
        {
            tinPath = this.folderBrowserDialog1.SelectedPath;
            fileInfo = new FileInfo(tinPath);
            if (tinWorkspaceFactory.IsWorkspace(fileInfo.DirectoryName))
            {
                tinWorkspace = tinWorkspaceFactory.OpenFromFile(fileInfo.Directory
                Name, 0) as ITinWorkspace ;
                //tinWorkspace.OpenTin(fileInfo.DirectoryName);
                tin = tinWorkspace.OpenTin(fileInfo.Name);
                tinLayer.Dataset = tin;
                tinLayer.Visible = false;
                pScene.AddLayer(tinLayer as ILayer, true);
                pSceneGraph.RefreshViewers();
            }
        }
    }
    //叠加TIN数
    private void button3_Click(object sender, EventArgs e)
    {
        ISceneGraph pSceneGraph = this.axSceneControl1.SceneGraph;
        IScene pScene = pSceneGraph.Scene;
        ILayer layer = pScene.get_Layer(0);
        ITinLayer tinLayer = layer as ITinLayer;
        layer = pScene.get_Layer(1);
        IRasterLayer rasterLayer = layer as IRasterLayer;
        ITinAdvanced tinAdvanced;
        ISurface surface;
        tinAdvanced =tinLayer.Dataset as ITinAdvanced ;
        surface = tinAdvanced.Surface;

        ILayerExtensions layerExtensions = (ILayerExtensions)rasterLayer ;
        I3DProperties i3dProperties = null;
```

```csharp
        for (int i = 0; i < layerExtensions.ExtensionCount; i++)
        {
            if (layerExtensions.get_Extension(i) is I3DProperties)
            {
                i3dProperties = (I3DProperties)layerExtensions.get_Extension(i);
            }
        }//get 3d properties from extension

        i3dProperties.BaseOption = esriBaseOption.esriBaseSurface;
        i3dProperties.BaseSurface = surface;
        i3dProperties.Apply3DProperties(rasterLayer);
    }
    //分层设色
    private void button4_Click(object sender, EventArgs e)
    {   ILayer layer;
        ISceneGraph pSceneGraph = this.axSceneControl1.SceneGraph;
        IScene pScene = pSceneGraph.Scene;

        for (int i = 0; i < pScene.LayerCount; i++)
        {
            layer = pScene.get_Layer(i);
            if (layer is ITinLayer)
            {
                ITinLayer tinLayer = layer as ITinLayer;
                //设置高程缩放因子
                I3DProperties i3DProperties=null ;
                ILayerExtensions layerExtensions = layer as ILayerExtensions;
                for (int j = 0; j < layerExtensions.ExtensionCount; j++)
                {
                    if (layerExtensions.get_Extension(j) is I3DProperties)
                    {
                        i3DProperties = layerExtensions.get_Extension(j) as I3DProperties;
                    }
                }
                if (i3DProperties != null)
                {
                    i3DProperties.ZFactor = 0.0005;
                    i3DProperties.RenderMode = esriRenderMode.esriRenderCache;
                    i3DProperties.Apply3DProperties(layer);
                }
                ITinRenderer tinRenderer = new TinElevationRendererClass() as ItinRenderer;
                if (tinRenderer is ITinColorRampRenderer)
                {
                    if (tinRenderer.Name == "Elevation")
                    {
                        ITinAdvanced tinAdvanced = tinLayer.Dataset as ITin Advanced;
                        double dZMin = tinAdvanced.Extent.ZMin;
                        double dZMax = tinAdvanced.Extent.ZMax;
                        double dInterval = (dZMax - dZMin) /30;
                        ITinColorRampRenderer tinColorRampRenderer=tinRenderer as ITinColorRampRenderer ;
                        IClassBreaksUIProperties classBreakUIProperties =tinRenderer as IClass BreaksUIProperties ;
                        INumberFormat numberFormat=classBreakUIProperties. NumberFormat ;
                        int lClasses=30;
                        tinColorRampRenderer.BreakCount =lClasses ;
                        double dLowBreak =dZMin ;
                        double dHighBreak=dLowBreak +dInterval ;
                        classBreakUIProperties.ColorRamp ="Custom";
                        IAlgorithmicColorRamp algorithmicColorRamp =Create
```

```csharp
                        AlgorithmicColorRamp ();
                    IEnumColors enumColor =algorithmicColorRamp.Colors ;
                    enumColor.Reset ();

                    for (int k =0;k<lClasses ;k++)
                    {
                        classBreakUIProperties.set_LowBreak (k,dLowBreak );
                        tinColorRampRenderer.set_Break (k,dHighBreak );
                        tinColorRampRenderer.set_Label(k, numberFormat.Value ToString
                        (dLowBreak) + " - " + numberFormat.ValueToString(dHigh Break));
                        dLowBreak = dHighBreak;
                        dHighBreak = dHighBreak + dInterval;
                        ISimpleFillSymbol  symbol =new SimpleFillSymbol Class ();
                        symbol.Color =enumColor.Next ();
                        tinColorRampRenderer.set_Symbol (k,symbol as ISymbol );
                    }
                    tinLayer.ClearRenderers ();
                    (tinColorRampRenderer as ITinRenderer ).Visible =true ;
                    tinLayer.InsertRenderer (tinColorRampRenderer as
                    ITinRenderer ,0);
                }
                pScene.SceneGraph.Invalidate (tinLayer ,true,false );
                this.axSceneControl1.SceneViewer.Redraw (true );
                pScene.SceneGraph.RefreshViewers ();
                this.axTOCControl1.ActiveView.PartialRefresh (esriView
                DrawPhase.esriViewForeground ,tinLayer ,pScene.Extent );
                this.axTOCControl1.Update ();

            }

        }
    }
}
private IAlgorithmicColorRamp CreateAlgorithmicColorRamp()
{
    //创建一个新AlgorithmicColorRampClass对象
    IAlgorithmicColorRamp algColorRamp = new AlgorithmicColorRampClass();
    IRgbColor fromColor = new RgbColorClass();
    IRgbColor toColor = new RgbColorClass();
    //创建起始颜色对象
    fromColor.Red = 255;
    fromColor.Green = 0;
    fromColor.Blue = 0;
    //创建终止颜色对象
    toColor.Red = 0;
    toColor.Green = 255;
    toColor.Blue = 0;
    //设置AlgorithmicColorRampClass的起止颜色属性
    algColorRamp.ToColor = fromColor;
    algColorRamp.FromColor = toColor;
    //设置梯度类型
    algColorRamp.Algorithm = esriColorRampAlgorithm.esriCIELabAlgorithm;
    //设置颜色带颜色数量
    algColorRamp.Size = 30;
    //创建颜色带
    bool bture = true;
    algColorRamp.CreateRamp(out bture);
    return algColorRamp;
}
```

9.3 三维分析

三维分析是指在数字高程模型的基础上,利用空间分析算法获取研究区域中与空间特征相关的一些信息的过程。三维分析一般实现如下功能:任意点的定位查询、获取三维坐标值、距离查询、面积查询、体积查询、填挖方分析、剖面图绘制、地质特征三维分析等。

ArcEngine 提供了支持 Raster 数据和 TIN 数据进行各种三维分析工作,本节介绍三维场景属性查询、坡度分析、通视分析及剖面图绘制等功能。

9.3.1 三维场景属性查询

三维场景属性查询,即在单击时返回该位置的坐标、高程等信息。通过 ISceneGraph 接口的方法 Locate,可以获取 IPoint 接口对象,利用 IPoint 接口获得单击处的地理坐标和高程值。

9.3.2 坡度分析

坡度即水平面与局部地表之间的正切值,它包含两个部分:斜度和坡向。斜度即高度变化的最大值比率,坡向即变化比率最大值的方向。坡度分析需要用到如下几个类:Point、RasterWorkspace、RasterBand、PixelBlock 和 DblPnt。

Point 组件类用于创建一个二纬点对象,通常有长度、高度和 ID 属性。

RasterWorkspace 类,Raseter 工作区对象,该对象不可创建,对该对象的引用必须通过其他对象接口函数获得。

RasterBand 类,在磁盘上的 Raster 数据集的某一个波段,该类对象不可创建,对该对象的引用必须通过其他对象接口函数获得。

PixelBlock 类,该类的对象是个不可创建对象,对该对象的引用必须通过其他对象接口函数获得,可以从 Raster 数据集或 Raster 数据的某个波段直接读写像素值。

9.3.3 通视分析

通视分析有着很强的应用背景,如观察哨所位置的选择,哨所的位置应该设置在能监视某一感兴趣的区域,视线不能被地形挡住;低空侦察飞机在飞行时选择雷达盲区等。根据实际问题输出维数的不同,可分为点通视、线通视、面通视等。由于通视分析使用的方法 Visibility 只适用于 Raster 表面,因此 ArcEngine 没有提供对 Tin 表面的分析方法。

9.3.4 剖面图绘制

剖面图可以反映地形表面沿着一条线的高程变化情况,可以帮助我们设计路线或评估一条给定路线是否适合进行铁轨的铺设等。剖面图的绘制通常使用以下几个类:SceneGraph、Polyline、Tin、RasterSurface、LineElement、RgbColor、SimpleLineSymbol、GraphicsLayer3D 和 DataGraph。

SceneGraph 类实现了 IDisplay3D 接口,提供了 6 个方法和两个属性,如 FlashLocation 用于在指定的位置产生闪光效果,AddFlashFeature 用于添加实体对象到动画列表。

Tin 和 RasterSurface 类实现了 ISurface 接口,提供了一些属性和方法,如 InterpolateShape 用于为一个实体内插高程值。RasterSurface 类还实现了 IRasterSurface 接口,提供了操作和分析一个 Raster 表面的方法和属性。

GraphicsLayer3D 类实现了 IGraphicsContainer3D 接口,提供了操作图形容器的方法和属性。该类还实现了 IGraphicsLayer 和 IGraphicsSelection 接口,IGraphicsLayer 接口提供了控制图形图层的

方法和属性，IGraphicsSelection 接口提供了选择三维图形的方法和属性。

　　DataGraph 类实现了 IDataGraph 和 IDataGraphProperties 接口，IDataGraph 接口提供了控制数据图表的方法和属性，IDataGraphProperties 接口提供了控制 DataGraph 对象属性的方法和属性，具体实现如图。

```
string m_FileName;
//添加 TIN 数
private void button1_Click(object sender, EventArgs e)
{
    ISceneGraph pSceneGraph = this.axSceneControl1.SceneGraph;
    IScene pScene = pSceneGraph.Scene;
    IRasterLayer pRasterLayer = new RasterLayerClass();
    ILayer pLayer;

    this.openFileDialog1.Title = "Raster layer";
    this.openFileDialog1.DefaultExt = ".TIF";
    this.openFileDialog1.Filter = "(*.tif)|*.tif";
    if (this.openFileDialog1.ShowDialog() == DialogResult.OK)
    {
        m_FileName = this.openFileDialog1.FileName;
        string pPathName = this.openFileDialog1.FileName;
        string pPath = pPathName.Substring(0, pPathName.LastIndexOf('\\'));
        string fileName = pPathName.Substring(pPath.Length + 1, pPath Name.Length
        - pPath.Length - 1);
        IWorkspaceFactory pwsf = new RasterWorkspaceFactoryClass();
        IRasterWorkspace pRasterWorkspace;
        if (pwsf.IsWorkspace(pPath))
        {
            pRasterWorkspace = pwsf.OpenFromFile(pPath, 0) as Iraster Workspace;
            IRasterDataset pRasterDataset = pRasterWorkspace.OpenRaster
            Dataset(fileName);
            //pRasterDataset.OpenFromFile(pPath);
            pRasterLayer.CreateFromDataset(pRasterDataset);
            pLayer = pRasterLayer as ILayer;
            pScene.AddLayer(pLayer, true);
            //pScene.ExaggerationFactor = 6;
            pSceneGraph.RefreshViewers();
        }
    }
}
//坡度分析
private void button2_Click(object sender, EventArgs e)
{
    IWorkspaceFactory workspaceFactory = new RasterWorkspaceFactoryClass();
    FileInfo fileInfo = new FileInfo(m_FileName);
    string filePath = fileInfo.DirectoryName;
    string fileName = fileInfo.Name;
    //filePath =filePath +@"\workspace1";
    IRasterWorkspace workspace = workspaceFactory.OpenFromFile(filePath, 0) as
    IRasterWorkspace;

    ILayer layer = this.axSceneControl1.Scene.get_Layer(0);
    IRasterLayer rasterLayer = layer as IRasterLayer;
    IRasterDataset rasterDataset = workspace.OpenRasterDataset(fileName);

    ISurfaceOp surfaceOp = new RasterSurfaceOpClass();
    IRasterAnalysisEnvironment rasterAnalysisEnveronment;
    rasterAnalysisEnveronment = surfaceOp as IRasterAnalysisEnvironment;
    rasterAnalysisEnveronment.OutWorkspace = workspace as IWorkspace;
```

```csharp
        object zFactor = new object();
        IGeoDataset geoDataset, rasterGeoDataset;
        rasterGeoDataset = rasterDataset as IGeoDataset;
        geoDataset = surfaceOp.Slope(rasterGeoDataset, esriGeoAnalysisSlope
        Enum.esriGeoAnalysisSlopePercentrise, ref  zFactor);

        IRasterBandCollection  rasterBandCollection  =  geoDataset  as  IrasterBandCollection;

        rasterBandCollection.SaveAs("podu.tif", workspace as IWorkspace, "TIFF");
    }

    private void axSceneControl1_OnMouseDown(object sender, ISceneControlEvents_OnMouseDownEvent e)
    {
        ISceneGraph pSceneGraph = this.axSceneControl1.SceneGraph;
        IScene pScene = pSceneGraph.Scene;
        int px = e.x;
        int py = e.y;
        IPoint point = null;
        object pOwner;
        object pObject;
        pSceneGraph.Locate(this.axSceneControl1.SceneViewer, px, py,
        esriScenePickMode.esriScenePickAll, true, out point, out pOwner, out pObject);
        if (point != null)
        {
            MessageBox.Show(point.X + "_" + point.Y + "_" + point.Z);
        }
    }
    //通视分析
    private void button3_Click(object sender, EventArgs e)
    {
        IWorkspaceFactory workspaceFactory = new RasterWorkspaceFactoryClass();
        FileInfo fileInfo = new FileInfo(m_FileName);
        string filePath = fileInfo.DirectoryName;
        string fileName = fileInfo.Name;

        IRasterWorkspace rasterWorkspace = workspaceFactory.Open FromFile (filePath,
        0) as IrasterWorkspace;

        IWorkspaceFactory pointWorkspaceFactory = new AccessWorkspaceFactory Class();
        IWorkspace featureWorkspace = pointWorkspaceFactory.OpenFromFile (filePath +
        @"\aaa.mdb", 0);
        IEnumDataset enumDataset = featureWorkspace.get_Datasets(esriDataset
        Type.esriDTAny);
        enumDataset.Reset();
        IGeoDataset featureDataset = enumDataset.Next() as IGeoDataset;


        ILayer layer = this.axSceneControl1.Scene.get_Layer(0);
        IRasterLayer rasterLayer = layer as IRasterLayer;
        IRasterDataset rasterDataset = rasterWorkspace.OpenRasterDataset (fileName);

        ISurfaceOp surfaceOp = new RasterSurfaceOpClass();
        IRasterAnalysisEnvironment rasterAnalysisEnveronment;
        rasterAnalysisEnveronment = surfaceOp as IRasterAnalysisEnvironment;
        rasterAnalysisEnveronment.OutWorkspace = rasterWorkspace as IWorkspace;
        object zFactor = new object();
        IGeoDataset geoDataset, rasterGeoDataset;
        rasterGeoDataset = rasterDataset as IGeoDataset;
        geoDataset = surfaceOp.Visibility(rasterGeoDataset, featureDataset, esriGeo
```

```
                AnalysisVisibilityEnum.esriGeoAnalysisVisibilityFrequency);
            IRasterBandCollection rasterBandCollection = geoDataset as IrasterBandCollection;
            rasterBandCollection.SaveAs("tongshi.tif", rasterWorkspace as IWorkspace, "TIFF");
        }
    }
}
```

9.4 本章小结

　　三维可视化是 GIS 应用的一个重要方面，本章讲述了三维数据的显示，通过三维数据加载、图层叠加、分层设色 3 个方面做介绍。三维分析是 GIS 应用系统中的高级分析功能，是衡量 GIS 功能的一个重要尺度，通过三维分析，用户可以得到更接近实际的分析结果，本章介绍了三维分析中常用的 3 个功能：坡度分析、通视分析、剖面图绘制。希望了解更多三维分析功能的读者可以参考专门的三维分析书籍或者查看 ArcEngine 在线帮助。

第二篇

应用提高篇

第 10 章　空间分析

第 10 章 空间分析

10.1 简介

空间分析是基于地理对象的位置和形态特征的空间数据分析技术，其目的在于提取和传输空间信息。通过空间分析，不但可以知道数据库中的数据，而且可以通过这些数据去揭示更深刻、更内在的规律和特征。空间分析根据使用的数据性质的不同，可以分为基于空间图形数据的分析运算、基于非空间属性的数据运算、空间和非空间数据的联合运算。其最终目的是解决人们涉及的地理空间实际问题，提取和传输地理空间信息。

10.2 空间查询

地图中包含了大量的信息，为了快速地了解所需信息，必须借助空间查询功能。空间查询主要有两类：基于属性查询；基于空间位置查询，也称为空间关系查询。

在 ArcEngine 平台中，查询使用的游标（Cursor）非常多，凡是和数据操作有关的地方几乎都会用到游标。游标是一个指向数据的指针，其本身不包含数据内容，只提供了一个连接 Row 对象或要素对象的桥梁。游标可分为 3 种类型：查询游标、插入游标、更新游标，每一种游标都有相应的方法来获取，如 Search、Insert、Update。更新和查询游标需要使用一个过滤器对象，筛选出需要进行操作的要素。

游标对象实现了 ICursor 接口，定义了一个 Row 集合或一个 Row 对象的属性和方法，如 ITable 的 Insert 方法用于获取一个插入型游标，该游标用于在表中插入一条记录，使用该方法插入记录比使用 CreateRow 和 Store 两个方法插入一条记录的速度更快；UpDate 方法用于获取一个更新型游标，该游标用于更新或删除一条记录，该方法比使用 Store 和 Delete 方法处理的速度更快；Search 方法用于对表进行查询，可以得到一个查询型 Cursor 对象，该对象指向一个或多个 Row 对象，通过使用 NextRow 方法来获取游标指向的下一个 Row 对象。

10.2.1 基于属性查询

基于属性查询是通过对要素的属性信息设定要求来查询定位空间位置。如查询所有的"真功夫"连锁店，此查询需要对要素名称设置限制条件，查询到结果后再利用图形和要素的对应关系，在地图上对符合条件的要素进行定位并渲染。根据属性字段的不同类型，主要有字符型字段查询、数值型字段查询、多条件查询等。

字符型字段查询通常使用 "LIKE" 进行模糊查询。数值型字段查询一般使用比较操作符和运算符来完成，如 ">"、">="、"<"、"<="、"+"、"-"、"*"、"/"。多条件查询一般需要组合多个条件，条件之间用 "AND"、"OR" 来连接，如 "名称='china' and city='beijing'"，还可以使用 IN 操作

符号，可同时查找一个字段内的多个字符串值，如"cityName in ('beijing','shanghai','nanjing')"。

QueryFilter 组件类指定了一个属性查询过滤器，利用该过滤器可从要素集中筛选出满足条件的行集。该组件类实现了 IQueryFilter 接口，提供了 1 个方法和 3 个属性，如 AddField 方法用于向输出字段集中增加一个字段，OutputSpatialReference 属性用于获取给定字段的输出几何图形的空间参考，SubFields 用于获取或设置输出几何图形的字段名称清单，WhereClause 用于获取或设置查询过滤条件。

使用 WhereClause 应注意两点：其一，由于 Engine 不支持 OrderBy，所以不能出现这个关键字，如果需要排序，则可通过 ITableSort 接口来完成；其二，SQL 语法在不同的数据源 Access、SDE、SHP、Coverages 之间存在差异，在 Access 中对字段值不区分大小写，而在 SDE、SHP 和 Coverages 中是区分大小写的，通配符在这些数据源中的表达形式不同，Access 使用"？"代表单个字符"*"表示一组字符，而在 SDE、SHP 和 Coverages 中则分别是"_"和"%"。

在 ITableSort 接口可以设置 TableSort 对象的属性并且执行排序。对一个 TableSort 而言，Table 和 Fields 属性是必须被设置的，而其他的属性是可选的，除了使用表、对象类、要素类外，选择集也可以作为一个需要排序的数据源。Fields 是一个用于排序的字段列表，当使用该接口的 Sort 方法排序时，可以先根据第 1 个字段排序，然后根据第 2 个字段排序，以此类推。该接口还提供了 Ascending 用于确定升序排序的字段，CaseSensitive 用于确定使用大小写敏感设置。

下面的代码片段演示了如何使用属性查询对象。

```
IQueryFilter queryFilter ;
queryFilter =new QueryFilterClass ();
//设置过滤器属性
queryFilter.WhereClause = "name ='california'";
ITableSort tableSort = new TableSortClass();
ILayer layer = this.axMapControl1.get_Layer(0);
IFeatureLayer featureLayer = layer as IFeatureLayer;
IFeatureClass featureClass = featureLayer.FeatureClass;
//设置排序表对象
tableSort.Table = featureClass as ITable;
tableSort.QueryFilter = queryFilter;
//设置排序字段
tableSort.Fields = "name ,address ";
tableSort.set_Ascending("name", false);
tableSort.set_CaseSensitive ("address", true);
tableSort.Sort(null);
ICursor sortedCursor = tableSort.Rows;
IRow row = sortedCursor.NextRow();
int namePosition =featureClass.FindField ("name");
int addressPosition =featureClass.FindField ("address");
while (row != null)
{
    Console.WriteLine("name is {0},address is {1}", row.get_Value(namePosition),
    row.get_Value(addressPosition));
}
```

10.2.2 基于空间位置查询

基于空间位置查询，是根据要素与要素之间的空间关系进行查询，主要有以下几种：相交（Intersect）、相接（Touch）、叠加（Overlap）、穿越（Crosses）、在内部（Within）和包含（Contains）。

相交有 3 种情况，其一，几何图形与目标几何图形相交；其二，查询几何图形的空间范围和目标几何图形的空间范围相交；其三，查询几何图形的空间范围和目标几何图形的索引范围相交。

相接是指两个几何图形只在它们的交界处相交，而两个几何图形内部的交集为空。如点和线的相接，那么点必须在线的端点处才会发生相接关系，点和点之间不会发生相接关系。

叠加是指两个几何图形的交集与这两个几何图形的维数相同，但其交集结果与这两个几何图形不同。

穿越是指查询几何图形与目标几何图形为跨越关系。穿越关系是低纬跨越于高纬,如铁路线穿越多个省份。

在内部是指当查询几何图形完全落入到目标几何图形内时,一个几何图形不能包含在另一个具有更低维数的几何图形中,如多边形不能落入线要素中。

包含是指一个查询几何图形包含了目标几何图形,如一个湖内有多个岛屿,则湖与岛屿为包含关系。注意:一个几何图形不能包含一个具有更高维数的几何图形。

空间几何查询主要用到 SpatialFilter 组件类,该类是 QueryFilter 组件类的子类,可以从属性和空间关系两方面查询条件来查询。QueryFilter 类实现了 ISpatialFilter 接口,该接口是给予空间关系和属性两方面规则来查询要素,使用该接口必须设置 Geometry、GeometryField 和 SpatialRel。该接口提供了 1 个方法和 10 个属性,如 AddField 方法用于在输出字段集合中增加一个字段;Geometry 属性用于设置或获取查询的几何图形,用此几何图形来限制空间位置;GeometryField 用于获取或设置过滤器中使用的被查询几何图形的空间域字段;SearchOrder 用于获取或设置空间查询顺序,即先进行属性查询还是先进行空间查询;SpatialRel 用于获取或设置查询时的空间关系,具体如表 10-1 所示。

表 10-1　　　　　　　　　　获取或设置查询时的空间关系

esriSpatialRelUndefined	关系未定义
esriSpatialRelIntersects	A 与 B 相交
esriSpatialRelEnvelopeIntersects	A 的包络与 B 的包络线相交
esriSpatialRelIndexIntersects	A 与 B 的索引相交
esriSpatialRelTouches	A 与 B 相接,即其边界处相接
esriSpatialRelOverlaps	A 与 B 相叠加,它们必须是同维对象,如都是多边形
esriSpatialRelCrosses	A 与 B 相交,如两条线交于一点,面和线交于一条线
esriSpatialRelWithin	A 在 B 的内部
esriSpatialRelContains	A 被 B 包含
esriSpatialRelRelation	Query geometry IBE(Interior-Boundary-Exterior) relationship with target geometry.

下面的代码片段演示了如何使用空间筛选器。

```
ISpatialFilter spatialFilter = new SpatialFilterClass();
//设置用于筛选几何对象,
spatialFilter.Geometry = point;
//设置筛选几何字段
spatialFilter.GeometryField = "shape";
//设置空间几何关系
spatialFilter.SpatialRel = esriSpatialRelEnum.esriSpatialRelContains;
ILayer layer = this.axMapControl1.get_Layer(0);
IFeatureLayer featureLayer = layer as IFeatureLayer;
IFeatureClass featureClass = featureLayer.FeatureClass;
IFeatureCursor featureCursor;
featureCursor = featureClass.Search(spatialFilter, false);
IFeature feature = featureCursor.NextFeature();
while (feature != null)
{
    Console.WriteLine("feature oid is {0}", feature.OID);
    feature = featureCursor.NextFeature();
}
```

10.2.3　要素选择集

要素选择集(SelectionSet)是被选择的行对象的集合,这些行只能来自单个表或要素类,但是一个表或要素类可以产生多个选择集对象。选择集有两种形式，一种是基于被选择的行对象的 OID 集合，另一种是 Row 对象本身，可以通过设置 SelectionType 属性来确定。SelectionType 有 3 种类型，其一，esriSelectionIDSet 代表选择集使用的是一个 OID 集合，这些 OID 值可以保存在一个物理表中，也可以保存在内存中；其二，esriSelectionTypeSnapShot，该类型表示选择集使用的是保存在内存中的实际 Row 对象；其三，esriSelectionTypeHybrid，该类型当选择数量少的时候选择集使用内存中的行对象，当数量多的时候使用 OID 集合。

使用标识集合（如 OID 集合）可以表示数目巨大的选择集，当数据源每一次被选择的对象还需要查询的时候，使用这种方式可以保证选择集是动态的，而且在数据源改变的时候会自动变化。Snapshot 类型的选择集速度最快，而且一旦选择集构造完成后就不再要求与数据源发生查询，当数据量少的时候该方式有效，当数据量大的时候占用内存大就会失去优势。Hybird 类型选择集包含了前两者的优点，可以依据选择数据的大小而自动选用不同的选择方式，但这个选择数据的大小并不能由程序来控制。

SelectionSet 对象主要实现 ISelectionSet 接口，提供了对选择集的管理和查询，如 Search 方法用于在选择集内进行再选择；Select 方法用于在目前的选择集中构造一个新的选择集对象；Add、AddList、RemoveList 等方法用于在选择集中添加和移除对象，这些方法是通过 OID 属性来完成的；Combine 方法则用于绑定两个选择集，注意这两个选择集必须来自同一个目标表或要素类。

下面的代码片段演示了如何使用选择集对象。

```
IQueryFilter queryFilter;
queryFilter = new QueryFilterClass();
//设置过滤器属性
queryFilter.WhereClause = "name ='california'";
ILayer layer = this.axMapControl1.get_Layer(0);
IFeatureLayer featureLayer = layer as IFeatureLayer;
IFeatureClass featureClass = featureLayer.FeatureClass;
ISelectionSet selectionSet;
IDataset dataset =featureClass as IDataset ;
selectionSet  =  featureClass.Select(queryFilter,  esriSelectionType.esriSelection
TypeHybrid, esri SelectionOption.esriSelectionOptionNormal, dataset.Workspace);
//获取 OID 集合
IEnumIDs  enumIDs=selectionSet.IDs ;
int namePosition =featureClass.FindField ("name");
int id =enumIDs.Next ();
IFeature feature ;
while (id!=-1)
{
    feature = featureClass.GetFeature(id);
    id = enumIDs.Next();
}
```

10.3 空间几何图形的集合运算

空间几何图形的集合运算可分为点与多边形集合运算、线与多边形集合运算、多边形与多边形集合运算等几种情况。

点与多边形集合运算，就是多边形对点的包含关系分析。利用此运算，可以找到哪些点在此多边形内。如银行布点时，先统计某个区域内有多少家营业点。

线与多边形集合运算，是比较线上坐标与多边形坐标的关系，判断线是否落入多边形内的分析。

多边形与多边形集合运算是将两个或多个多边形进行叠加产生一个新多边形的操作。

对于矢量数据图层，集合运算主要是叠置求交（Interset）和叠置求和（Union）操作。叠置求交是将两个图层中共同的区域内的要素和属性组合到第三图层，其输出结果要素类型与输入图层几何类型一致。

10.4 空间拓扑运算

空间拓扑描述的是自然界地理对象的空间位置关系，即相邻、重合、连通等。拓扑（Topology）是在同一个要素集（FeatureDataset）下的要素类（Feature Class）之间的关系的集合，所以要参与一个拓扑的所有要素类，必须在同一个要素集内（也就是具有同一个空间参考），这样进行的拓扑检查才是精确的。一个要素集可以有多个拓扑，但每个要素类最多只能参与一个拓扑，一个拓扑中可以定义多个规则，是地理对象空间属性的一部分，在目前 ESRI 提供的数据存储方式中，Coverage 和 GeoDatabase 能够建立拓扑，而 Shape 格式的数据不能建立拓扑。

ESRI 提供了 27 种拓扑关系，如表 10-2 所示。

表 10-2　　　　　　　　　　　　ESRI 提供的拓扑关系

esriTRTAny	任何拓扑规则，查询拓扑的时候使用
esriTRTFeatureLargerThanClusterTolerance	地理要素小于聚类容限被删除
esriTRTAreaNoGaps	面是封闭的
esriTRTAreaNoOverlap	面不相交
esriTRTAreaCoveredByAreaClass	The rule is an area covered by area class rule.
esriTRTAreaAreaCoverEachOther	两个区域完全重合
esriTRTAreaCoveredByArea	一个区域被另一个区域覆盖
esriTRTAreaNoOverlapArea	一个面没有与之相交的其他面
esriTRTLineCoveredByAreaBoundary	线被区域的边线覆盖
esriTRTPointCoveredByAreaBoundary	点在面的边界上
esriTRTPointProperlyInsideArea	点完全在面内
esriTRTLineNoOverlap	无重合的线
esriTRTLineNoIntersection	无相交的线
esriTRTLineNoDangles	无摇摆的线
esriTRTLineNoPseudos	线不存在伪节点
esriTRTLineCoveredByLineClass	The rule is a line covered by line class rule.
esriTRTLineNoOverlapLine	The rule is a line-no overlap line rule.
esriTRTPointCoveredByLine	点被线覆盖
esriTRTPointCoveredByLineEndpoint	点被线的尾节点覆盖
esriTRTAreaBoundaryCoveredByLine	一个面的边界被线覆盖
esriTRTAreaBoundaryCoveredByAreaBoundary	一个面的边界被另一个面的边界覆盖
esriTRTLineNoSelfOverlap	不存在自重合的线
esriTRTLineNoSelfIntersect	不存在自相交的线
esriTRTLineNoIntersectOrInteriorTouch	The rule is a line-no intersect or interior touch rule.
esriTRTLineEndpointCoveredByPoint	线的尾节点被点覆盖
esriTRTAreaContainPoint	面包含点
esriTRTLineNoMultipart	The rule is a line cannot be multipart rule.

ArcEngine 中提供了 ITopologicalOperator 接口用于拓扑运算，ITopologicalOperator 接口用来通过对已存在的几何对象做空间拓扑运算，以产生新的结合对象。实现该接口的类有 Point、Multipoint、Polyline、Polygon 和 MultiPatch 等，这些都是高级几何对象。另外 GeometryBag 也实现了该接口，低级的构建几何对象，如 Segments(Line、Circular Arc、Elliptic Arc、Bezier Curve)、Paths 或者 Rings 如果想使用该接口，则需包装成高级几何对象。

ITopologicalOperator 接口在 GIS 开发中的使用非常广泛，通常 GIS 系统中的缓冲区分析、裁剪几何图形、几何图形差分操作、几何图形合并操作等都需要使用此接口。下面介绍 ITopologicalOperator 接口主要的方法。

Boundary 方法用于几何图形对象的边界。如图 10-1 所示，Polygon 几何对象的 Boundary 是组成它的 Polyline 几何对象，Polyline 几何对象的 Boundary 是组成它的顶点 Point 几何对象，而 Point 几何对象的 Boundary 是空对象。

Buffer 用于集合对象缓冲区拓扑操作，Buffer 方法用于给一个高级几何对象产生一个缓冲区，无论是 Polygon、Polyline 还是 Point，它们的缓冲区都是具有面积的几何对象，如图 10-2 所示。

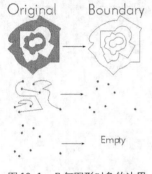
▲图 10-1　几何图形对象的边界

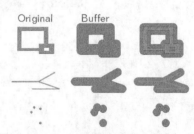
▲图 10-2　有面积的几何对象

Clip 用于对几何对象进行裁剪空间拓扑操作。Clip 方法可用一个 Envelope 对象对一个几何对象进行裁剪，结果是几何对象被 Envelope 对象所包围的部分，如图 10-3 所示。

ConstructUion 用于将多个枚举对象与单个几何对象合并为单个几何对象。

ConvexHull 用于构建几何对象的凸多边形。ConvexHull 方法用于产生一个几何图形的最小的边框凸多边形（没有凹面包含几何图形的最小多边形），如图 10-4 所示。

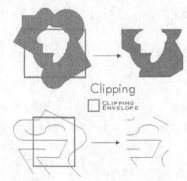
▲图 10-3　Envelope 对象所包围的部分

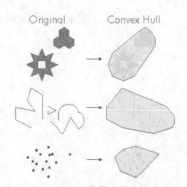
▲图 10-4　几何图形的最小的边框凸多边形

Cut 用于切割几何对象。Cut 方法不支持 GeometryBags 几何对象，它可以指定一条切割曲线和一个几何图形，经过切割运算后把几何图形分为左右两部分，左右两部分是相对曲线的方向而言。

点与多点不能被切割，Polyline 和 Polygon 只有与切割曲线相交时才能执行 Cut 方法，如图 10-5 所示。

Difference 用于从一个几何图形中减去其与另一个几何图形相交的部分，产生两个几何对象的差集，如图 10-6 所示。

▲图 10-5　Cut 用于切割几何对象　　　　▲图 10-6　两个几何对象的差集

Intersection 方法用于两个同维度几何对象的交集部分，如图 10-7 所示。

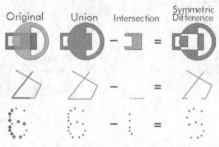

▲图 10-7　几何对象的交集部分

Simplify 用于使几何对象拓扑一致，如图 10-8 所示。

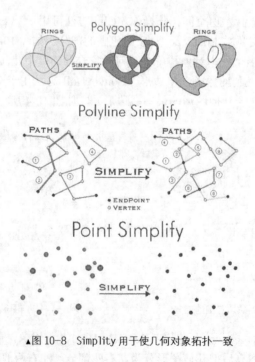

▲图 10-8　Simplity 用于使几何对象拓扑一致

SymmetricDifference 对称差分可将两个几何图形的并集部分减去两个几何图形交集的部分，如图 10-9 所示。

Union 用于合并两个同维度的几何对象为单个几何对象，如图 10-10 所示。Union 方法和 ConstructUnion 都用于合并几何对象，所不同的是前者合并两个同维度的几何对象为单个几何对象，而后者是高效地将多个枚举几何对象与单个几何对象合并为单个几何对象，这对于大量几何对象的合并是非常高效的。

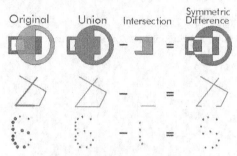

▲图 10-9　几何图形交集的部分

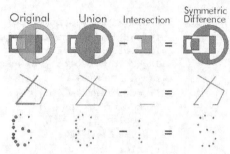

▲图 10-10　Union 合并几何对象

下面代码演示如何使用这些运算：

```csharp
//***********************************************************************
//功能演示用到的辅助方法
//***********************************************************************
//获取颜色对象
private IRgbColor getRGB(int r, int g, int b)
{
    IRgbColor pColor;
    pColor = new RgbColorClass();
    pColor.Red = r;
    pColor.Green = g;
    pColor.Blue = b;
    return pColor;
}
//获取简单线符号
private ISymbol getSimpleLineSymbol()
{
    ISimpleLineSymbol simpleLineSymbol = new SimpleLineSymbolClass();
    simpleLineSymbol.Style = esriSimpleLineStyle.esriSLSDashDotDot;
    simpleLineSymbol.Width = 10;
    IRgbColor rgbColor = getRGB(255, 0, 0);
    simpleLineSymbol.Color = rgbColor;
    ISymbol symbol = simpleLineSymbol as ISymbol;
    symbol.ROP2 = esriRasterOpCode.esriROPNotXOrPen;
    return symbol;
}
//获取简单面填充符号
private ISimpleFillSymbol  getSimpleFillSymbol(int fillColor,int lineColor)
{
    ISimpleFillSymbol simpleFillSymbol = new SimpleFillSymbolClass();
    simpleFillSymbol.Style = esriSimpleFillStyle.esriSFSSolid;
    simpleFillSymbol.Color = getRGB(fillColor, 0, 0);
    //创建边线符号
    ISimpleLineSymbol simpleLineSymbol = new SimpleLineSymbolClass();
    simpleLineSymbol.Style = esriSimpleLineStyle.esriSLSDashDotDot;
    simpleLineSymbol.Color = getRGB(0, lineColor, 0);
```

```csharp
            simpleLineSymbol.Width = 5;
            ISymbol symbol = simpleLineSymbol as ISymbol;
            symbol.ROP2 = esriRasterOpCode.esriROPNotXOrPen;

            simpleFillSymbol.Outline = simpleLineSymbol;
            return simpleFillSymbol;


        }
        //创建颜色带
        private IColorRamp CreateAlgorithmicColorRamp(int count)
        {
            //创建一个新AlgorithmicColorRampClass对象
            IAlgorithmicColorRamp algColorRamp = new AlgorithmicColorRampClass();
            IRgbColor fromColor = new RgbColorClass();
            IRgbColor toColor = new RgbColorClass();
            //创建起始颜色对象
            fromColor.Red = 255;
            fromColor.Green = 0;
            fromColor.Blue = 0;
            //创建终止颜色对象
            toColor.Red = 0;
            toColor.Green = 0;
            toColor.Blue = 255;
            //设置AlgorithmicColorRampClass的起止颜色属性
            algColorRamp.ToColor = fromColor;
            algColorRamp.FromColor = toColor;
            //设置梯度类型
            algColorRamp.Algorithm = esriColorRampAlgorithm.esriCIELabAlgorithm;
            //设置颜色带颜色数量
            algColorRamp.Size = count;
            //创建颜色带
            bool bture = true;
            algColorRamp.CreateRamp(out bture);
            return algColorRamp;
        }
        //转换像素到地图单位
        private double ConvertPixelsToMapUnits(IActiveView pActiveView, double pixelUnits)
        {
            tagRECT pRect = pActiveView.ScreenDisplay.DisplayTransformation.get_DeviceFrame();
            int pixelExtent = pRect.right - pRect.left;

            double realWorldDisplayExtent = pActiveView.ScreenDisplay.DisplayTransformation.VisibleBounds.Width;
            double sizeOfOnePixel = realWorldDisplayExtent / pixelExtent;
            return pixelUnits * sizeOfOnePixel;
        }
//*******************************************************************************
//功能函数
//*******************************************************************************
        //Boundary
        private void button1_Click(object sender, EventArgs e)
        {

            IFeatureLayer featureLayer;
            IFeatureCursor featureCursor;
            featureLayer = this.axMapControl1.Map.get_Layer(1) as IFeatureLayer ;
            featureCursor = featureLayer.FeatureClass.Search(null, true);
            IFeature feature;
```

```csharp
            feature = featureCursor.NextFeature();
            IGeometry geometry ;
            ISymbol symbol;
            IActiveView activeView = this.axMapControl1.ActiveView;
            if (feature != null)
            {
                IPolygon polygon;
                polygon = feature.Shape as IPolygon;
                ITopologicalOperator topo = polygon as ITopologicalOperator;
                topo.Simplify();
                geometry = topo.Boundary;
                if (geometry != null)
                {
                    if (geometry.GeometryType == esriGeometryType.esriGeometry Polyline)
                    {
                        symbol = getSimpleLineSymbol() as ISymbol ;
                        activeView.ScreenDisplay.StartDrawing(activeView.
                        ScreenDisplay.hDC, (short)esriScreenCache.esriNoScreenCache);
                        activeView.ScreenDisplay.SetSymbol(symbol);
                        activeView.ScreenDisplay.DrawPolyline (geometry);
                        activeView.ScreenDisplay.FinishDrawing();
                        activeView.ScreenDisplay.FinishDrawing();
                    }
                }
            }

}
//Buffer
private void button2_Click(object sender, EventArgs e)
{
    IFeatureLayer featureLayer =this.axMapControl1.Map.get_Layer (1) as IfeatureLayer ;
    IFeatureCursor featureCursor ;
    IFeature feature;
    IGeometry geometry;
    IActiveView activeView=this.axMapControl1.ActiveView ;
    ITopologicalOperator topo;
    featureCursor =featureLayer.FeatureClass .Search (null,false );
    feature =featureCursor.NextFeature ();
    ISpatialFilter spatialFilter =new SpatialFilterClass ();
    IFeatureSelection featureSelection;
    if(feature !=null)
    {
        topo = feature.Shape as ITopologicalOperator;
        double bufferLength =ConvertPixelsToMapUnits (activeView ,2);
        geometry = topo.Buffer(bufferLength);

        spatialFilter.SpatialRel =esriSpatialRelEnum.esriSpatial RelIntersects ;
        spatialFilter.Geometry = geometry;
        spatialFilter.GeometryField = featureLayer.FeatureClass.Shape FieldName;
        spatialFilter.SubFields = "continent";
        spatialFilter.WhereClause = "";
        featureSelection = featureLayer as IFeatureSelection;
        featureSelection.SelectFeatures (spatialFilter ,esriSelectionResultEnum .esriSelectionResultNew ,false );
        ISelectionSet selectionSet =featureSelection.SelectionSet ;
        ICursor cursor;
        selectionSet.Search (null,true ,out cursor );
        featureCursor=cursor as IFeatureCursor ;
        feature =featureCursor.NextFeature ();
```

```csharp
            while (feature !=null)
            {
                this.axMapControl1.Map.SelectFeature(featureLayer, feature);
                feature=featureCursor.NextFeature ();
            }
            activeView.PartialRefresh(esriViewDrawPhase.esriViewGeoSelection, null,
            null);
        }
    }
    //Clip
    private void button3_Click(object sender, EventArgs e)
    {
        IFeatureLayer featureLayer =this.axMapControl1.Map.get_Layer (1) as Ifeature
        Layer ;
        IFeatureCursor featureCursor ;
        IFeature feature;
        IGeometry geometry;
        IActiveView activeView=this.axMapControl1.ActiveView ;
        ITopologicalOperator topo;
        featureCursor =featureLayer.FeatureClass .Search (null,false );
        feature =featureCursor.NextFeature ();
        ISpatialFilter spatialFilter =new SpatialFilterClass ();
        IFeatureSelection featureSelection;
        IEnvelope env = new EnvelopeClass();
        if (feature != null)
        {
            topo = feature.Shape as ITopologicalOperator;
            env = feature.Shape.Envelope;
            double width,height;
            width = env.XMax - env.XMin;
            height = env.YMax - env.YMin;
            env.XMin = env.XMin + width / 3;
            env.XMax = env.XMax - width / 3;
            env.YMin = env.YMin + height / 3;
            env.YMax = env.YMax - height / 3;
            geometry = new PolygonClass();
            topo.QueryClipped(env, geometry);
            ISymbol symbol = getSimpleFillSymbol(255,255) as ISymbol ;

            activeView.ScreenDisplay.StartDrawing(activeView.ScreenDisplay. hDC,
            (short)esriScreenCache.esriNoScreenCache);
            activeView.ScreenDisplay.SetSymbol(symbol);
            activeView.ScreenDisplay.DrawPolygon(geometry);
            activeView.ScreenDisplay.FinishDrawing();
            activeView.ScreenDisplay.FinishDrawing();
        }
    }
    //ConstructUion
    private void button4_Click(object sender, EventArgs e)
    {
        IFeatureLayer featureLayer =this.axMapControl1.Map.get_Layer (1) as
        IFeatureLayer ;
        IFeature feature;
        IActiveView activeView=this.axMapControl1.ActiveView ;
        ITopologicalOperator topo;

        object missing=Type.Missing ;
        IGeometryCollection geometryCollection = new GeometryBagClass();
        for (int i = 0; i < 3; i++)
        {
            feature=featureLayer.FeatureClass.GetFeature(i);
```

```csharp
            geometryCollection.AddGeometry(feature.Shape, ref missing, ref missing);
        }
        IPolygon newPolygon = new PolygonClass();
        topo = newPolygon as ITopologicalOperator;
        topo.ConstructUnion(geometryCollection as IEnumGeometry);
        ISymbol symbol = getSimpleFillSymbol(255,255) as ISymbol ;

        activeView.ScreenDisplay.StartDrawing(activeView.ScreenDisplay.hDC,
            (short)esriScreenCache.esriNoScreenCache);
        activeView.ScreenDisplay.SetSymbol(symbol);
        activeView.ScreenDisplay.DrawPolygon(newPolygon as IGeometry);
        activeView.ScreenDisplay.FinishDrawing();
        activeView.ScreenDisplay.FinishDrawing();
}
//ConvexHull
private void button5_Click(object sender, EventArgs e)
{
    IFeatureLayer featureLayer = this.axMapControl1.Map.get_Layer(1) as IfeatureLayer;
    IFeature feature;
    IActiveView activeView = this.axMapControl1.ActiveView;
    ITopologicalOperator topo;

    IGeometry geometry;
    feature = featureLayer.FeatureClass.GetFeature(0);
    topo = feature.Shape as ITopologicalOperator ;
    geometry=topo.ConvexHull();


    ISymbol symbol = getSimpleLineSymbol ();
    activeView.ScreenDisplay.StartDrawing(activeView.ScreenDisplay.hDC,
        (short)esriScreenCache.esriNoScreenCache);
    activeView.ScreenDisplay.SetSymbol(symbol);
    activeView.ScreenDisplay.DrawPolyline(geometry);
    activeView.ScreenDisplay.FinishDrawing();
    activeView.ScreenDisplay.FinishDrawing();
}
//Cut
private void button6_Click(object sender, EventArgs e)
{
    IFeatureLayer featureLayer = this.axMapControl1.Map.get_Layer(1) as IfeatureLayer;
    IFeature feature;
    IActiveView activeView = this.axMapControl1.ActiveView;
    ITopologicalOperator topo;

    IGeometry leftGeometry,rightGeometry;
    feature = featureLayer.FeatureClass.GetFeature(0);
    topo = feature.Shape as ITopologicalOperator;
    IEnvelope env = feature.Shape.Envelope;
    IPolyline polyline = new PolylineClass();
    IPoint point = new PointClass();
    point.PutCoords(env.XMin, env.YMin);
    polyline.FromPoint = point;
    point.PutCoords(env.XMax, env.YMax);
    polyline.ToPoint = point;
    topo.Cut(polyline, out leftGeometry, out rightGeometry);

    ISymbol symbol =getSimpleFillSymbol (255,255) as ISymbol ;

    activeView.ScreenDisplay.StartDrawing(activeView.ScreenDisplay.hDC,
        (short)esriScreenCache.esriNoScreenCache);
```

```csharp
            activeView.ScreenDisplay.SetSymbol(symbol);
            activeView.ScreenDisplay.DrawPolygon(leftGeometry);

            activeView.ScreenDisplay.FinishDrawing();
            activeView.ScreenDisplay.FinishDrawing();
        }
        //Difference
        private void button7_Click(object sender, EventArgs e)
        {
            IFeatureLayer featureLayer = this.axMapControl1.Map.get_Layer(1) as IfeatureLayer;
            IFeature feature;
            IActiveView activeView = this.axMapControl1.ActiveView;
            ITopologicalOperator topo;

            IGeometry geometry;
            feature = featureLayer.FeatureClass.GetFeature(0);
            topo = feature.Shape as ITopologicalOperator;
            IEnvelope env = feature.Shape.Envelope;
            double width, height;
            width = env.XMax - env.XMin;
            height = env.YMax - env.YMin;
            object Missing = Type.Missing;
            IPolygon polygon = new PolygonClass();
            IPointCollection pointCollection = polygon as IPointCollection;
            IPoint point = new PointClass();
            point.PutCoords(env.XMin +width/3,env.YMin +height/3);
            pointCollection.AddPoint(point, ref Missing, ref Missing);
            point.PutCoords(env.XMin +width/3, env.YMax -height /3);
            pointCollection.AddPoint(point, ref Missing, ref Missing);
            point.PutCoords(env.XMax -width /3, env.YMax -height /3);
            pointCollection.AddPoint(point, ref Missing, ref Missing);
            point.PutCoords(env.XMax -width /3,env.YMin +height /3);
            pointCollection.AddPoint(point, ref Missing, ref Missing);
            polygon.SimplifyPreserveFromTo();
            geometry = topo.Difference(polygon as IGeometry );
            ISymbol symbol = getSimpleFillSymbol(255, 255) as ISymbol ;

            activeView.ScreenDisplay.StartDrawing(activeView.ScreenDisplay.hDC,
            (short)esriScreenCache.esriNoScreenCache);
            activeView.ScreenDisplay.SetSymbol(symbol);
            activeView.ScreenDisplay.DrawPolygon(geometry);

            activeView.ScreenDisplay.FinishDrawing();
            activeView.ScreenDisplay.FinishDrawing();
        }
        //Intersect
        private void button8_Click(object sender, EventArgs e)
        {
            IFeatureLayer featureLayer = this.axMapControl1.Map.get_Layer(1) as IfeatureLayer;
            IFeature feature;
            IActiveView activeView = this.axMapControl1.ActiveView;
            ITopologicalOperator topo;

            IGeometry geometry;
            feature = featureLayer.FeatureClass.GetFeature(0);
            topo = feature.Shape as ITopologicalOperator;
            IEnvelope env = feature.Shape.Envelope;
            double width, height;
            width = env.XMax - env.XMin;
            height = env.YMax - env.YMin;
```

```csharp
            object Missing = Type.Missing;
            IPolygon polygon = new PolygonClass();
            IPointCollection pointCollection = polygon as IPointCollection;
            IPoint point = new PointClass();
            point.PutCoords(env.XMin + width / 3, env.YMin + height / 3);
            pointCollection.AddPoint(point, ref Missing, ref Missing);
            point.PutCoords(env.XMin + width / 3, env.YMax - height / 3);
            pointCollection.AddPoint(point, ref Missing, ref Missing);
            point.PutCoords(env.XMax - width / 3, env.YMax - height / 3);
            pointCollection.AddPoint(point, ref Missing, ref Missing);
            point.PutCoords(env.XMax - width / 3, env.YMin + height / 3);
            pointCollection.AddPoint(point, ref Missing, ref Missing);
            polygon.SimplifyPreserveFromTo();
            geometry = topo.Intersect (polygon as IGeometry,esriGeometryDimension. esri
            Geometry2Dimension );
            ISymbol symbol = getSimpleFillSymbol(255, 255 ) as ISymbol ;

            activeView.ScreenDisplay.StartDrawing(activeView.ScreenDisplay.hDC,
            (short)esriScreenCache.esriNoScreenCache);
            activeView.ScreenDisplay.SetSymbol(symbol);
            activeView.ScreenDisplay.DrawPolygon(geometry);

            activeView.ScreenDisplay.FinishDrawing();
            activeView.ScreenDisplay.FinishDrawing();
        }
        //SymmetricDifference
        private void button9_Click(object sender, EventArgs e)
        {
            IFeatureLayer featureLayer = this.axMapControl1.Map.get_Layer(1) as Ifeature
            Layer;
            IFeature feature;
            IActiveView activeView = this.axMapControl1.ActiveView;
            ITopologicalOperator topo;

            IGeometry geometry;
            feature = featureLayer.FeatureClass.GetFeature(0);
            topo = feature.Shape as ITopologicalOperator;
            IEnvelope env = feature.Shape.Envelope;
            double width, height;
            width = env.XMax - env.XMin;
            height = env.YMax - env.YMin;
            object Missing = Type.Missing;
            IPolygon polygon = new PolygonClass();
            IPointCollection pointCollection = polygon as IPointCollection;
            IPoint point = new PointClass();
            point.PutCoords(env.XMin + width / 3, env.YMin + height / 3);
            pointCollection.AddPoint(point, ref Missing, ref Missing);
            point.PutCoords(env.XMin + width / 3, env.YMax - height / 3);
            pointCollection.AddPoint(point, ref Missing, ref Missing);
            point.PutCoords(env.XMax + width / 3, env.YMax - height / 3);
            pointCollection.AddPoint(point, ref Missing, ref Missing);
            point.PutCoords(env.XMax + width / 3, env.YMin + height / 3);
            pointCollection.AddPoint(point, ref Missing, ref Missing);
            polygon.SimplifyPreserveFromTo();
            geometry = topo.SymmetricDifference (polygon as IGeometry);
            ISymbol symbol = getSimpleFillSymbol(255, 255) as ISymbol ;

            activeView.ScreenDisplay.StartDrawing(activeView.ScreenDisplay.hDC,
            (short)esriScreenCache.esriNoScreenCache);
            activeView.ScreenDisplay.SetSymbol(symbol);
            activeView.ScreenDisplay.DrawPolygon(geometry);
```

```csharp
        activeView.ScreenDisplay.FinishDrawing();
        activeView.ScreenDisplay.FinishDrawing();
    }
    //Union
    private void button10_Click(object sender, EventArgs e)
    {
        IFeatureLayer featureLayer = this.axMapControl1.Map.get_Layer(1) as Ifeature
        Layer;
        IFeature feature;
        IActiveView activeView = this.axMapControl1.ActiveView;
        ITopologicalOperator topo;

        IGeometry geometry;
        feature = featureLayer.FeatureClass.GetFeature(0);
        topo = feature.Shape as ITopologicalOperator;
        IEnvelope env = feature.Shape.Envelope;
        double width, height;
        width = env.XMax - env.XMin;
        height = env.YMax - env.YMin;
        object Missing = Type.Missing;
        IPolygon polygon = new PolygonClass();
        IPointCollection pointCollection = polygon as IPointCollection;
        IPoint point = new PointClass();
        point.PutCoords(env.XMin + width / 3, env.YMin + height / 3);
        pointCollection.AddPoint(point, ref Missing, ref Missing);
        point.PutCoords(env.XMin + width / 3, env.YMax - height / 3);
        pointCollection.AddPoint(point, ref Missing, ref Missing);
        point.PutCoords(env.XMax + width / 3, env.YMax - height / 3);
        pointCollection.AddPoint(point, ref Missing, ref Missing);
        point.PutCoords(env.XMax + width / 3, env.YMin + height / 3);
        pointCollection.AddPoint(point, ref Missing, ref Missing);
        polygon.SimplifyPreserveFromTo();
        geometry = topo.Union (polygon as IGeometry);

        ISimpleFillSymbol simpleFillSymbol = getSimpleFillSymbol(255,255) ;

        activeView.ScreenDisplay.StartDrawing(activeView.ScreenDisplay.hDC,
        (short)esriScreenCache.esriNoScreenCache);
        activeView.ScreenDisplay.SetSymbol(simpleFillSymbol as ISymbol);
        activeView.ScreenDisplay.DrawPolygon(geometry as IGeometry);
        activeView.ScreenDisplay.FinishDrawing();
    }
}
}
```

10.5 空间关系运算

空间关系运算主要用到 Irelationaloperator 和 IProximityOperator 接口，通过该接口的方法获取两个几何对象之间的关系是否存在。

10.5.1 IRelationalOperator 接口

几何对象之间都存在着某种关联关系，如包含、相等、在内部、相交、叠加等。这些关联关系都可以通过 IRelationalOperator 接口来获得。关系运算是在两个几何对象之间进行的，通过 IRelationalOperator 的某一个方法返回一个布尔值来说明这两个几何对象是否有这种关系。所有支持 ITopologicaloperator 的几何对象的类也实现了 IRelationalOperator 接口,其中包括 Envelope 对象，这意味着还可以对两个几何对象的 Envelope 进行关联关系检查，如表 10-3 所示。

表 10-3 关联关系

Contain	检查两个几何图形，几何图形1是否包含几何图形2
Cross	用于检测两个几何图形是否相交
Equal	用于检测两个几何图形是否相等
Touch	用于检测两个几何图形是否相连
Disjoint	用于检测两个几何图形是否不相交
Overlap	用于检测两个几何图形是否有重叠
Relation	用于检测是否存在定义 relationship
Within	检查几何图形1是否被包含于几何图形2

各种关系效果图如图10-11、图10-12、图10-13、图10-14、图10-15、图10-16和图10-17所示。

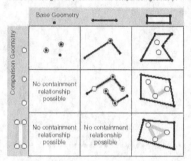

▲图 10-11 Contain 关系效果   ▲图 10-12 Corss 关系效果

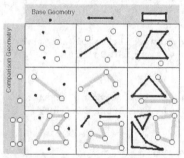

▲图 10-13 Disjoint 关系效果   ▲图 10-14 Equal 关系效果

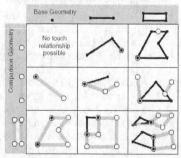

▲图 10-15 Overlap 关系效果   ▲图 10-16 Touch 关系效果

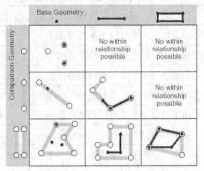

▲图10-17 Within 关系效果

```
//**************************************************************************
//功能函数用到的辅助方法
//**************************************************************************
//获取颜色对象
private IRgbColor getRGB(int r, int g, int b)
{
    IRgbColor pColor;
    pColor = new RgbColorClass();
    pColor.Red = r;
    pColor.Green = g;
    pColor.Blue = b;
    return pColor;
}
//**************************************************************************
//空间关系运算功能函数
//**************************************************************************
//Contains
private void button1_Click(object sender, EventArgs e)
{
    //简单填充符号
    ISimpleFillSymbol simpleFillSymbol = new SimpleFillSymbolClass();
    simpleFillSymbol.Style = esriSimpleFillStyle.esriSFSSolid;
    simpleFillSymbol.Color = getRGB(255, 0, 0);
    //创建边线符号
    ISimpleLineSymbol simpleLineSymbol = new SimpleLineSymbolClass();
    simpleLineSymbol.Style = esriSimpleLineStyle.esriSLSDashDotDot;
    simpleLineSymbol.Color = getRGB(0, 255, 0);
    simpleLineSymbol.Width = 2;
    ISymbol symbol = simpleLineSymbol as ISymbol;
    //symbol.ROP2 = esriRasterOpCode.esriROPNotXOrPen;
    simpleFillSymbol.Outline = simpleLineSymbol;
    //创建面对象
    object Missing = Type.Missing;
    IPolygon polygon1 = new PolygonClass();
    IPointCollection pointCollection = polygon1 as IPointCollection;
    IPoint point = new PointClass();
    point.PutCoords(20, 20);
    pointCollection.AddPoint(point, ref Missing, ref Missing);
    point.PutCoords(20, 60);
    pointCollection.AddPoint(point, ref Missing, ref Missing);
    point.PutCoords(60, 60);
    pointCollection.AddPoint(point, ref Missing, ref Missing);
    point.PutCoords(60, 20);
    pointCollection.AddPoint(point, ref Missing, ref Missing);
    polygon1.SimplifyPreserveFromTo();
```

```csharp
            IPolygon polygon2 = new PolygonClass();
            pointCollection = polygon2 as IPointCollection;
            point = new PointClass();
            point.PutCoords(10, 10);
            pointCollection.AddPoint(point, ref Missing, ref Missing);
            point.PutCoords(10, 80);
            pointCollection.AddPoint(point, ref Missing, ref Missing);
            point.PutCoords(80, 80);
            pointCollection.AddPoint(point, ref Missing, ref Missing);
            point.PutCoords(80, 10);
            pointCollection.AddPoint(point, ref Missing, ref Missing);
            polygon2.SimplifyPreserveFromTo();

            IActiveView activeView = this.axMapControl1.ActiveView;
            activeView.ScreenDisplay.StartDrawing(activeView.ScreenDisplay.hDC,
            (short)esriScreenCache.esriNoScreenCache);
            activeView.ScreenDisplay.SetSymbol(simpleFillSymbol as ISymbol);
            activeView.ScreenDisplay.DrawPolygon(polygon2 as IGeometry);
            simpleFillSymbol.Color = getRGB(0, 0, 255);
            activeView.ScreenDisplay.SetSymbol(simpleFillSymbol as ISymbol);
            activeView.ScreenDisplay.DrawPolygon(polygon1 as IGeometry);
            activeView.ScreenDisplay.FinishDrawing();

            IRelationalOperator relationOperator = polygon2 as IRelationalOperator;
            bool isContains = false;
            isContains = relationOperator.Contains(polygon1 as IGeometry);
            if (isContains == true)
            {
                this.Text = "红色面实体包含蓝色面实体";
            }

        }

        //Crosses
        private void button2_Click(object sender, EventArgs e)
        {
            //简单填充符号
            ISimpleFillSymbol simpleFillSymbol = new SimpleFillSymbolClass();
            simpleFillSymbol.Style = esriSimpleFillStyle.esriSFSSolid;
            simpleFillSymbol.Color = getRGB(255, 0, 0);
            //创建边线符号
            ISimpleLineSymbol simpleLineSymbol = new SimpleLineSymbolClass();
            simpleLineSymbol.Style = esriSimpleLineStyle.esriSLSDashDotDot;
            simpleLineSymbol.Color = getRGB(0, 255, 0);
            simpleLineSymbol.Width = 2;
            ISymbol symbol = simpleLineSymbol as ISymbol;
            //symbol.ROP2 = esriRasterOpCode.esriROPNotXOrPen;
            simpleFillSymbol.Outline = simpleLineSymbol;

            //创建简单线符号
            ISimpleLineSymbol simpleLineSymbol2 = new SimpleLineSymbolClass();
            simpleLineSymbol2.Style = esriSimpleLineStyle.esriSLSDashDotDot;
            IPolyline polyline = new PolylineClass();
            IPoint point = new PointClass();
            point.PutCoords(10, 10);
            polyline.FromPoint = point;
            point.PutCoords(90, 90);
            polyline.ToPoint = point;
            simpleLineSymbol.Width = 10;
            IRgbColor rgbColor = getRGB(0, 0, 255);
```

```csharp
            simpleLineSymbol2.Color = rgbColor;
            //创建面对象
            object Missing = Type.Missing;
            IPolygon polygon1 = new PolygonClass();
            IPointCollection pointCollection = polygon1 as IPointCollection;
            point = new PointClass();
            point.PutCoords(20, 20);
            pointCollection.AddPoint(point, ref Missing, ref Missing);
            point.PutCoords(20, 60);
            pointCollection.AddPoint(point, ref Missing, ref Missing);
            point.PutCoords(60, 60);
            pointCollection.AddPoint(point, ref Missing, ref Missing);
            point.PutCoords(60, 20);
            pointCollection.AddPoint(point, ref Missing, ref Missing);
            polygon1.SimplifyPreserveFromTo();
            IActiveView activeView = this.axMapControl1.ActiveView;
            activeView.ScreenDisplay.StartDrawing(activeView.ScreenDisplay.hDC,
            (short)esriScreenCache.esriNoScreenCache);
            activeView.ScreenDisplay.SetSymbol(simpleFillSymbol as ISymbol);
            activeView.ScreenDisplay.DrawPolygon(polygon1 as IGeometry);

            activeView.ScreenDisplay.SetSymbol(simpleLineSymbol2 as ISymbol);
            activeView.ScreenDisplay.DrawPolygon(polyline as IGeometry);
            activeView.ScreenDisplay.FinishDrawing();

            IRelationalOperator relationOperator = polygon1 as IRelationalOperator;
            bool isCrosses = false;
            isCrosses = relationOperator.Crosses(polyline as IGeometry);
            if (isCrosses == true)
            {
                this.Text = "红色面实体与蓝色线实体相交";
            }
        }
        //Equal
        private void button3_Click(object sender, EventArgs e)
        {
            //简单填充符号
            ISimpleFillSymbol simpleFillSymbol = new SimpleFillSymbolClass();
            simpleFillSymbol.Style = esriSimpleFillStyle.esriSFSSolid;
            simpleFillSymbol.Color = getRGB(255, 0, 0);
            //创建边线符号
            ISimpleLineSymbol simpleLineSymbol = new SimpleLineSymbolClass();
            simpleLineSymbol.Style = esriSimpleLineStyle.esriSLSDashDotDot;
            simpleLineSymbol.Color = getRGB(0, 255, 0);
            simpleLineSymbol.Width = 2;
            ISymbol symbol = simpleLineSymbol as ISymbol;
            //symbol.ROP2 = esriRasterOpCode.esriROPNotXOrPen;
            simpleFillSymbol.Outline = simpleLineSymbol;
            //创建面对象
            object Missing = Type.Missing;
            IPolygon polygon1 = new PolygonClass();
            IPointCollection pointCollection = polygon1 as IPointCollection;
            IPoint point = new PointClass();
            point.PutCoords(20, 20);
            pointCollection.AddPoint(point, ref Missing, ref Missing);
            point.PutCoords(20, 60);
            pointCollection.AddPoint(point, ref Missing, ref Missing);
            point.PutCoords(60, 60);
            pointCollection.AddPoint(point, ref Missing, ref Missing);
            point.PutCoords(60, 20);
            pointCollection.AddPoint(point, ref Missing, ref Missing);
```

```csharp
        polygon1.SimplifyPreserveFromTo();


        IPolygon polygon2 = new PolygonClass();
        pointCollection = polygon2 as IPointCollection;
        point = new PointClass();
        point.PutCoords(20, 20);
        pointCollection.AddPoint(point, ref Missing, ref Missing);
        point.PutCoords(20, 60);
        pointCollection.AddPoint(point, ref Missing, ref Missing);
        point.PutCoords(60, 60);
        pointCollection.AddPoint(point, ref Missing, ref Missing);
        point.PutCoords(60, 20);
        pointCollection.AddPoint(point, ref Missing, ref Missing);
        polygon2.SimplifyPreserveFromTo();

        IActiveView activeView = this.axMapControl1.ActiveView;
        activeView.ScreenDisplay.StartDrawing(activeView.ScreenDisplay.hDC,
        (short)esriScreenCache.esriNoScreenCache);
        activeView.ScreenDisplay.SetSymbol(simpleFillSymbol as ISymbol);
        activeView.ScreenDisplay.DrawPolygon(polygon2 as IGeometry);
        simpleFillSymbol.Color = getRGB(0, 0, 255);
        activeView.ScreenDisplay.SetSymbol(simpleFillSymbol as ISymbol);
        activeView.ScreenDisplay.DrawPolygon(polygon1 as IGeometry);
        activeView.ScreenDisplay.FinishDrawing();

        IRelationalOperator relationOperator = polygon2 as IRelationalOperator;
        bool isEqual = false;
        isEqual = relationOperator.Equals(polygon1 as IGeometry);
        if (isEqual == true)
        {
            this.Text = "红色面实体与蓝色面实体相等";
        }
    }
    //Touches
    private void button4_Click(object sender, EventArgs e)
    {
        //简单填充符号
        ISimpleFillSymbol simpleFillSymbol = new SimpleFillSymbolClass();
        simpleFillSymbol.Style = esriSimpleFillStyle.esriSFSSolid;
        simpleFillSymbol.Color = getRGB(255, 0, 0);
        //创建边线符号
        ISimpleLineSymbol simpleLineSymbol = new SimpleLineSymbolClass();
        simpleLineSymbol.Style = esriSimpleLineStyle.esriSLSDashDotDot;
        simpleLineSymbol.Color = getRGB(0, 255, 0);
        simpleLineSymbol.Width = 2;
        ISymbol symbol = simpleLineSymbol as ISymbol;
        //symbol.ROP2 = esriRasterOpCode.esriROPNotXOrPen;
        simpleFillSymbol.Outline = simpleLineSymbol;
        //创建面对象
        object Missing = Type.Missing;
        IPolygon polygon1 = new PolygonClass();
        IPointCollection pointCollection = polygon1 as IPointCollection;
        IPoint point = new PointClass();
        point.PutCoords(20, 20);
        pointCollection.AddPoint(point, ref Missing, ref Missing);
        point.PutCoords(20, 60);
        pointCollection.AddPoint(point, ref Missing, ref Missing);
        point.PutCoords(60, 60);
        pointCollection.AddPoint(point, ref Missing, ref Missing);
        point.PutCoords(60, 20);
        pointCollection.AddPoint(point, ref Missing, ref Missing);
```

```csharp
            polygon1.SimplifyPreserveFromTo();


            IPolygon polygon2 = new PolygonClass();
            pointCollection = polygon2 as IPointCollection;
            point = new PointClass();
            point.PutCoords(60, 20);
            pointCollection.AddPoint(point, ref Missing, ref Missing);
            point.PutCoords(60, 80);
            pointCollection.AddPoint(point, ref Missing, ref Missing);
            point.PutCoords(80, 80);
            pointCollection.AddPoint(point, ref Missing, ref Missing);
            point.PutCoords(80, 20);
            pointCollection.AddPoint(point, ref Missing, ref Missing);
            polygon2.SimplifyPreserveFromTo();

            IActiveView activeView = this.axMapControl1.ActiveView;
            activeView.ScreenDisplay.StartDrawing(activeView.ScreenDisplay.hDC,
            (short)esriScreenCache.esriNoScreenCache);
            activeView.ScreenDisplay.SetSymbol(simpleFillSymbol as ISymbol);
            activeView.ScreenDisplay.DrawPolygon(polygon2 as IGeometry);
            simpleFillSymbol.Color = getRGB(0, 0, 255);
            activeView.ScreenDisplay.SetSymbol(simpleFillSymbol as ISymbol);
            activeView.ScreenDisplay.DrawPolygon(polygon1 as IGeometry);
            activeView.ScreenDisplay.FinishDrawing();

            IRelationalOperator relationOperator = polygon2 as IRelationalOperator;
            bool isTouches = false;
            isTouches = relationOperator.Touches(polygon1 as IGeometry);
            if (isTouches == true)
            {
                this.Text = "红色面实体与蓝色面实体相连";
            }
        }
        //Disjoint
        private void button5_Click(object sender, EventArgs e)
        {
            //简单填充符号
            ISimpleFillSymbol simpleFillSymbol = new SimpleFillSymbolClass();
            simpleFillSymbol.Style = esriSimpleFillStyle.esriSFSSolid;
            simpleFillSymbol.Color = getRGB(255, 0, 0);
            //创建边线符号
            ISimpleLineSymbol simpleLineSymbol = new SimpleLineSymbolClass();
            simpleLineSymbol.Style = esriSimpleLineStyle.esriSLSDashDotDot;
            simpleLineSymbol.Color = getRGB(0, 255, 0);
            simpleLineSymbol.Width = 2;
            ISymbol symbol = simpleLineSymbol as ISymbol;
            //symbol.ROP2 = esriRasterOpCode.esriROPNotXOrPen;
            simpleFillSymbol.Outline = simpleLineSymbol;
            //创建面对象
            object Missing = Type.Missing;
            IPolygon polygon1 = new PolygonClass();
            IPointCollection pointCollection = polygon1 as IPointCollection;
            IPoint point = new PointClass();
            point.PutCoords(20, 20);
            pointCollection.AddPoint(point, ref Missing, ref Missing);
            point.PutCoords(20, 60);
            pointCollection.AddPoint(point, ref Missing, ref Missing);
            point.PutCoords(60, 60);
            pointCollection.AddPoint(point, ref Missing, ref Missing);
            point.PutCoords(60, 20);
            pointCollection.AddPoint(point, ref Missing, ref Missing);
```

```csharp
            polygon1.SimplifyPreserveFromTo();


            IPolygon polygon2 = new PolygonClass();
            pointCollection = polygon2 as IPointCollection;
            point = new PointClass();
            point.PutCoords(70, 20);
            pointCollection.AddPoint(point, ref Missing, ref Missing);
            point.PutCoords(70, 80);
            pointCollection.AddPoint(point, ref Missing, ref Missing);
            point.PutCoords(80, 80);
            pointCollection.AddPoint(point, ref Missing, ref Missing);
            point.PutCoords(80, 20);
            pointCollection.AddPoint(point, ref Missing, ref Missing);
            polygon2.SimplifyPreserveFromTo();

            IActiveView activeView = this.axMapControl1.ActiveView;
            activeView.ScreenDisplay.StartDrawing(activeView.ScreenDisplay.hDC,
            (short)esriScreenCache.esriNoScreenCache);
            activeView.ScreenDisplay.SetSymbol(simpleFillSymbol as ISymbol);
            activeView.ScreenDisplay.DrawPolygon(polygon2 as IGeometry);
            simpleFillSymbol.Color = getRGB(0, 0, 255);
            activeView.ScreenDisplay.SetSymbol(simpleFillSymbol as ISymbol);
            activeView.ScreenDisplay.DrawPolygon(polygon1 as IGeometry);
            activeView.ScreenDisplay.FinishDrawing();

            IRelationalOperator relationOperator = polygon2 as IRelationalOperator;
            bool isDisjoint = false;
            isDisjoint = relationOperator.Disjoint(polygon1 as IGeometry);
            if (isDisjoint == true)
            {
                this.Text = "红色面实体与蓝色面实体不相交";
            }
        }
        //Overlaps
        private void button6_Click(object sender, EventArgs e)
        {
            //简单填充符号
            ISimpleFillSymbol simpleFillSymbol = new SimpleFillSymbolClass();
            simpleFillSymbol.Style = esriSimpleFillStyle.esriSFSSolid;
            simpleFillSymbol.Color = getRGB(255, 0, 0);
            //创建边线符号
            ISimpleLineSymbol simpleLineSymbol = new SimpleLineSymbolClass();
            simpleLineSymbol.Style = esriSimpleLineStyle.esriSLSDashDotDot;
            simpleLineSymbol.Color = getRGB(0, 255, 0);
            simpleLineSymbol.Width = 2;
            ISymbol symbol = simpleLineSymbol as ISymbol;
            //symbol.ROP2 = esriRasterOpCode.esriROPNotXOrPen;
            simpleFillSymbol.Outline = simpleLineSymbol;
            //创建面对象
            object Missing = Type.Missing;
            IPolygon polygon1 = new PolygonClass();
            IPointCollection pointCollection = polygon1 as IPointCollection;
            IPoint point = new PointClass();
            point.PutCoords(20, 20);
            pointCollection.AddPoint(point, ref Missing, ref Missing);
            point.PutCoords(20, 60);
            pointCollection.AddPoint(point, ref Missing, ref Missing);
            point.PutCoords(60, 60);
            pointCollection.AddPoint(point, ref Missing, ref Missing);
            point.PutCoords(60, 20);
            pointCollection.AddPoint(point, ref Missing, ref Missing);
```

```csharp
            polygon1.SimplifyPreserveFromTo();


    IPolygon polygon2 = new PolygonClass();
    pointCollection = polygon2 as IPointCollection;
    point = new PointClass();
    point.PutCoords(50, 50);
    pointCollection.AddPoint(point, ref Missing, ref Missing);
    point.PutCoords(50, 80);
    pointCollection.AddPoint(point, ref Missing, ref Missing);
    point.PutCoords(80, 80);
    pointCollection.AddPoint(point, ref Missing, ref Missing);
    point.PutCoords(80, 50);
    pointCollection.AddPoint(point, ref Missing, ref Missing);
    polygon2.SimplifyPreserveFromTo();

    IActiveView activeView = this.axMapControl1.ActiveView;
    activeView.ScreenDisplay.StartDrawing(activeView.ScreenDisplay.hDC,
    (short)esriScreenCache.esriNoScreenCache);
    activeView.ScreenDisplay.SetSymbol(simpleFillSymbol as ISymbol);
    activeView.ScreenDisplay.DrawPolygon(polygon2 as IGeometry);
    simpleFillSymbol.Color = getRGB(0, 0, 255);
    activeView.ScreenDisplay.SetSymbol(simpleFillSymbol as ISymbol);
    activeView.ScreenDisplay.DrawPolygon(polygon1 as IGeometry);
    activeView.ScreenDisplay.FinishDrawing();

    IRelationalOperator relationOperator = polygon2 as IRelationalOperator;
    bool isOverlaps = false;
    isOverlaps = relationOperator.Overlaps(polygon1 as IGeometry);
    if (isOverlaps == true)
    {
        this.Text = "红色面实体与蓝色面实体有重叠";
    }
}
//Within
private void button7_Click(object sender, EventArgs e)
{
    //简单填充符号
    ISimpleFillSymbol simpleFillSymbol = new SimpleFillSymbolClass();
    simpleFillSymbol.Style = esriSimpleFillStyle.esriSFSSolid;
    simpleFillSymbol.Color = getRGB(255, 0, 0);
    //创建边线符号
    ISimpleLineSymbol simpleLineSymbol = new SimpleLineSymbolClass();
    simpleLineSymbol.Style = esriSimpleLineStyle.esriSLSDashDotDot;
    simpleLineSymbol.Color = getRGB(0, 255, 0);
    simpleLineSymbol.Width = 2;
    ISymbol symbol = simpleLineSymbol as ISymbol;
    //symbol.ROP2 = esriRasterOpCode.esriROPNotXOrPen;
    simpleFillSymbol.Outline = simpleLineSymbol;
    //创建面对象
    object Missing = Type.Missing;
    IPolygon polygon1 = new PolygonClass();
    IPointCollection pointCollection = polygon1 as IPointCollection;
    IPoint point = new PointClass();
    point.PutCoords(20, 20);
    pointCollection.AddPoint(point, ref Missing, ref Missing);
    point.PutCoords(20, 60);
    pointCollection.AddPoint(point, ref Missing, ref Missing);
    point.PutCoords(60, 60);
    pointCollection.AddPoint(point, ref Missing, ref Missing);
    point.PutCoords(60, 20);
    pointCollection.AddPoint(point, ref Missing, ref Missing);
```

```
            polygon1.SimplifyPreserveFromTo();

            IPolygon polygon2 = new PolygonClass();
            pointCollection = polygon2 as IPointCollection;
            point = new PointClass();
            point.PutCoords(10, 10);
            pointCollection.AddPoint(point, ref Missing, ref Missing);
            point.PutCoords(10, 80);
            pointCollection.AddPoint(point, ref Missing, ref Missing);
            point.PutCoords(80, 80);
            pointCollection.AddPoint(point, ref Missing, ref Missing);
            point.PutCoords(80, 10);
            pointCollection.AddPoint(point, ref Missing, ref Missing);
            polygon2.SimplifyPreserveFromTo();

            IActiveView activeView = this.axMapControl1.ActiveView;
            activeView.ScreenDisplay.StartDrawing(activeView.ScreenDisplay.hDC,
(short)esriScreenCache.esriNoScreenCache);
            activeView.ScreenDisplay.SetSymbol(simpleFillSymbol as ISymbol);
            activeView.ScreenDisplay.DrawPolygon(polygon2 as IGeometry);
            simpleFillSymbol.Color = getRGB(0, 0, 255);
            activeView.ScreenDisplay.SetSymbol(simpleFillSymbol as ISymbol);
            activeView.ScreenDisplay.DrawPolygon(polygon1 as IGeometry);
            activeView.ScreenDisplay.FinishDrawing();

            IRelationalOperator relationOperator = polygon1 as IRelationalOperator;
            bool isWithin = false;
            isWithin = relationOperator.Within(polygon2 as IGeometry);
            if (isWithin == true)
            {
                this.Text = "蓝色面实体被包含于红色面实体";
            }
        }
    }
}
```

10.5.2　IProximityOperator 接口

　　IProximityOperator 接口用于获取两个图形的距离，以及给定一个点，求另一个几何图形上离给定点最近的点。该接口的主要方法有：QueryNearestPoint、ReturnDistance 和 ReturnNearestPoint。ReturnDistance 方法用于返回两个几何对象间的最短距离，QueryNearestPoint 方法用于查询获取几何对象上离给定点最近距离的点的引用，ReturnNearestPoint 方法用于创建并返回几何对象上给定输入点的最近距离的点。QueryNearestPoint 与 ReturnNearesPoint 的效果一样，如图 10-18～图 10-20 所示。

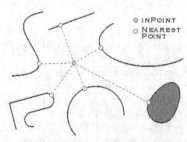

▲图 10-18　QueryNearest Point 效果

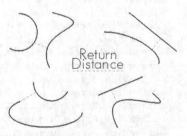

▲图 10-19　Return Distance 效果

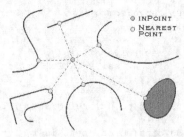

▲图 10-20 ReturnNearest Point 效果

```csharp
//QueryNearestPoint
    private void button8_Click(object sender, EventArgs e)
    {
        //创建简单线符号
        ISimpleLineSymbol simpleLineSymbol = new SimpleLineSymbolClass();
        simpleLineSymbol.Style = esriSimpleLineStyle.esriSLSDash ;

        //创建线
        IPolyline polyline = new PolylineClass();
        IPoint point = new PointClass();
        point.PutCoords(10, 10);
        polyline.FromPoint = point;
        point.PutCoords(90, 90);
        polyline.ToPoint = point;
        simpleLineSymbol.Width = 10;
        IRgbColor rgbColor = getRGB(255, 0, 0);
        simpleLineSymbol.Color = rgbColor;
        //设置指定点和符号
        point.PutCoords (70,30);
        ISimpleMarkerSymbol simpleMarkerSymbol=new SimpleMarkerSymbolClass ();
        simpleMarkerSymbol.Style = esriSimpleMarkerStyle.esriSMSCircle   ;
        simpleMarkerSymbol.Size = 20;
        rgbColor = getRGB(0, 255, 0);
        simpleMarkerSymbol.Color = rgbColor;
        //求最近点
        IPoint nearPoint =new PointClass ();
        IProximityOperator proximityOperator = polyline as IProximityOperator;
        proximityOperator.QueryNearestPoint(point, esriSegmentExtension.esriExtendAtFrom,
        nearPoint);
        //画出指定点到最近点的连线
        ISimpleLineSymbol simpleLineSymbol2 = new SimpleLineSymbolClass();
        simpleLineSymbol2.Style = esriSimpleLineStyle.esriSLSSolid ;
        simpleLineSymbol2.Width = 5;
        IPolyline polyline2 = new PolylineClass();
        polyline2.FromPoint = point;
        polyline2.ToPoint = nearPoint;
        rgbColor = getRGB(0, 0, 255);
        simpleLineSymbol2.Color = rgbColor;

        IActiveView activeView = this.axMapControl1.ActiveView;
        activeView.ScreenDisplay.StartDrawing(activeView.ScreenDisplay.hDC,
        (short)esriScreenCache.esriNoScreenCache);
        activeView.ScreenDisplay.SetSymbol(simpleLineSymbol as ISymbol);
        activeView.ScreenDisplay.DrawPolyline(polyline as IGeometry);
        activeView.ScreenDisplay.SetSymbol(simpleMarkerSymbol as ISymbol);
        activeView.ScreenDisplay.DrawPoint (point  as IGeometry);
        activeView.ScreenDisplay.SetSymbol(simpleLineSymbol2 as ISymbol);
        activeView.ScreenDisplay.DrawPolyline(polyline2 as IGeometry);
        activeView.ScreenDisplay.FinishDrawing();
```

```csharp
}
//ReturnDistance
private void button9_Click(object sender, EventArgs e)
{
    //创建简单线符号
    ISimpleLineSymbol simpleLineSymbol = new SimpleLineSymbolClass();
    simpleLineSymbol.Style = esriSimpleLineStyle.esriSLSDash;
    ISimpleLineSymbol simpleLineSymbol2 = new SimpleLineSymbolClass();
    simpleLineSymbol2.Style = esriSimpleLineStyle.esriSLSDashDotDot;
    //创建线
    IPolyline polyline = new PolylineClass();
    IPoint point = new PointClass();
    point.PutCoords(10, 10);
    polyline.FromPoint = point;
    point.PutCoords(90, 90);
    polyline.ToPoint = point;
    simpleLineSymbol.Width = 10;
    IRgbColor rgbColor = getRGB(255, 0, 0);
    simpleLineSymbol.Color = rgbColor;

    IPolyline polyline2 = new PolylineClass();
    point = new PointClass();
    point.PutCoords(50, 20);
    polyline2.FromPoint = point;
    point.PutCoords(70, 30);
    polyline2.ToPoint = point;
    simpleLineSymbol2.Width = 10;
    rgbColor = getRGB(0, 0, 255);
    simpleLineSymbol2.Color = rgbColor;

    double distance = 0.0;
    IProximityOperator proximityOperator = polyline as IProximityOperator;
    distance= proximityOperator.ReturnDistance (polyline2  as IGeometry );


    IActiveView activeView = this.axMapControl1.ActiveView;
    activeView.ScreenDisplay.StartDrawing(activeView.ScreenDisplay.hDC,
    (short)esriScreenCache.esriNoScreenCache);
    activeView.ScreenDisplay.SetSymbol(simpleLineSymbol as ISymbol);
    activeView.ScreenDisplay.DrawPolyline(polyline as IGeometry);

    activeView.ScreenDisplay.SetSymbol(simpleLineSymbol2 as ISymbol);
    activeView.ScreenDisplay.DrawPolyline(polyline2 as IGeometry);
    activeView.ScreenDisplay.FinishDrawing();

    MessageBox.Show("两线之间的距离为: " + distance+"单位");
}
//ReturnNearestPoint
private void button10_Click(object sender, EventArgs e)
{
    //创建简单线符号
    ISimpleLineSymbol simpleLineSymbol = new SimpleLineSymbolClass();
    simpleLineSymbol.Style = esriSimpleLineStyle.esriSLSDash;

    //创建线
    IPolyline polyline = new PolylineClass();
    IPoint point = new PointClass();
    point.PutCoords(10, 10);
    polyline.FromPoint = point;
    point.PutCoords(90, 90);
```

```
            polyline.ToPoint = point;
            simpleLineSymbol.Width = 10;
            IRgbColor rgbColor = getRGB(255, 0, 0);
            simpleLineSymbol.Color = rgbColor;
            //设置指定点和符号
            point.PutCoords(70, 30);
            ISimpleMarkerSymbol simpleMarkerSymbol = new SimpleMarkerSymbolClass();
            simpleMarkerSymbol.Style = esriSimpleMarkerStyle.esriSMSCircle;
            simpleMarkerSymbol.Size = 20;
            rgbColor = getRGB(0, 255, 0);
            simpleMarkerSymbol.Color = rgbColor;
            //求最近点
            IPoint nearPoint = new PointClass();
            IProximityOperator proximityOperator = polyline as IProximityOperator;
            nearPoint= proximityOperator.ReturnNearestPoint (point, esriSegment Extension.
            esriExtendAtFrom);
            //画出指定点到最近点的连线
            ISimpleLineSymbol simpleLineSymbol2 = new SimpleLineSymbolClass();
            simpleLineSymbol2.Style = esriSimpleLineStyle.esriSLSSolid;
            simpleLineSymbol2.Width = 5;
            IPolyline polyline2 = new PolylineClass();
            polyline2.FromPoint = point;
            polyline2.ToPoint = nearPoint;
            rgbColor = getRGB(0, 0, 255);
            simpleLineSymbol2.Color = rgbColor;

            IActiveView activeView = this.axMapControl1.ActiveView;
            activeView.ScreenDisplay.StartDrawing(activeView.ScreenDisplay.hDC,
            (short)esriScreenCache.esriNoScreenCache);
            activeView.ScreenDisplay.SetSymbol(simpleLineSymbol as ISymbol);
            activeView.ScreenDisplay.DrawPolyline(polyline as IGeometry);
            activeView.ScreenDisplay.SetSymbol(simpleMarkerSymbol as ISymbol);
            activeView.ScreenDisplay.DrawPoint(point as IGeometry);
            activeView.ScreenDisplay.SetSymbol(simpleLineSymbol2 as ISymbol);
            activeView.ScreenDisplay.DrawPolyline(polyline2 as IGeometry);
            activeView.ScreenDisplay.FinishDrawing();
        }
```

10.6 网络及网络分析

网络分析是地理信息空间分析的一个重要组成部分，它依据网络拓扑关系，并通过考察网络元素的空间、属性数据，对网络的性能特征进行多方面的分析计算。网络由两个基本部分组成，其一是边线（Edge），其二是交汇点（Junction）。边线与边线之间通过交汇点连接。网络在 Geodatabase 中有两种描述：几何网络（Geometric NetWork）和逻辑网络（Logical NetWork）。几何网络是组成线性网络的要素的集合，这些要素称为网络要素（NetWork Feature）。逻辑网络主要是用特定的属性表存储网络的连通性信息，不存储坐标。

网络分析主要用到的接口有 ITraceFlowSolver、IPointToEID、IEIDHelper、INetWorkCollection、IgeometricNetwork 和 INetSchema 等。

ItraceFlowSolver 接口提供了网络中基本的追踪方法，组件类 TraceFlowSolver 封装了该接口，该接口封装了 7 个方法和一个属性。如 FindCircuits 方法用于寻找网络中所有可到达的网络元素，且这些元素组成闭合环；FindCommonAncestors 方法用于寻找网络中位于起始点上游的所有公共网络要素；FindFlowElements 方法用于根据指定流的方法寻找所有可到达的网络要素；FindFlowEndElements 方法用于根据指定流的方法寻找所有可到达的网络终止元素；FindPath 方法用于寻找网络中给定点之间的路径。

IEIDHelper 接口提供了从网络元素 ID 枚举中获取要素或几何图形的方法和属性。EIDHelper 组件类封装了该接口，该接口提供了 3 个方法和 6 个属性，如 CreateEnumEIDInfo 方法用于创建一个 EIDInfo 枚举，GeometricNetwork 属性用于设置网络元素 ID 的源几何网络。

用到的接口主要有 InetworkCollection、IgeometricNetwork、IPointToEID、ITraceFlowSolverGEN（它实现了 ITraceFlowSolver 的接口）、InetSchema 和 IEIDHelper 等。

主要步骤如下。

（1）获取几何网络工作空间。
（2）定义一个边线旗数组，把离点串最近的网络元素添加进数组。
（3）设置开始和结束边线的权重。
（4）进行路径分析。
（5）得到路径分析的结果。

10.6.1 主要对象类

（1）NetworkLoader 组件对象类。

主要作用：由 FeatureDataset 生成 NetWork。

（2）TraceFlowSolver 组件对象类。

主要作用：指定 Network 和 barriers，定义权重，进行网络分析操作等。

TraceFlowSolver 类实现了以下接口。

① InetSolver 接口：指定 Network 和 barriers。

InetSolve.SourceNetwork：设定 Network（网络）。

InetSolve.ElementBarriers：设定 NetElementBarriers（障碍元素）。

② InetSolverWeights 接口：定义权重。

InetSolverWeights.ToFromEdgeFilterWeight：设定边从终点到起点的权重。

InetSolverWeights.FromToEdgeFilterWeight：设定边从起点到终点的权重。

InetSolverWeights.JunctionFilterWeight：设定节点的权重。

③ ITraceFlowSolver 接口：网络分析。

ItraceFlowSolver.PutEdgeOrigins 设置最短路径标记。

ItraceFlowSolver.PutJunctionOrigins 在指定的标志间寻找最短路径。

ItraceFlowSolver.FindPath：在指定的标志间寻找最短路径。

（3）EdgeFlag 组件对象类。

主要作用：指定路径的起点（点在边上）或路径算法。

EdgeFlag 组件对象类实现了以下接口。

① INetFlag 接口：指定边上作为 Flag 的元素。

② IedgeFlag 接口：指定路径的起点（点在边上）或路径算法。

IedgeFlag.Position：点在边上的位置。

IedgeFlag.TwoWay：是否是可以双向搜索。

（4）JunctionFlag 组件对象类。

主要作用：指定路径的起点（点在节点上）。

JunctionFlag 组件对象类实现了以下接口。

① INetFlg 接口：指定点上作为 Flag 的元素。

② IedgeFlag 接口：指定边上作为 Flag 的元素。

（5）JunctionFlagDisplay 组件对象类。

主要作用：显示节点标志。

JunctionFlagDisplay 组件对象类实现了以下接口。

① IJunctionFlagDisplay 接口：

② IflagDisplay 接口：显示节点标志，设定 FeatureClass、ID and SubID、Symbol。

IflagDisplay.FeatureClassID：显示实体类的 ID。

IflagDisplay.FID：显示实体的 ID。

IflagDisplay.SubID：显示实体单个要素的 ID。

（6）EdgeFlagDisplay 组件对象类。

主要作用：显示边节点标志。

EdgeFlagDisplay 组件对象类实现了以下接口。

① IEdgeFlagDisplay 接口：显示边节点标记。

IEdgeFlagDisplay.Percentage 接口：沿边线元素位置标记。

② IflagDisplay 接口：显示节点标志，设定 FeatureClass、ID and SubID、Symbol。

IflagDisplay.FeatureClassID：显示实体类 ID。

IflagDisplay.FID：显示实体的 ID。

IflagDisplay.SubID：显示实体单个要素的 ID。

（7）NetElementBarriers 组件对象类。

主要作用：设置障碍。

NetElementBarriers 组件对象类实现了以下接口。

① INetElementBarriers 接口：设置障碍。

InetElementBarriers.ElementType：障碍 ElementType。

InetElementBarriers.SetBarriers：设置障碍物（a set of network feature）。

② INetElementBarriers2 接口：

INetElementBarriers2.SetBarriersByEID：通过 element EID 设置障碍。

（8）PointToEID 组件对象类。

主要作用：搜索最近的节点或边点。

PointToEID 组件对象类实现了以下接口。

IPointToEID 接口：搜索最近的节点或边点。

IPointToEID.GetNearestEdge：搜索最近的边点。

IPointToEID.GetNearestJunction：搜索最近的节点。

10.6.2 类之间的相互关系

（1）TraceFlowSolver 组件对象类与 NetElementBarriers 组件对象类之间的关系。

NetElementBarriers 组件对象类的主要作用是生成障碍，TraceFlowSolver 组件对象类则是通过 InetSolver 接口的 InetSolve.ElementBarriers 方法设置障碍。

（2）TraceFlowSolver 组件对象类与 EdgeFlag 组件对象类之间的关系。

EdgeFlag 组件对象类的主要作用是生成 EdgeFlag，TraceFlowSolver 组件对象类则是通过 ITraceFlowSolver 接口的 ItraceFlowSolver.PutEdgeOrigins 方法设置边标志（多个）。

（3）TraceFlowSolver 组件对象类与 JunctionFlag 组件对象类之间的关系。

JunctionFlag 组件对象类的主要作用是生成 JunctionFlag，TraceFlowSolver 组件对象类则是通过 ITraceFlowSolver 接口的 ItraceFlowSolver.PutJunctionOrigins 方法设置节点标志（多个）。

10.7 本章小结

空间分析是 GIS 系统中的一个高级功能，通过空间对象之间的关系挖掘出空间对象的联系，在现实生活中具有重要的意义，如银行选址、区域人口统计分析、传染病控制等。本章介绍空间对象的集合运算、空间对象的拓扑运算、空间对象的关系运算等。读者应掌握 ITopologicalOperator 等接口的应用。

第三篇

综合实例篇

第 11 章　符号库管理系统的开发
第 12 章　空间数据管理系统

第 11 章 符号库管理系统的开发

11.1 简介

地图是由符号构筑的"大厦",而符号是地图的基本元素。地图中的符号是地图语言中最重要的部分,要表达成千上万的物体和现象,就必须设计和制作相应的图像符号。地图使用这些符号表现复杂的自然或社会现象,它与"见物绘物"的风景画和对客观实体的机械缩影的航片、卫片截然不同。地图上使用分门别类的地图符号对复杂的事物进行抽象概括,使实体很小的物体仍得以清晰地表示。地面上受遮盖的物体(隧道、涵洞等)和许多自然及社会现象,如工农业产值、行政界线、人口数、太阳辐射等无形的现象,仍能通过地图符号或注记表达出来。因此地图上浓缩存储了大量有关地点、状况、相互关系、自然和经济等动态现象,详细记录了对象的空间分布、组合、联系及随时间的变化,凝聚了极丰富的空间信息,从而使地图成为人们认识和研究客观世界的重要工具。

近年来由于专题地图的迅速发展,地图的应用不断扩大,地图符号的设计制作成了一个重要而繁重的任务。它不仅关系到地图表示的质量,而且也影响地图的成图速度和自动化制图的发展。由此可见地图符号的设计和制作,在地图的制作中占据着十分重要的位置。

随着地理信息系统的深入发展,地图种类和内容的不断增加,新地图的表示法和彩色印刷提供的有利条件,使地图符号的数量上升很快。原先的 GIS 软件中自带的符号库已经不能满足使用者的需求,因此世界各 GIS 软件生产商又在软件中嵌入了能够根据用户自己的要求进行二次开发的符号编辑模块。这样一来,用户就可以根据地图表达的实际需要来制作形式各样、大小不一、颜色多变的符号。

11.2 系统设计

ArcMap 中用来制作和管理符号的模块是 "Styles",它提供了一套完整的工具以帮助使用者创建一幅地图,每一种 style 包括了一系列符号及地图元素,提供符号的特性、标记的确定、颜色的选择、图例、线形比例尺特征以及其他的信息,因此它可以帮助用户维护符号的形状、大小、颜色等。用户可以剪切、复制、粘贴、重命名任何样式,还可以删除一些 ArcMap 提供的,而又不需要的符号和地图元素。

我们常把地图符号按其几何性质的不同分为点状符号、线状符号、面状符号等 3 类,这也符合图形设计软件中数据组织的技术特征,在 ArcEngine 中提供了 ServerStyle 的符号格式文件。本章在参考 ArcMap 符号库的基础上用 ArcEngine +C# 开发 ServerStyle 符号库管理系统。该系统功能比较简单,只简单介绍了各种类型符号中的某一种符号的创建,其他符号的创建,读者可以参考第 5 章的实例,在本案例的基础上进行扩展。

下面介绍各种类型符号的设计。

11.2.1 主程序界面设计

在主程序界面打开一个 ServerStyle 格式文件，初始化符号库列表，右边的窗体部分显示了当前选择的符号类中所有的符号，右边窗体提供了"大图标"、"小图标"、"详细列表"3 种浏览符号视图，"新建样式符号"根据当前视图的符号类型，显示点、线、面的符号新建窗口，左下角显示了当前选择的符号的大样图，如图 11-1 所示。

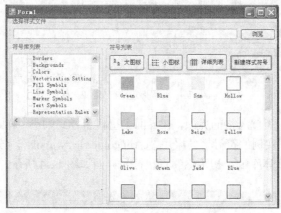

▲图 11-1 主程序界面

11.2.2 点状符号

点状符号常用来表示在当前的比例尺和表示方式下，呈点状分布的地理实体和现象，不论符号大小，实际上以点的概念定位，而符号的面积不具有实地的面积意义。这时，符号的大小与地图比例尺无关，且具有定位特征。它在图中的位置由一个点来确定，即符号的定位点，通常为符号的几何中心点或符号底部的中心点，例如控制点、居民点及其他独立地物点等符号。

在 ArcMap 中，可以按下面的步骤打开点符号。

（1）启动 ArcMap，如果未创建符号库，则需要创建符号库；如果已经创建符号库，则需要添加符号库。

（2）单击符号库名，接着再单击 marker symbols 符号文件夹，在右侧窗口的空白处单击右键，指向 new，单击 marker symbol，弹出"symbol property editor"对话框，如图 11-2 和图 11-3 所示，从中可以对符号的各种属性进行设置，单击"OK"按钮保存退出。

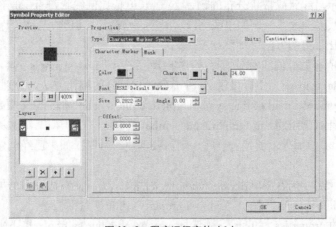

▲图 11-2 程序运行窗体（1）

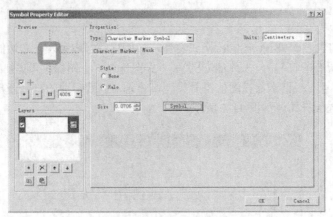

▲图 11-3 程序运行窗体（2）

本实例点状符号采用类似风格的界面，在界面上提供了样式大小、样式角度、样式颜色、偏移量等参数来新建点符号，实例的符号字体采用 "esri geometric symbols"，该字体在安装 ARCGIS 的时候已经添加进系统的字体库中，初始化的点状符号界面如图 11-4 所示。

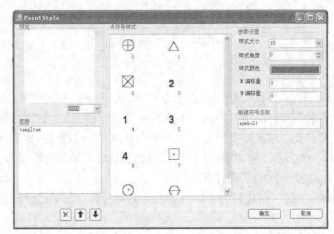

▲图 11-4 初始化的点符号界面

11.2.3 线状符号

线状符号是表示呈线状或带状分布的物体。对于长度依比例线状符号，符号沿着某个方向延伸，且长度与地图比例尺发生关系，例如单线河流、渠道、水涯线、道路、航线等符号。制作线状符号时要特别注意数字化采集的方向，如陡坎符号。

在 ArcMap 中所有做好的线符号均存放在符号库下属的 line symbols 符号文件夹中。ArcMap 符号样式管理（style manage）中提供了 5 种类型线状符号的制作方法，分别是 cartographic line symbol、hash line symbol、marker line symbol、picture line symbol 和 simple line symbol。同样，线状符号的制作也是针对常用的 cartographic line symbol 展开。

（1）启动 ArcMap，如果未创建符号库，则需要创建符号库；如果已经创建符号库，则需要添加符号库。

（2）单击符号库名，接着再单击 line symbols 文件夹，然后在右边空白处单击鼠标右键，在弹出菜单中单击 new\line symbol，弹出 "symbol property editor" 对话框。

（3）在对话框的 properties 栏的 type 项选择 cartographic line symbol。接下来与点状符号一样对

各属性项进行设置,如图 11-5 所示。

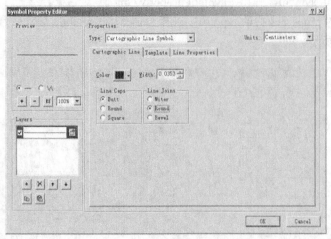

▲图 11-5　线状符号

本实例线状符号采用类似风格的界面,在界面上提供了 3 种线:简单线、制图线、哈希线。初始化的线状符号界面如图 11-6 所示。

▲图 11-6　初始化的线状符号界面

11.2.4　面状符号

面状符号具有实际的二维特征,它们以面定位,其形状与其所代表对象的实际形状一致。这时,符号所处的范围同地图比例尺发生关系,且不论这种范围是明显的还是隐喻的,是精确的还是模糊的。用这种地图符号表示的有水部范围、林地范围、土地利用分类范围、各种区划范围、动植物和矿藏资源分布范围等。

在 ArcMap 中所有做好的面符号均存放在样式库下属的 fill symbols 符号文件夹中。ArcMap 的符号样式管理(style manage)中提供了 5 种类型面状符号的制作方法,分别是 gradient fill symbol、line fill symbol、marker fill symbol、picture fill symbol 和 simple fill symbol。下面根据 marker fill symbol 展开。

(1) 启动 ArcMap,如果未创建符号库,则需要创建符号库;如果已经创建符号库,则需要添加符号库。

（2）单击符号库名，接着再单击 fill symbols 文件夹，然后在右边空白处单击鼠标右键，在弹出菜单中单击 new\fill symbol，弹出"symbol property editor"对话框。

（3）在对话框的 properties 栏的 type 项中选择 marker fill symbol。剩下的属性项设置同前面所述类似，如图 11-7 所示。

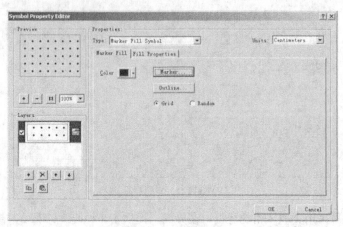

▲图 11-7　面状符号

本实例面状符号采用类似风格的界面，在界面上提供了简单填充。初始化的面状符号界面如图 11-8 所示。

▲图 11-8　初始化的面状符号界面

11.3　符号管理工具实现

主窗口界面用于加载样式文件，并展示样式文件管理接口的使用，将选定的样式文件加载在界面上，遍历样式文件中的所有样式。

符号管理工具实现程序如下所示：

```
using System;
using System.Collections.Generic;
using System.ComponentModel;
using System.Data;
```

```csharp
using System.Drawing;
using System.Text;
using System.Windows.Forms;
using ESRI.ArcGIS.Display;
using ESRI.ArcGIS.Carto;
using ESRI.ArcGIS.Controls;

namespace StyleManager
{
    public partial class Mainfrm : Form
    {
        //样式集
        IStyleGallery styleGallery;
        //样式仓库
        IStyleGalleryStorage styleGalleryStorage;
        //当前样式类
        private string CurrentStyleGalleryClass = "";
        //当前样式类索引
        private int CurrentStyleGalleryClassIndex = -1;
        //当前样式仓库名
        private string CurrentCategoryName;
        //当前样式符号
        private string CurrentStyleFile;
        //当前listview中样式项
        private int CurrentListViewIndex = -1;
        //样式转图片类
        ConvertClass convertClass;
        //样式文件名
        public string fileName;
        //当前选择的节点
        TreeNode CurNode;
        public Mainfrm()
        {
            InitializeComponent();
            convertClass = new ConvertClass();
        }

        //打开样式文件
        private void btOpenStyle_Click(object sender, EventArgs e)
        {
            this.openFileDialog1.Title = "样式文件";
            this.openFileDialog1.DefaultExt = ".Serverstyle";
            this.openFileDialog1.Filter = "(*.Serverstyle)|*.Serverstyle";
            if (this.openFileDialog1.ShowDialog() == DialogResult.OK)
            {
                styleGallery = new ServerStyleGalleryClass();
                styleGalleryStorage = styleGallery as IStyleGalleryStorage;
                AddFile(this.openFileDialog1.FileName);
                this.textBox1.Text = this.openFileDialog1.FileName;
                fileName = this.openFileDialog1.FileName;
            }
        }
        //添加样式文件到树形控件
        private void AddFile(string filename)
        {
            //判断是否存在样式文件
            if (!System.IO.File.Exists(filename))
            {
                return;
            }
```

```csharp
        //判断是否已经加载到树形控件
        bool isExist = false;
        foreach (TreeNode node in this.tvStyleNode.Nodes)
        {
            if (node.Tag.ToString() == filename)
            {
                isExist = true;
            }
        }
        if (!isExist)
        {
            //设置根节点
            System.IO.FileInfo fileinfo = new System.IO.FileInfo(filename);
            TreeNode rootNode = new TreeNode(fileinfo.Name);
            rootNode.Tag = fileinfo.FullName;
            this.tvStyleNode.Nodes.Add(rootNode);
            //加载各种类型的样式目录到树控件
            for (int i = 0; i < this.styleGallery.ClassCount; i++)
            {
                TreeNode node = new TreeNode(this.styleGallery.get_Class(i).Name);
                node.Tag = i;
                rootNode.Nodes.Add(node);
            }
            this.tvStyleNode.ExpandAll();
        }
    }

    //获取颜色对象
    private IRgbColor getRGB(int r, int g, int b)
    {
        IRgbColor pColor;
        pColor = new RgbColorClass();
        pColor.Red = r;
        pColor.Green = g;
        pColor.Blue = b;
        return pColor;
    }

    //大图标视图
    private void tbtLargeIcon_Click(object sender, EventArgs e)
    {
        this.lvSymbolView.View = System.Windows.Forms.View.LargeIcon;


    }
    //小图标视图
    private void tbtSmallIcon_Click(object sender, EventArgs e)
    {
        this.lvSymbolView.View = System.Windows.Forms.View.SmallIcon ;
    }
    //详细列表视图
    private void tbtDetail_Click(object sender, EventArgs e)
    {
        this.lvSymbolView.View = System.Windows.Forms.View.Details ;
    }
    //在listview中预览样式符号
    private void PreviewSymbols(string StyleFile, string StyleGalleryClass)
    {
        //符号枚举接口
        ESRI.ArcGIS.Display.IEnumStyleGalleryItem enumStyleItem;
        //单个符号对象
        ESRI.ArcGIS.Display.IStyleGalleryItem styleItem;
```

```csharp
//判断是初次加载样式文件,或新的样式文件
if (StyleFile != this.CurrentStyleFile)
{
    //设置当前样式文件
    this.CurrentStyleFile = StyleFile;
    //设置当前样式类型
    this.CurrentStyleGalleryClass = StyleGalleryClass;
    //设置符号样式文件位置
    this.styleGalleryStorage = styleGallery as ESRI.ArcGIS.Display. IstyleGalleryStorage;
    this.styleGalleryStorage.AddFile(this.CurrentStyleFile);
    //获取枚举的符号
    enumStyleItem = this.styleGallery.get_Items(this.CurrentStyle GalleryClass, this.CurrentStyleFile, "");
}
else
{
    //设置当前选择的符号类型
    CurrentStyleGalleryClass = StyleGalleryClass;
    //获取枚举的符号
    enumStyleItem = this.styleGallery.get_Items(CurrentStyleGallery Class, this. CurrentStyleFile, CurrentCategoryName);
}
//设置 LISTVIEW 的大图标、小图标
System.Windows.Forms.ImageList largeImage = new ImageList();
System.Windows.Forms.ImageList smallImage = new ImageList();
largeImage.ImageSize = new Size(32, 32);
smallImage.ImageSize = new Size(16, 16);
System.Drawing.Bitmap bmpB, bmpS;
System.Windows.Forms.ListViewItem lvItem;
this.lvSymbolView.Items.Clear();
this.lvSymbolView.Columns.Clear();
//设置 LISTVIEW 的列样式
this.lvSymbolView.LargeImageList = largeImage;
this.lvSymbolView.SmallImageList = smallImage;
this.lvSymbolView.Columns.Add("Name", 180, System.Windows.Forms. HorizontalAlignment.Left);
this.lvSymbolView.Columns.Add("Index", 50, System.Windows.Forms. HorizontalAlignment.Left);
this.lvSymbolView.Columns.Add("Category", 120, System.Windows.Forms. HorizontalAlignment.Left);
//重设枚举的位置
enumStyleItem.Reset();
//获取第 1 个符号
styleItem = enumStyleItem.Next();
int ImageIndex = 0;
//获取当前样式类
ESRI.ArcGIS.Display.IStyleGalleryClass styleClass = styleGallery.get_ Class(CurrentStyleGalleryClassIndex);
while (styleItem != null)
{
    //生成符号的图片预览
    bmpB = convertClass.StyleGalleryItemToBmp(32, 32, styleClass, styleItem);
    bmpS = convertClass.StyleGalleryItemToBmp(16, 16, styleClass, styleItem);
    //添加到控件的图标列表
    largeImage.Images.Add(bmpB);
    smallImage.Images.Add(bmpS);
    //初始化 LISTVIEW 的子项
    lvItem = new ListViewItem(new string[] { styleItem.Name, styleItem. ID.ToString(), styleItem.Category }, ImageIndex);
```

```csharp
            this.lvSymbolView.Items.Add(lvItem);
            styleItem = enumStyleItem.Next();
            ImageIndex++;
        }
        //释放符号枚举,注意,如果不释放有可能在下次加载的时候出错,无法加载
        System.Runtime.InteropServices.Marshal.ReleaseComObject (enumStyleItem);
}

//选择样式类型
private void tvStyleNode_AfterSelect(object sender, TreeViewEventArgs e)
{
    TreeNode node = this.tvStyleNode .SelectedNode;
    //判断节点不为空,并且不是根节点
    if (node != null && node.Parent != null)
    {
        //设置当前样式类型索引
        this.CurrentStyleGalleryClassIndex = int.Parse(node.Tag.ToString ());
        //将符号加载在 LISTVIEW 中预览
        PreviewSymbols(node.Parent.Tag.ToString(), node.Text);
        CurNode = node;
    }
}


//选择符号在图像控件中预览
private void lvSymbolView_ItemSelectionChanged(object sender, ListView Item SelectionChangedEventArgs e)
{
    IStyleGalleryItem styleItem=null;
    IEnumStyleGalleryItem enumStyleItem;
    //获取枚举的符号
    enumStyleItem = this.styleGallery.get_Items(this.CurrentStyleGallery Class, this. CurrentStyleFile, "");
    enumStyleItem.Reset ();

    styleItem = enumStyleItem.Next();
    while (styleItem != null)
    {
        if (styleItem.Name == this.lvSymbolView.Items[e.ItemIndex].Text )
        {
            CurrentListViewIndex = e.ItemIndex;
            break;
        }
        styleItem = enumStyleItem.Next();
    }
    IStyleGalleryClass styleGalleryClass = styleGallery.get_Class(Current StyleGallery ClassIndex);
    Bitmap image = convertClass.StyleGalleryItemToBmp(this.pictureBox1. Width, this.pictureBox1.Height, styleGalleryClass, styleItem);
    this.pictureBox1.Image = image;
    //释放符号枚举,注意,如果不释放有可能在下次加载的时候出错,无法加载
    System.Runtime.InteropServices.Marshal.ReleaseComObject (enumStyleItem);
}
private void addNewStyle_Click(object sender, EventArgs e)
{
    //点符号
    if (this.CurrentStyleGalleryClass == "Marker Symbols")
    {
        PointStyle pointStyle = new PointStyle();
        pointStyle.fileName = fileName;
```

```csharp
                pointStyle.ShowDialog();
                //将符号加载到LISTVIEW中预览
                PreviewSymbols(CurNode.Parent.Tag.ToString(), CurNode.Text);
            }
            //线符号
            else if (this.CurrentStyleGalleryClass == "Line Symbols")
            {
                LineStyle lineStyle = new LineStyle();
                lineStyle.fileName = fileName;
                lineStyle.CurrentStyleGalleryClassIndex = CurrentStyleGallery ClassIndex;
                lineStyle.ShowDialog();
                //将符号加载到LISTVIEW中预览
                PreviewSymbols(CurNode.Parent.Tag.ToString(), CurNode.Text);
            }
            //面符号
            else if (this.CurrentStyleGalleryClass == "Fill Symbols")
            {
                PolygonStyle polygonStyle = new PolygonStyle();
                polygonStyle.fileName = fileName;
                polygonStyle.CurrentStyleGalleryClassIndex = CurrentStyleGallery ClassIndex;
                polygonStyle.ShowDialog();
                //将符号加载到LISTVIEW中预览
                PreviewSymbols(CurNode.Parent.Tag.ToString(), CurNode.Text);
            }
        }
    }
}
```

GListBox是本示例创建的组件类，该类继承Listbox，通过扩展，可以在该组件上显示图片，完整代码如下：

```csharp
using System;
using System.Collections.Generic;
using System.Text;
using System.Windows.Forms ;
using System.Drawing ;

namespace StyleManager
{

    // GListBox 类对系统自带的listbox进行扩展，可以显示图片
    //读者可以在此上进行再扩展
    public class GListBox : ListBox
    {
        //显示图像列表
        private ImageList _myImageList;
        public ImageList ImageList
        {
            get { return _myImageList; }
            set { _myImageList = value; }
        }
        public GListBox()
        {
            this.DrawMode = DrawMode.OwnerDrawFixed;
        }
        //重写子项的绘制
        protected override void OnDrawItem(System.Windows.Forms.DrawItemEventArgs e)
        {
            e.DrawBackground();
            e.DrawFocusRectangle();
            GListBoxItem item;
            Rectangle bounds = e.Bounds;
            Size imageSize = _myImageList.ImageSize;
```

```csharp
    try
    {
        item = (GListBoxItem)Items[e.Index];
        if (item.ImageIndex != -1)
        {
            ImageList.Draw(e.Graphics, bounds.Left, bounds.Top, item.ImageIndex);
            e.Graphics.DrawString(item.Text, e.Font, new SolidBrush (e.ForeColor),
                bounds.Left + imageSize.Width, bounds.Top);
        }
        else
        {
            e.Graphics.DrawString(item.Text, e.Font, new SolidBrush (e.ForeColor),
                bounds.Left, bounds.Top);
        }
    }
    catch
    {
        if (e.Index != -1)
        {
            e.Graphics.DrawString(Items[e.Index].ToString(), e.Font,
                new SolidBrush(e.ForeColor), bounds.Left, bounds.Top);
        }
        else
        {
            e.Graphics.DrawString(Text, e.Font, new SolidBrush(e.ForeColor),
                bounds.Left, bounds.Top);
        }
    }
    base.OnDrawItem(e);
}
// GListBoxItem 类
public class GListBoxItem
{
    private string _myText;
    private int _myImageIndex;
    private CheckBox _myCheck;
    // 属性
    public string Text
    {
        get { return _myText; }
        set { _myText = value; }
    }
    public int ImageIndex
    {
        get { return _myImageIndex; }
        set { _myImageIndex = value; }
    }

    //构造函数
    public GListBoxItem(string text, int index)
    {
        _myText = text;
        _myImageIndex = index;
    }
    public GListBoxItem(string text) : this(text, -1) { }
    public GListBoxItem() : this("") { }
    public GListBoxItem(int index)
    {
        _myCheck = new CheckBox();
        _myImageIndex = index;
    }
```

```csharp
        public override string ToString()
        {
            return _myText;
        }
    }
    class ListBoxICon
    {
    }
}
```

PointStyle 窗体用于点状样式的创建,点状样式的创建参照 ArcMap 的创建,完整代码如下:

```csharp
namespace StyleManager
{
    partial class PointStyle
    {
        private GListBox gListBox ;


//
            // gListBox
            //
            this.gListBox.AllowDrop = true;
            this.gListBox.DrawMode = System.Windows.Forms.DrawMode.OwnerDrawFixed;
            this.gListBox.FormattingEnabled = true;
            this.gListBox.ImageList = this.listBoxImage;
            gListBoxItem1.ImageIndex = -1;
            gListBoxItem1.Text = "tempItem";
            this.gListBox.Items.AddRange(new object[] {
            gListBoxItem1});
            this.gListBox.Location = new System.Drawing.Point(4, 14);
            this.gListBox.Name = "gListBox";
            this.gListBox.Size = new System.Drawing.Size(190, 160);
            this.gListBox.TabIndex = 0;
            this.gListBox.SelectedIndexChanged += new System.EventHandler(this.gListBox_SelectedIndexChanged);
```

```csharp
using System;
using System.Collections.Generic;
using System.Text;
using System.Drawing;
using ESRI.ArcGIS.Display;

namespace StyleManager
{
    class ConvertClass
    {
        //通过指定的高度和宽度生成图像
        public System.Drawing.Bitmap StyleGalleryItemToBmp(
            int iWidth,
            int iHeight,
            IStyleGalleryClass styleGalleryClass,
            IStyleGalleryItem styleGalleryItem)
        {
            Bitmap bmp = new Bitmap(iWidth, iHeight);
            Graphics gImage = Graphics.FromImage(bmp);

            ESRI.ArcGIS.Display.tagRECT rect = new ESRI.ArcGIS.Display.tagRECT();
            rect.right = bmp.Width;
```

```csharp
            rect.bottom = bmp.Height;
            //生成预览
            System.IntPtr hdc = new IntPtr();
            hdc = gImage.GetHdc();
            styleGalleryClass.Preview(styleGalleryItem.Item, hdc.ToInt32(), ref rect);
            gImage.ReleaseHdc(hdc);
            gImage.Dispose();
            return bmp;
        }
    }
}

using System;
using System.Collections.Generic;
using System.Collections;
using System.ComponentModel;
using System.Data;
using System.Drawing;
using System.Drawing.Text;
using System.Drawing.Imaging;
using System.Drawing.Drawing2D;
using System.Text;
using System.Windows.Forms;
using System.IO;
using ESRI.ArcGIS.Display;

using ESRI.ArcGIS.Geometry;
namespace StyleManager
{
    public partial class PointStyle : Form
    {
        public PointStyle()
        {
            InitializeComponent();
        }
        //当前图层颜色
        Color curColor ;
        //符号图片列表
        System.Windows.Forms.ImageList largeImage;
        //图像列表选择项
        int selectItem = -1;
        //listbox选择项索引
        int listBoxSelectIndex = -1;

        //字符家族
        FontFamily fontFamily;
        //样式文件名
        public string fileName = "";
        //保存选择的符号
        ArrayList arrayList = new ArrayList();
        //保存符号的偏移量
        ArrayList arrayXOffset = new ArrayList();
        ArrayList arrayYOffset = new ArrayList();
        //保存符号的大小
        ArrayList arraySize = new ArrayList();
        //保存符号的颜色
        ArrayList arrayColor = new ArrayList();
        //保存符号的角度
        ArrayList arrayAngle= new ArrayList();
```

```csharp
//缩放比率
float  Radio=1;
//将 ESRI GEOMETRIC SYMBOLS 头 10 个字符添加到当前窗体的 listview 控件中
//在这里可以根据项目的需要添加其他字符
private void PointStyle_Load(object sender, EventArgs e)
{
    FileInfo fileInfo = new FileInfo(fileName);
    string path = fileInfo.DirectoryName;
    //获取字体集
    InstalledFontCollection MyFont = new InstalledFontCollection();
    FontFamily[] MyFontFamilies = MyFont.Families;
    int Count = MyFontFamilies.Length;
    int index = -1;
    int key=0;
    for (int i = 0; i < Count; i++)
    {
        //读取 ESRI GEOMETRIC SYMBOLS 字体
        if (MyFontFamilies[i].Name.ToLower() == "esri geometric symbols")
        {
            index = i;
        }
    }
    if (index >=0)
    {
        fontFamily = MyFontFamilies[index];
        //创建字体
         System.Drawing.Font font = new Font(fontFamily, 36, GraphicsUnit. Pixel);
        //创建 LISTVIEW 子项
        ListViewItem listViewItem;
        largeImage = new ImageList();
        largeImage.ImageSize = new Size(60, 36);

        for (int i = 0; i <= 9; i++)
        {
            //根据字体创建图片
            Bitmap bitmap = new Bitmap(60,36);
            Graphics graphics = Graphics.FromImage(bitmap);
            System.Drawing.Color  curColor =new System.Drawing.Color  ();
            curColor = System.Drawing.Color.FromArgb(0, 0, 0);
            graphics.DrawString(i.ToString(), font, new SolidBrush (curColor), new PointF(0, 0));
            graphics.Save();
            graphics.Dispose();
            bitmap.Save(path + @"\" + i.ToString() + ".bmp");
            largeImage.Images.Add(bitmap);
            listViewItem = new ListViewItem();
            listViewItem.ImageIndex = i;
            listViewItem.Text = i.ToString();
            this.listView1.Items.Add(listViewItem);
        }
        this.listView1.LargeImageList = largeImage;
        this.gListBox.ItemHeight = largeImage.ImageSize.Height;
        this.listBoxImage.ImageSize = largeImage.ImageSize;
        this.curColor = this.btColorDialog.BackColor;
    }
}

//获取颜色对象将 Color 装为 Arcengine 的 RGB 颜色
private IRgbColor getRGB(int r, int g, int b)
{
    IRgbColor pColor;
```

```csharp
            pColor = new RgbColorClass();
            pColor.Red = r;
            pColor.Green = g;
            pColor.Blue = b;
            return pColor;
        }
        //选取符号颜色
        private void btColorDialog_Click(object sender, EventArgs e)
        {
            if (this.colorDialog1.ShowDialog() == DialogResult.OK)
            {
                btColorDialog.BackColor = this.colorDialog1.Color;
                curColor = this.colorDialog1.Color;

            }
        }
        //根据listview选择项创建新子样式
        private void listView1_DoubleClick(object sender, EventArgs e)
        {
            //在GLISTBOX里添加符号
            GListBoxItem listBoxItem = new GListBoxItem();
            this.listBoxImage.Images.Add(largeImage.Images[selectItem]);
            ListViewItem listViewItem = new ListViewItem();
            listViewItem = this.listView1.Items[selectItem];
            arrayList.Add(listViewItem.Text );
            //偏移量
            float xOffset = float.Parse(this.txtXOffset.Text.ToString());
            arrayXOffset.Add(xOffset);
            float yOffset = float.Parse(this.txtYOffset.Text.ToString());
            arrayYOffset.Add(yOffset);
            //符号大小
            int symbolSize = int.Parse(this.cbSize.Text.ToString());
            arraySize.Add(symbolSize);
            //符号角度
            int symbolAngle = int.Parse(this.nupAngle.Value.ToString());
            arrayAngle.Add(symbolAngle);

            arrayColor.Add(curColor);
            if (this.gListBox.Items.Count > 0)
            {
                GListBoxItem tempItem =(GListBoxItem ) this.gListBox.Items[0];
                if (tempItem.Text == "tempItem")
                {
                    this.gListBox.Items.RemoveAt(0);
                }
            }
            listBoxItem.ImageIndex = this.listBoxImage.Images.Count - 1;
            listBoxItem.Text = listViewItem.Text;
            this.gListBox.Items.Add(listBoxItem);
            //在PICTUREBOX 中画出新建符号
            RePainPictureBox(1);


        }
        //在PICTUREBOX 中画出新建符号
        private void RePainPictureBox(float ratio)
        {
            //在pictureBox 中预览图像
            Bitmap bitmap = new Bitmap(pbView.Width, pbView.Height);
            Graphics graphics = Graphics.FromImage(bitmap);
            System.Drawing.Color color = System.Drawing.Color.FromArgb(0, 0, 0);
            System.Drawing.Pen pen = new Pen(new SolidBrush(color), 2);
```

```csharp
        pen.DashStyle = DashStyle.DashDot;
        graphics.DrawLine(pen, new PointF(0, pbView.Height / 2), new PointF (pbView.
        Width, pbView.Height / 2));
        graphics.DrawLine(pen, new PointF(pbView.Width / 2, 0), new PointF (pbView.
        Width / 2, pbView.Height));

        PointF pointF = new PointF();
        float symbolSize;
        float xOffset;
        float yOffset;
        float symbolAngle;
        System.Drawing.Color  symbolColor;
        //在画图中画出各个子样式
        for (int i = 0; i < arrayList.Count; i++)
        {
            pointF = new PointF();
            symbolSize =float.Parse ( arraySize[i].ToString()) *ratio ;
            xOffset = float.Parse(arrayXOffset[i].ToString());
            yOffset =float.Parse(arrayYOffset[i].ToString());
            symbolAngle = float.Parse(arrayAngle[i].ToString());
            pointF.X = pbView.Width / 2 - symbolSize / 2 + xOffset;
            pointF.Y = pbView.Height / 2 - symbolSize / 2 + yOffset;
            System.Drawing.Font font = new Font(fontFamily, symbolSize, Graphics
            Unit.Pixel);
            //根据旋转角度，旋转当前子样式
            Matrix matrix = new Matrix();
            matrix.RotateAt(symbolAngle, new PointF(pbView.Width / 2, pbView. Height /
            2), MatrixOrder.Append);
            graphics.Transform = matrix;
            symbolColor = (System.Drawing.Color)arrayColor[i];
            graphics.DrawString(arrayList[i].ToString(), font, new SolidBrush (symbol
            Color), pointF);
            graphics.Save();
        }
        graphics.Dispose();
        pbView.Image = bitmap;
}
private void listView1_ItemSelectionChanged(object sender, ListViewItem Selection
ChangedEventArgs e)
{
        selectItem = e.ItemIndex;
}
//关闭创建符号窗体
private void btCancel_Click(object sender, EventArgs e)
{
        this.Close();
        this.Dispose();
}

private void cbPercent_SelectedIndexChanged(object sender, EventArgs e)
{
        string strRadio = this.cbPercent.Text;
        strRadio.Substring(0, strRadio.Length - 1);
        Radio = float.Parse(strRadio);
        RePainPictureBox(Radio );
}
//符号的属性数组维护
private void UpdateArrayList(string updateType)
{
        int tempIndex=0;
        if (updateType.ToLower() == "del")
```

```csharp
        {
            arrayList.RemoveAt(listBoxSelectIndex );
            arrayXOffset.RemoveAt(listBoxSelectIndex );
            arrayYOffset.RemoveAt(listBoxSelectIndex);
            arraySize.RemoveAt(listBoxSelectIndex );
            arrayColor.RemoveAt(listBoxSelectIndex );
            arrayAngle.RemoveAt(listBoxSelectIndex );
        }
        else if (updateType.ToLower() == "up")
        {
            //判断是第1个
            if (listBoxSelectIndex == 0)
            {
            }
            else
            {
                tempIndex = -1;
                moveArray(tempIndex);
            }
        }
        else if (updateType.ToLower() == "down")
        {
            //判断是最后一个
            if (listBoxSelectIndex ==this.gListBox.Items.Count -1)
            {
            }
            else
            {
                tempIndex = 1;
                moveArray(tempIndex);
            }
        }
    }

    //移动数组
    private void moveArray(int tempIndex)
    {
        string tempStr;
        float tempFloat;
        int  tempInt;
        object  color;
        Image  image;

        tempStr = arrayList[listBoxSelectIndex].ToString();
        arrayList[listBoxSelectIndex] = arrayList[listBoxSelectIndex + temp Index];
        arrayList[listBoxSelectIndex + tempIndex]= tempStr;

        tempFloat =float .Parse ( arrayXOffset [listBoxSelectIndex].ToString ());
        arrayXOffset [listBoxSelectIndex ]=arrayXOffset [listBoxSelectIndex+temp Index ];
        arrayXOffset [listBoxSelectIndex +tempIndex ]=tempFloat ;

        tempFloat =float .Parse ( arrayYOffset [listBoxSelectIndex ].ToString ());
        arrayYOffset [listBoxSelectIndex ]=arrayYOffset [listBoxSelectIndex+temp Index ];
        arrayYOffset [listBoxSelectIndex +tempIndex ]=tempFloat ;

        tempInt =int.Parse ( arraySize [listBoxSelectIndex ].ToString ());
        arraySize [listBoxSelectIndex ]=arraySize [listBoxSelectIndex +temp- Index ];
        arraySize[listBoxSelectIndex + tempIndex] = tempInt;

        color  = arrayColor [listBoxSelectIndex ];
```

```csharp
        arrayColor [listBoxSelectIndex ]=arrayColor [listBoxSelectIndex +temp- Index ];
        arrayColor[listBoxSelectIndex + tempIndex] = color;

        tempFloat =float .Parse ( arrayAngle [listBoxSelectIndex ].ToString ());
        arrayAngle [listBoxSelectIndex ]=arrayAngle [listBoxSelectIndex +temp- Index ];
        arrayAngle [listBoxSelectIndex +tempIndex ]=tempFloat ;

        GListBoxItem tempListItem =(GListBoxItem ) this.gListBox.Items[listBox
SelectIndex];
        this.gListBox.Items[listBoxSelectIndex] = this.gListBox.Items
[listBoxSelectIndex + tempIndex];
        this.gListBox.Items[listBoxSelectIndex + tempIndex] = tempListItem;
}
//删除 gListBox 中的符号
private void btDel_Click(object sender, EventArgs e)
{
    if (listBoxSelectIndex >= 0)
    {
        //this.listBoxImage.Images.RemoveAt(listBoxSelectIndex);
        UpdateArrayList("del");
        this.gListBox.Items.RemoveAt(listBoxSelectIndex);
        RePainPictureBox(Radio);
    }
}
//选择 gListBox 中的符号
private void gListBox_SelectedIndexChanged(object sender, EventArgs e)
{
    listBoxSelectIndex = this.gListBox.SelectedIndex;
}
//将 gListBox 中的符号上移
private void btUp_Click(object sender, EventArgs e)
{
    UpdateArrayList("up");
    RePainPictureBox(Radio);
    if (listBoxSelectIndex <= 0)
        this.gListBox.SelectedIndex = 0;
    else
        this.gListBox.SelectedIndex = listBoxSelectIndex - 1;
}
//将 gListBox 中的符号下移
private void btDown_Click(object sender, EventArgs e)
{
    UpdateArrayList("down");
    RePainPictureBox(Radio);
    if (listBoxSelectIndex >= this.gListBox.Items.Count-1)
        this.gListBox.SelectedIndex = this.gListBox.Items.Count - 1;
    else
        this.gListBox.SelectedIndex = listBoxSelectIndex + 1;
}
//将符号保存到样式文件
private void btSave_Click(object sender, EventArgs e)
{
    //构建图片存储临时目录
    FileInfo fileInfo = new FileInfo(fileName);
    string path = fileInfo.DirectoryName;
    string bitmapFileName = path + @"\" + this.txtSymbolName.Text+".bmp";
    PointF pointF;
    //创建新的画图,并将底色清为白色
    Bitmap bitmap = new Bitmap(100, 100, PixelFormat.Format24bppRgb);
    Graphics graphics = Graphics.FromImage(bitmap);
    graphics.Clear(Color.White   );
```

```csharp
//将各个子样式在画图上绘制
for (int i = 0; i < arrayList.Count; i++)
{
    pointF = new PointF();
    float symbolSize = float.Parse(arraySize[i].ToString());
    float xOffset = float.Parse(arrayXOffset[i].ToString());
    float yOffset = float.Parse(arrayYOffset[i].ToString());
    float symbolAngle = float.Parse(arrayAngle[i].ToString());
    pointF.X = 50 - symbolSize / 2 + xOffset;
    pointF.Y = 50 - symbolSize / 2 + yOffset;
    System.Drawing.Font font = new Font(fontFamily, symbolSize, GraphicsUnit.Pixel);
    Matrix matrix = new Matrix();
    matrix.RotateAt(symbolAngle, new PointF(50, 50), MatrixOrder.Append);

    graphics.Transform = matrix;
    System.Drawing.Color symbolColor = (System.Drawing.Color)arrayColor[i];
    graphics.DrawString(arrayList[i].ToString(), font, new SolidBrush(symbolColor), pointF);
    graphics.Save();
}
graphics.SmoothingMode = SmoothingMode.HighQuality;
graphics.Dispose();
//将图片保存到临时目录
bitmap.Save(bitmapFileName, System.Drawing.Imaging.ImageFormat.Bmp);
//创建图片样式
IPictureMarkerSymbol pictureMarkerSymbol = new PictureMarkerSymbolClass();
pictureMarkerSymbol.CreateMarkerSymbolFromFile(esriIPictureType.esriIPictureBitmap, bitmapFileName);
pictureMarkerSymbol.Angle = 0;

pictureMarkerSymbol.Size = 30;
pictureMarkerSymbol.XOffset = 0;
pictureMarkerSymbol.YOffset = 0;

IStyleGallery styleGallery;
IStyleGalleryItem styleGalleryItem;
IStyleGalleryStorage styleGalleryStorge;
//创建新的样式
styleGalleryItem = new ServerStyleGalleryItemClass();
styleGalleryItem.Name = this.txtSymbolName.Text;
styleGalleryItem.Category = "default";
object objSymbol = pictureMarkerSymbol;
styleGalleryItem.Item = objSymbol;

styleGallery = new ServerStyleGalleryClass();
styleGalleryStorge = styleGallery as IStyleGalleryStorage;
styleGalleryStorge.TargetFile = fileName;
//添加新样式
styleGallery.AddItem(styleGalleryItem);
//保存新样式
styleGallery.SaveStyle(fileName, fileInfo.Name, "marker Symbols");

this.Close();
this.Dispose();
    }
  }
}
```

LineStyle 窗体用于创建线状样式，参照 ArcMap 的创建，完整代码如下：

```csharp
using System;
using System.Collections.Generic;
using System.Collections;
using System.ComponentModel;
using System.Data;
using System.Drawing;
using System.Drawing.Text;
using System.Drawing.Imaging;
using System.Drawing.Drawing2D;
using System.Text;
using System.Windows.Forms;
using System.IO;
using ESRI.ArcGIS.Display;

using ESRI.ArcGIS.Geometry;

namespace StyleManager
{
    public partial class LineStyle : Form
    {
        //当前listbox选择项索引
        int listBoxSelectIndex = -1;
        //符号库文件名
        public string fileName = "";
        //当前样式类索引
        public int CurrentStyleGalleryClassIndex;
        //缩放比例
        float Radio = 1;

        //制图线\哈希线的CAP\JOIN选择
        string curCatCap = "Butt";
        string curCatJoin = "Round";
        string curHashCap = "Butt";
        string curHashJoin = "Round";
        public LineStyle()
        {
            InitializeComponent();
            //初始化listbox图片大小
            Size size = new Size();
            size.Width = this.listBoxImage.ImageSize.Width - 30;
            size.Height = this.listBoxImage.ImageSize.Height;
            this.gListBox.ImageList.ImageSize = size;
            this.gListBox.ItemHeight = this.listBoxImage.ImageSize.Height;
        }
        //添加新的子样式
        private void btNew_Click(object sender, EventArgs e)
        {
            //删除默认listbox项
            if (this.gListBox.Items.Count > 0)
            {
                GListBoxItem tempItem = (GListBoxItem)this.gListBox.Items[0];
                if (tempItem.Text == "tempItem")
                {
                    this.gListBox.Items.RemoveAt(0);
                }
            }
            //简单线
            if (tcType.SelectedIndex == 0)
            {
                ISimpleLineSymbol simpleLineSymbol = createSimpleLine(btSimpleColor.
```

```csharp
            BackColor, cbSimpleStyle.Text,double.Parse ( cbSimpleWidth.Text.ToString ()) );
            addListBoxItem(simpleLineSymbol as ISymbol );
        }
        //制图线
        else if (tcType.SelectedIndex == 1)
        {
            ICartographicLineSymbol cartographicLineSymbol = createCartogra phicLine
            (btCatColor.BackColor,double.Parse ( nupCatWidth.Value.ToString ()) ,
            double.Parse (txtCatOffset.Text), curCatCap, curCatJoin);
            addListBoxItem(cartographicLineSymbol as ISymbol );
        }
        //哈希线
        else if (tcType.SelectedIndex == 2)
        {
            IHashLineSymbol hashLineSymbol = createHashLine(btHashColor.
            BackColor,double.Parse( nupHashWidth.Value.ToString ()),double.Parse
            ( nupHashAngle. Value.ToString ()),double.Parse ( txtHashOffset.Text),
            curHashCap, curHashJoin);
            addListBoxItem(hashLineSymbol as ISymbol );
        }
        RePainPictureBox(Radio);
    }
    //添加 GLISTBOX 项
    private void addListBoxItem(ISymbol symbol)
    {
        IStyleGallery styleGallery;
        IStyleGalleryItem styleGalleryItem;
        IStyleGalleryStorage styleGalleryStorge;

        styleGalleryItem = new ServerStyleGalleryItemClass();
        styleGalleryItem.Name = this.txtSymbolName.Text;
        styleGalleryItem.Category = "default";
        object objSymbol = symbol ;
        styleGalleryItem.Item = objSymbol;

        styleGallery = new ServerStyleGalleryClass();
        styleGalleryStorge = styleGallery as IStyleGalleryStorage;
        styleGalleryStorge.TargetFile = fileName;

        IStyleGalleryClass styleGalleryClass = styleGallery.get_Class(Current
        StyleGallery ClassIndex);
        ConvertClass convertClass = new ConvertClass();
        Bitmap bitmap = convertClass.StyleGalleryItemToBmp(190, 36, style GalleryClass,
        styleGalleryItem);

        GListBoxItem listBoxItem = new GListBoxItem();

        this.listBoxImage.Images.Add(bitmap );
        listBoxItem.ImageIndex = this.listBoxImage.Images.Count - 1;
        listBoxItem.Text = "sss";
        this.gListBox.Items.Add(listBoxItem);

    }
    //符号的属性数组维护
    private void UpdateArrayList(string updateType)
    {
        int tempIndex = 0;
        if (updateType.ToLower() == "del")
        {
```

```csharp
        }
        else if (updateType.ToLower() == "up")
        {
            //判断是第 1 个
            if (listBoxSelectIndex == 0)
            {
            }
            else
            {
                tempIndex = -1;
                moveListBoxItem(tempIndex);
            }
        }
        else if (updateType.ToLower() == "down")
        {
            //判断是最后一个
            if (listBoxSelectIndex == this.gListBox.Items.Count - 1)
            {
            }
            else
            {
                tempIndex = 1;
                moveListBoxItem(tempIndex);
            }
        }
    }
    //移动 listbox 项顺序
    private void moveListBoxItem(int tempIndex)
    {
        GListBoxItem tempListItem = (GListBoxItem)this.gListBox.Items[listBoxSelect
        Index];
        this.gListBox.Items[listBoxSelectIndex] = this.gListBox.Items [list
        BoxSelectIndex + tempIndex];
        this.gListBox.Items[listBoxSelectIndex + tempIndex] = tempListItem;
    }
    //获取 CAP\JOIN 的选择
    private void checkRadioButton()
    {
        if (rbCatButt.Checked == true)
        {
            curCatCap = "Butt";
        }
        else if (rbCatRoundC.Checked == true)
        {
            curCatCap = "Round";
        }
        else if (rbCatSquare.Checked == true)
        {
            curCatCap = "Square";
        }

        if (rbCatMitre.Checked == true)
        {
            curCatJoin = "Mitre";
        }
        else if (rbCatRoundJ.Checked == true)
        {
            curCatJoin = "Round";
        }
        else if (rbCatBevel.Checked == true)
        {
            curCatJoin = "Bevel";
```

```csharp
        }
        if (rbHashButt.Checked == true)
        {
            curHashCap = "Butt";
        }
        else if (rbHashRoundC.Checked == true)
        {
            curHashCap = "Round";
        }
        else if (rbHashSquare.Checked == true)
        {
            curHashCap = "Square";
        }

        if (rbHashMitre.Checked == true)
        {
            curHashJoin = "Mitre";
        }
        else if (rbHashRoundJ.Checked == true)
        {
            curHashJoin = "Round";
        }
        else if (rbHashBevel.Checked == true)
        {
            curHashJoin = "Bevel";
        }
    }
    //创建简单符号
    private ISimpleLineSymbol createSimpleLine(Color color,string style, double width)
    {
        ISimpleLineSymbol simpleLineSymbol = new SimpleLineSymbolClass();
        //设置子样式类型
        switch (style )
        {
            case "Solid":
                simpleLineSymbol.Style = esriSimpleLineStyle.esriSLSSolid ;
                break ;
            case "Dashed":
                simpleLineSymbol.Style = esriSimpleLineStyle.esriSLSDash;
                break ;
            case "Dotted":
                simpleLineSymbol.Style = esriSimpleLineStyle.esriSLSDot ;
                break ;
            case "Dash-Dot":
                simpleLineSymbol.Style = esriSimpleLineStyle.esriSLSDashDot;
                break ;
            case "Dash-Dot-Dot":
                simpleLineSymbol.Style = esriSimpleLineStyle.esriSLSDashDotDot;
                break ;
            case "null":
                simpleLineSymbol.Style = esriSimpleLineStyle.esriSLSNull;
                break ;
        }
        simpleLineSymbol.Width = width ;
        IRgbColor rgbColor = getRGB(color.R ,color.G , color.B );
        simpleLineSymbol.Color = rgbColor;
        return simpleLineSymbol ;
    }
    //获取颜色对象将Color装为Arcengine的RGB颜色
    private IRgbColor getRGB(int r, int g, int b)
```

```csharp
{
    IRgbColor pColor;
    pColor = new RgbColorClass();
    pColor.Red = r;
    pColor.Green = g;
    pColor.Blue = b;
    return pColor;
}
//创建哈希符号
private IHashLineSymbol createHashLine(Color color, double width ,double angle,
double offset,string cap,string join)
{
    IHashLineSymbol hashLineSymbol = new HashLineSymbolClass();
    ILineProperties lineProperties = hashLineSymbol as ILineProperties;
    lineProperties.Offset = offset;
    double[] dob = new double[6];
    dob[0] = 0;
    dob[1] = 1;
    dob[2] = 2;
    dob[3] = 3;
    dob[4] = 4;
    dob[5] = 5;
    ITemplate template = new TemplateClass();
    //间隔
    template.Interval = 1;
    for (int i = 0; i < dob.Length; i += 2)
    {
        template.AddPatternElement(dob[i], dob[i + 1]);
    }
    lineProperties.Template = template;
    hashLineSymbol.Width = width;
    hashLineSymbol.Angle = angle;
    hashLineSymbol.HashSymbol = createCartographicLine(color, width, offset, cap,
    join);
    IRgbColor hashColor = new RgbColor();
    hashColor = getRGB(color.R, color.G, color.B);
    hashLineSymbol.Color = hashColor;
    return hashLineSymbol;
}
//创建制图符号
private ICartographicLineSymbol createCartographicLine(Color color, double width,
double offset, string cap, string join)
{
    ICartographicLineSymbol cartographicLineSymbol = new CartographicLine
    SymbolClass();
    switch (cap)
    {
        case "Butt":
            cartographicLineSymbol.Cap = esriLineCapStyle.esriLCSButt;
            break;
        case "Round":
            cartographicLineSymbol.Cap = esriLineCapStyle.esriLCSRound;
            break;
        case "Square":
            cartographicLineSymbol.Cap = esriLineCapStyle.esriLCSSquare;
            break;
    }

    switch (join)
    {
        case "Mitre":
            cartographicLineSymbol.Join = esriLineJoinStyle.esriLJSMitre ;
```

```csharp
                break;
            case "Round":
                cartographicLineSymbol.Join = esriLineJoinStyle.esriLJSRound;
                break;
            case "Bevel":
                cartographicLineSymbol.Join = esriLineJoinStyle.esriLJSBevel;
                break;
    }
    cartographicLineSymbol.Width = width ;
    cartographicLineSymbol.MiterLimit = 4;
    ILineProperties lineProperties;
    lineProperties = cartographicLineSymbol as ILineProperties;
    lineProperties.Offset = offset ;
    double[] dob = new double[6];
    dob[0] = 0;
    dob[1] = 1;
    dob[2] = 2;
    dob[3] = 3;
    dob[4] = 4;
    dob[5] = 5;
    ITemplate template = new TemplateClass();
    template.Interval = 1;
    for (int i = 0; i < dob.Length; i += 2)
    {
        template.AddPatternElement(dob[i], dob[i + 1]);
    }
    lineProperties.Template = template;
    IRgbColor rgbColor = getRGB(color.R, color.G, color.B);
    cartographicLineSymbol.Color = rgbColor;
    return cartographicLineSymbol;
}

private void btDel_Click(object sender, EventArgs e)
{
    if (listBoxSelectIndex >= 0)
    {
        UpdateArrayList("del");
        this.gListBox.Items.RemoveAt(listBoxSelectIndex);
        //this.listBoxImage.Images.RemoveAt(listBoxSelectIndex);
        RePainPictureBox(Radio);
    }
}

//选择 gListBox 中的符号
private void gListBox_SelectedIndexChanged(object sender, EventArgs e)
{
    listBoxSelectIndex = this.gListBox.SelectedIndex;
}
//将 gListBox 中的符号上移
private void btUp_Click(object sender, EventArgs e)
{
    UpdateArrayList("up");
    RePainPictureBox(Radio);
    if (listBoxSelectIndex <= 0)
        this.gListBox.SelectedIndex = 0;
    else
        this.gListBox.SelectedIndex = listBoxSelectIndex - 1;

}
//将 gListBox 中的符号下移
private void btDown_Click(object sender, EventArgs e)
{
```

```csharp
        UpdateArrayList("down");
        RePainPictureBox(Radio);
         if (listBoxSelectIndex >= this.gListBox.Items.Count - 1)
            this.gListBox.SelectedIndex = this.gListBox.Items.Count - 1;
        else
            this.gListBox.SelectedIndex = listBoxSelectIndex + 1;

}
//简单线颜色
private void btSimpleColor_Click(object sender, EventArgs e)
{
    if (this.colorDialog1.ShowDialog() == DialogResult.OK)
    {
        btSimpleColor.BackColor = this.colorDialog1.Color;
    }
}
//制图线颜色
private void btCatColor_Click(object sender, EventArgs e)
{
    if (this.colorDialog1.ShowDialog() == DialogResult.OK)
    {
        btCatColor.BackColor = this.colorDialog1.Color;
    }
}
//哈希线颜色
private void btHashColor_Click(object sender, EventArgs e)
{
    if (this.colorDialog1.ShowDialog() == DialogResult.OK)
    {
        btHashColor.BackColor = this.colorDialog1.Color;
    }
}
//ListBox 中的选择项
private void gListBox_SelectedIndexChanged_1(object sender, EventArgs e)
{
    listBoxSelectIndex = this.gListBox.SelectedIndex;
}
//在 PictureBox 中画出新建符号
private void RePainPictureBox(float ratio)
{
    //在 PictureBox 中预览图像
    Bitmap bitmap = new Bitmap(pbView.Width, pbView.Height);
    Graphics graphics = Graphics.FromImage(bitmap);
    System.Drawing.Color color = System.Drawing.Color.FromArgb(0, 0, 0);

    Bitmap image;
    int startX;
    int startY;
    GListBoxItem glistItem ;
    for (int i = 0; i < this.gListBox.Items.Count ; i++)
    {
        glistItem =(GListBoxItem)this.gListBox.Items [i];
        image =(Bitmap ) listBoxImage.Images[glistItem.ImageIndex ];
        startX = pbView.Width / 2 - listBoxImage.ImageSize.Width / 2;
        startY = pbView.Height / 2 - listBoxImage.ImageSize.Height / 2;

        image.MakeTransparent(image.GetPixel(0, 0));
        System.Drawing.Rectangle rectangle = new Rectangle(startX, startY, list
BoxImage.ImageSize.Width, listBoxImage.ImageSize.Height);
        graphics.DrawImage(image, rectangle);
        graphics.Save ();
```

```csharp
    }
    graphics.Dispose();
    pbView.Image = bitmap;
}
//选择缩放比率
private void cbPercent_SelectedIndexChanged(object sender, EventArgs e)
{
    string strRadio = this.cbPercent.Text;
    strRadio.Substring(0, strRadio.Length - 1);
    Radio = float.Parse(strRadio);
    RePainPictureBox(Radio);
}
//关闭窗口
private void btCancel_Click(object sender, EventArgs e)
{
    this.Close();
    this.Dispose();
}
//保存样式
private void btSave_Click(object sender, EventArgs e)
{
    //指定临时图片保存位置
    FileInfo fileInfo = new FileInfo(fileName);
    string path = fileInfo.DirectoryName;
    string bitmapFileName = path + @"\" + this.txtSymbolName.Text + ".bmp";
    //创建新的画图,并将底色清为白色
    Bitmap bitmap = new Bitmap(listBoxImage.ImageSize.Width, listBoxImage.ImageSize.Height, PixelFormat.Format24bppRgb);
    Graphics graphics = Graphics.FromImage(bitmap);
    graphics.Clear(Color.White);

    Bitmap image;
    int startX;
    int startY;
    GListBoxItem glistItem;
    //将各个子样式在画图上绘制
    for (int i = 0; i < this.gListBox.Items.Count; i++)
    {
        glistItem = (GListBoxItem)this.gListBox.Items[i];
        image = (Bitmap)listBoxImage.Images[glistItem.ImageIndex];
        startX = 50 - listBoxImage.ImageSize.Width / 2;
        startY = 50 - listBoxImage.ImageSize.Height / 2;

        image.MakeTransparent(image.GetPixel(0, 0));
        System.Drawing.Rectangle rectangle = new Rectangle(startX, startY, listBoxImage.ImageSize.Width, listBoxImage.ImageSize.Height);
        graphics.DrawImage(image, rectangle);
        graphics.Save();
    }
    graphics.Dispose();
    //保存画图
    bitmap.Save(bitmapFileName, System.Drawing.Imaging.ImageFormat.Bmp);

    //创建图片类型,在这里可以参考第5章的例子,用其他类型转换
    IPictureLineSymbol pictureLineSymbol = new PictureLineSymbolClass();
    pictureLineSymbol.CreateLineSymbolFromFile(esriIPictureType.esriIPictureBitmap, bitmapFileName);

    pictureLineSymbol.Offset = 0;
    pictureLineSymbol.Width = 10;
    pictureLineSymbol.Rotate = false;
```

```csharp
            IStyleGallery styleGallery;
            IStyleGalleryItem styleGalleryItem;
            IStyleGalleryStorage styleGalleryStorge;
            //添加新样式
            styleGalleryItem = new ServerStyleGalleryItemClass();
            styleGalleryItem.Name = this.txtSymbolName.Text;
            styleGalleryItem.Category = "default";
            object objSymbol = pictureLineSymbol;
            styleGalleryItem.Item = objSymbol;

            styleGallery = new ServerStyleGalleryClass();
            styleGalleryStorge = styleGallery as IStyleGalleryStorage;
            styleGalleryStorge.TargetFile = fileName;
            styleGallery.AddItem(styleGalleryItem);
            //保存新样式
            styleGallery.SaveStyle(fileName, fileInfo.Name, "Line Symbols");

            this.Close();
            this.Dispose();
        }
    }
}
```

PolygonStyle 窗体用于面状样式，本示例演示了简单填充样式，完整代码如下：

```csharp
using System;
using System.Collections.Generic;
using System.ComponentModel;
using System.Data;
using System.IO;
using System.Drawing;
using System.Drawing.Imaging;
using System.Text;
using System.Windows.Forms;
using ESRI.ArcGIS.Display;


namespace StyleManager
{
    /// <summary>
    /// 该类为创建面符号类，由于篇幅的关系，这里只编写了简单填充类，其他类可以参考第 5 章的例子进行补充
完善
    /// </summary>
    public partial class PolygonStyle : Form
    {

        //ListBox 当前选中项索引
        int listBoxSelectIndex = -1;
        //符号文件名
        public string fileName = "";
        //当前符号类索引
        public int CurrentStyleGalleryClassIndex;
        //缩放比例
        float Radio = 1;
        //边线的宽度
        double curLineWidth = 1;
        //边线颜色
        Color lineColor;
        //填充颜色
```

```csharp
        Color fillColor;

        public PolygonStyle()
        {
            InitializeComponent();
            //初始化 ListBox 图像的大小
            Size size = new Size();
            size.Width = this.listBoxImage.ImageSize.Width - 30;
            size.Height = this.listBoxImage.ImageSize.Height;
            this.gListBox.ImageList.ImageSize = size;
            this.gListBox.ItemHeight = this.listBoxImage.ImageSize.Height;
            //初始化填充颜色和边线颜色
            fillColor = this.btSimpleColor.BackColor;
            lineColor = this.btLineColor.BackColor;
        }

        //增加新子样式
        private void btNew_Click(object sender, EventArgs e)
        {
            //删除默认的 ListBox 的项
            if (this.gListBox.Items.Count > 0)
            {
                GListBoxItem tempItem = (GListBoxItem)this.gListBox.Items[0];
                if (tempItem.Text == "tempItem")
                {
                    this.gListBox.Items.RemoveAt(0);
                }
            }
            //简单填充符号
            ISimpleFillSymbol simpleFillSymbol = new SimpleFillSymbolClass();
            simpleFillSymbol.Style = esriSimpleFillStyle.esriSFSSolid;
            simpleFillSymbol.Color = getRGB(fillColor.R ,fillColor.G ,fillColor.B );
            //创建边线符号
            ISimpleLineSymbol simpleLineSymbol = new SimpleLineSymbolClass();
            simpleLineSymbol.Style = esriSimpleLineStyle.esriSLSDashDotDot;
            simpleLineSymbol.Color = getRGB(lineColor.R , lineColor.G , lineColor.B );
            simpleLineSymbol.Width = curLineWidth;
            simpleFillSymbol.Outline = simpleLineSymbol;
            //将新增的子样式添加到列表中
            addListBoxItem(simpleFillSymbol);
            RePainPictureBox(Radio);

        }
        //获取颜色对象将 Color 装为 Arcengine 的 RGB 颜色
        private IRgbColor getRGB(int r, int g, int b)
        {
            IRgbColor pColor;
            pColor = new RgbColorClass();
            pColor.Red = r;
            pColor.Green = g;
            pColor.Blue = b;
            return pColor;
        }
        //删除子样式
        private void btDel_Click(object sender, EventArgs e)
        {
            if (listBoxSelectIndex >= 0)
            {
                UpdateArrayList("del");
                this.gListBox.Items.RemoveAt(listBoxSelectIndex);
                //this.listBoxImage.Images.RemoveAt(listBoxSelectIndex);
```

```csharp
            RePainPictureBox(Radio);
        }
    }
    //上移子样式
    private void btUp_Click(object sender, EventArgs e)
    {
        UpdateArrayList("up");
        RePainPictureBox(Radio);
        if (listBoxSelectIndex <= 0)
            this.gListBox.SelectedIndex = 0;
        else
            this.gListBox.SelectedIndex = listBoxSelectIndex - 1;
    }
    //下移子样式
    private void btDown_Click(object sender, EventArgs e)
    {
        UpdateArrayList("down");
        RePainPictureBox(Radio);
        if (listBoxSelectIndex >= this.gListBox.Items.Count - 1)
            this.gListBox.SelectedIndex = this.gListBox.Items.Count - 1;
        else
            this.gListBox.SelectedIndex = listBoxSelectIndex + 1;
    }
    //移动数组
    private void moveListBoxItem(int tempIndex)
    {
        GListBoxItem tempListItem = (GListBoxItem)this.gListBox.Items[listBoxSelectIndex];
        this.gListBox.Items[listBoxSelectIndex] = this.gListBox.Items[listBoxSelectIndex + tempIndex];
        this.gListBox.Items[listBoxSelectIndex + tempIndex] = tempListItem;
    }
    //更新 ListBox 的顺序
    private void UpdateArrayList(string updateType)
    {
        int tempIndex = 0;
        if (updateType.ToLower() == "del")
        {

        }
        else if (updateType.ToLower() == "up")
        {
            //判断是第 1 个
            if (listBoxSelectIndex == 0)
            {
            }
            else
            {
                tempIndex = -1;
                moveListBoxItem(tempIndex);
            }
        }
        else if (updateType.ToLower() == "down")
        {
            //判断是最后一个
            if (listBoxSelectIndex == this.gListBox.Items.Count - 1)
            {
            }
            else
            {
                tempIndex = 1;
```

```csharp
            moveListBoxItem(tempIndex);
        }
    }
}
//在 PictureBox 中画出新建符号
private void RePainPictureBox(float ratio)
{
    //在 PictureBox 中预览图像
    Bitmap bitmap = new Bitmap(pbView.Width, pbView.Height);
    Graphics graphics = Graphics.FromImage(bitmap);
    System.Drawing.Color color = System.Drawing.Color.FromArgb(0, 0, 0);

    Bitmap image;
    int startX;
    int startY;
    GListBoxItem glistItem;
    //将子样式在绘图纸上绘制
    for (int i = 0; i < this.gListBox.Items.Count; i++)
    {
        glistItem = (GListBoxItem)this.gListBox.Items[i];
        image = (Bitmap)listBoxImage.Images[glistItem.ImageIndex];
        startX = pbView.Width / 2 - listBoxImage.ImageSize.Width / 2;
        startY = pbView.Height / 2 - listBoxImage.ImageSize.Height / 2;
        //指定透明色颜色
        image.MakeTransparent(image.GetPixel(0, 0));
        System.Drawing.Rectangle rectangle = new Rectangle(startX, startY,
        listBoxImage.ImageSize.Width, listBoxImage.ImageSize.Height);
        graphics.DrawImage(image, rectangle);
        graphics.Save();
    }
    graphics.Dispose();
    pbView.Image = bitmap;
}
//选择缩放
private void cbPercent_SelectedIndexChanged(object sender, EventArgs e)
{
    string strRadio = this.cbPercent.Text;
    strRadio.Substring(0, strRadio.Length - 1);
    Radio = float.Parse(strRadio);
    RePainPictureBox(Radio);
}
//填充颜色
private void btSimpleColor_Click(object sender, EventArgs e)
{
    if (this.colorDialog1.ShowDialog() == DialogResult.OK)
    {
        btSimpleColor.BackColor = this.colorDialog1.Color;
        fillColor = this.colorDialog1.Color;
    }
}
//边框颜色
private void btLineColor_Click(object sender, EventArgs e)
{
    if (this.colorDialog1.ShowDialog() == DialogResult.OK)
    {
        btLineColor.BackColor = this.colorDialog1.Color;
        lineColor = this.colorDialog1.Color;
    }
}
//边框宽度
private void nupLineWidth_ValueChanged(object sender, EventArgs e)
```

```csharp
{
    this.curLineWidth = (double ) nupLineWidth.Value ;
}
//选择的子样式
private void gListBox_SelectedIndexChanged(object sender, EventArgs e)
{
    listBoxSelectIndex = this.gListBox.SelectedIndex;
}
//添加 gListBox 项
private void addListBoxItem(ISimpleFillSymbol simpleFillSymbol)
{

    IStyleGallery styleGallery;
    IStyleGalleryItem styleGalleryItem;
    IStyleGalleryStorage styleGalleryStorge;

    styleGalleryItem = new ServerStyleGalleryItemClass();
    styleGalleryItem.Name = this.txtSymbolName.Text;
    styleGalleryItem.Category = "default";
    object objSymbol = simpleFillSymbol;
    styleGalleryItem.Item = objSymbol;

    styleGallery = new ServerStyleGalleryClass();
    styleGalleryStorge = styleGallery as IStyleGalleryStorage;
    styleGalleryStorge.TargetFile = fileName;

    IStyleGalleryClass styleGalleryClass = styleGallery.get_Class(Current
    StyleGallery ClassIndex);
    ConvertClass convertClass = new ConvertClass();
    Bitmap bitmap = convertClass.StyleGalleryItemToBmp(100, 30, styleGalleryClass,
    styleGalleryItem);

    GListBoxItem listBoxItem = new GListBoxItem();

    this.listBoxImage.Images.Add(bitmap);
    listBoxItem.ImageIndex = this.listBoxImage.Images.Count - 1;
    listBoxItem.Text = "sss";
    this.gListBox.Items.Add(listBoxItem);


}
//保存符号
private void btSave_Click(object sender, EventArgs e)
{
    //指定临时图片保存位置
    FileInfo fileInfo = new FileInfo(fileName);
    string path = fileInfo.DirectoryName;
    string bitmapFileName = path + @"\" + this.txtSymbolName.Text + ".bmp";
    //创建新的画图,并将底色清为白色
    Bitmap bitmap = new Bitmap(listBoxImage.ImageSize.Width, listBoxImage.
    ImageSize.Height, PixelFormat.Format24bppRgb);
    Graphics graphics = Graphics.FromImage(bitmap);
    graphics.Clear(Color.White);

    Bitmap image;
    int startX;
    int startY;
    GListBoxItem glistItem;
    //将各个子样式在画图上绘制
    for (int i = 0; i < this.gListBox.Items.Count; i++)
    {
```

```csharp
            glistItem = (GListBoxItem)this.gListBox.Items[i];
            image = (Bitmap)listBoxImage.Images[glistItem.ImageIndex];

            startX = 0;
            startY = 0;
            System.Drawing.Rectangle rectangle = new Rectangle(startX, startY, image.Width, image.Height);
            graphics.DrawImage(image, rectangle);
            graphics.Save();

        }
        graphics.Dispose();
        //保存画图
        bitmap.Save(bitmapFileName, System.Drawing.Imaging.ImageFormat.Bmp);

        //创建图片填充类型,在这里可以参考第5章的例子,用其他类型转换
        IPictureFillSymbol pictureFillSymbol = new PictureFillSymbolClass();
        pictureFillSymbol.CreateFillSymbolFromFile(esriIPictureType.esriIPictureBitmap, bitmapFileName);

        IStyleGallery styleGallery;
        IStyleGalleryItem styleGalleryItem;
        IStyleGalleryStorage styleGalleryStorge;
        //添加新样式
        styleGalleryItem = new ServerStyleGalleryItemClass();
        styleGalleryItem.Name = this.txtSymbolName.Text;
        styleGalleryItem.Category = "default";
        object objSymbol = pictureFillSymbol;
        styleGalleryItem.Item = objSymbol;

        styleGallery = new ServerStyleGalleryClass();
        styleGalleryStorge = styleGallery as IStyleGalleryStorage;
        styleGalleryStorge.TargetFile = fileName;
        styleGallery.AddItem(styleGalleryItem);
        //保存新样式
        styleGallery.SaveStyle(fileName, fileInfo.Name, "Fill Symbols");

        this.Close();
        this.Dispose();
    }
    //关闭窗口
    private void btCancel_Click(object sender, EventArgs e)
    {
        this.Close();
        this.Dispose();
    }
}
```

11.4 本章小结

本章通过设计一个类似 ArcMap 的符号库编辑器的示例,综合演练符号对象相关的接口、类的使用。

第 12 章 空间数据管理系统

12.1 简介

空间数据管理系统通过一系列功能模块组成的集成化应用向用户提供空间数据的管理工具,所有的模块以地理信息数据库为基础。空间数据管理系统主要包括以下功能。

(1) 数据质量检查模块:主要依据数据库建库标准以及相应的国家标准和行业规范,对空间数据的位置精度、拓扑关系以及属性数据的完整性和逻辑一致性进行检查,生成检查报告,从而严格控制数据质量。

(2) 数据入库模块:将检验合格的空间数据导入空间数据库。

(3) 版本管理模块:由于空间数据具有时间的特性,不同时期具有不同的形状和属性,因此需要将过去的数据保留于历史库中,以便用户对历史情况进行查询。

(4) 数据浏览查询模块:该模块使用户能够对地理信息数据进行图形浏览,完成图形和属性之间的双向查询与检索,并根据用户的要求对数据进行汇总统计。

(5) 制图输出模块:该模块向用户提供以标准图幅或自定义的方式打印地图数据的工具,并根据地理数据库中的数据制作各种专题图。

(6) 空间数据编辑模块,该模块提供了空间数据编辑功能,通过简单的交互实现对数据库要素的增加和删除,以及修改图形特征和属性。

(7) 元数据管理:提供对入库地理信息元数据的集中管理,包括元数据字段的增加、删除,以及数据类型修改等。

(8) 符号库管理:依据各类图件的图式规范制作一套地图符号库。

12.2 空间数据管理框架设计

根据空间数据管理系统功能,设计确定系统总体框架结构如图 12-1 所示。

12.3 空间数据管理实现

本章的例子采用个人数据库。在上一章的例子中已经实现了"符号库管理"、"数据浏览查询"和"制图"等功能,本章着重实现"打开数据库"、"创建数据库"、"元数据管理"、"数据入库"和"数据编辑"等功能。

在空间数据管理系统主界面打开数据库后,在左侧会显示当前数据库中的数据集,单击数据集,在右侧的 MapControl 中则会显示当前数据集中的内容,如图 12-2 所示示例的主界面。

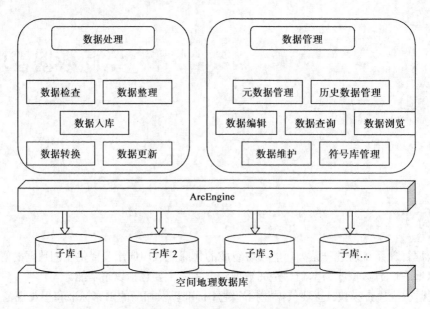

▲图 12-1　空间数据管理框架结构

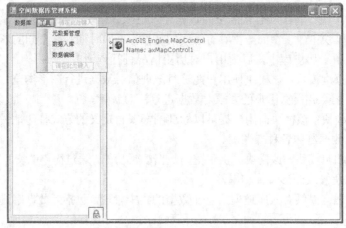

▲图 12-2　空间数据管理字流打开数据库界面

实现程序如下所示：

```
using System;
using System.IO;
using System.Collections;
using System.Collections.Generic;
using System.ComponentModel;
using System.Data;
using System.Drawing;
using System.Text;
using System.Windows.Forms;
using ESRI.ArcGIS.Geometry;
using ESRI.ArcGIS.Controls;
using ESRI.ArcGIS.Carto;
using ESRI.ArcGIS.esriSystem;
using ESRI.ArcGIS.DataSourcesFile;
using ESRI.ArcGIS.DataSourcesGDB;
using ESRI.ArcGIS.DataSourcesRaster;
using ESRI.ArcGIS.Geodatabase;
```

```csharp
namespace GeoManage
{
    public partial class frmMain : Form
    {
        //全局个人数据库目录
        string m_FullPath;
        //全局数据库名
        string m_FullDatasetName;
        //当前的工作控件
        IWorkspace  m_Workspace;
        //工作空间列表
        ArrayList m_WorkspaceList = new ArrayList();
        //当前选择节点
        TreeNode m_CurrentNode;
        //当前图层
        IFeatureLayer m_curFeatureLayer;
        public frmMain()
        {
            InitializeComponent();
            m_Workspace = null;
            refreshTree();
        }
        //创建新数据库
        private void 创建新数据库ToolStripMenuItem_Click(object sender, EventArgs e)
        {
            CreateDatabase createDatabase = new CreateDatabase();
            createDatabase.ShowDialog();
            m_FullDatasetName = createDatabase.FileName;
            m_FullPath = createDatabase.FilePath;
            m_Workspace = createDatabase.NewWorkspace;
            m_WorkspaceList.Add(m_Workspace);
            refreshTree();
        }
        //刷新工作空间树
        private void refreshTree()
        {
            this.treeView1.Nodes.Clear();
            TreeNode rootNode;
            for (int i = 0; i < m_WorkspaceList.Count; i++)
            {
                IWorkspace workspace = m_WorkspaceList[i] as IWorkspace ;
                FileInfo fileInfo = new FileInfo(workspace.PathName);
                string fileName =fileInfo.Name ;
                rootNode = new TreeNode();
                rootNode.Tag = workspace.PathName;
                rootNode.Name = fileName.Substring(0, fileName.LastIndexOf('.'));
                rootNode.Text = fileName.Substring(0, fileName.LastIndexOf('.'));
                IEnumDatasetName enumDatasetName= workspace.get_DatasetNames (esriDatasetType.esriDTFeatureClass);
                enumDatasetName.Reset();
                IDatasetName datasetName = enumDatasetName.Next();
                while (datasetName != null)
                {
                    TreeNode childNode = new TreeNode();
                    childNode.Text = datasetName.Name;
                    rootNode.Nodes.Add(childNode);
                    datasetName = enumDatasetName.Next();
                }
                this.treeView1.Nodes.Add(rootNode);
                this.treeView1.ExpandAll();
            }
        }
```

```csharp
//打开数据库
private void 打开ToolStripMenuItem_Click(object sender, EventArgs e)
{
    this.openFileDialog1.Title = "个人空间数据库";
    this.openFileDialog1.Filter = "geodatabase(*.mdb)|*.mdb";
    if (this.openFileDialog1.ShowDialog() == DialogResult.OK)
    {
        string fileName =this.openFileDialog1.FileName ;
        FileInfo fileInfo =new FileInfo (fileName );
        m_FullPath =fileInfo.DirectoryName ;
        m_FullDatasetName =fileInfo.Name ;
        IWorkspaceFactory workspaceFactory = new AccessWorkspaceFactory Class();
        IWorkspace workspace = workspaceFactory.OpenFromFile(fileName, 0);
        if (workspace != null)
        {
            m_Workspace = workspace;
            m_WorkspaceList.Add(workspace);
        }
    }
    refreshTree();
}
//选择树控件节点后，在地图控件中显示当前的图层
private void treeView1_AfterSelect(object sender, TreeViewEventArgs e)
{
    TreeNode tempNode = new TreeNode();
    IWorkspace workspace=null;
    IEnumDataset enumDataset = null;
    IDataset dataset = null;
    IFeatureLayer featureLayer = new FeatureLayerClass();
    int i;
    m_CurrentNode = e.Node;
    if (e.Node.Parent != null)
    {
        for ( i = 0; i < m_WorkspaceList.Count; i++)
        {
            workspace = m_WorkspaceList[i] as IWorkspace ;
            if (workspace.PathName.ToString () == e.Node.Parent.Tag.ToString ())
            {
                break ;
            }
        }
        if (i < m_WorkspaceList.Count)
        {
            enumDataset = workspace.get_Datasets(esriDatasetType.esriDTFeature Class);
            enumDataset.Reset();
            dataset = enumDataset.Next();
            while (dataset!=null)
            {
                if (dataset.Name == e.Node.Text)
                {
                    featureLayer.FeatureClass = dataset as IFeatureClass;
                    m_curFeatureLayer = featureLayer;
                    break;
                }
                dataset = enumDataset.Next();
            }
        }
    }

    this.axMapControl1.ClearLayers();
    this.axMapControl1.AddLayer(featureLayer as ILayer);
    this.axMapControl1.Extent = this.axMapControl1.FullExtent;
}
```

```
private void 元数据管理ToolStripMenuItem_Click(object sender, EventArgs e)
{
    schema fSchema = new schema();
    fSchema.CurNode = m_CurrentNode;
    fSchema.CurWorkspace = m_Workspace;
    fSchema.CurFeatureLayer = m_curFeatureLayer;
    fSchema.ShowDialog();
}

private void 数据入库ToolStripMenuItem_Click(object sender, EventArgs e)
{
    importdata fImportdata = new importdata();
    fImportdata.CurWorkspaceList = m_WorkspaceList;
    fImportdata.ShowDialog();
    refreshTree();
}

private void 数据编辑ToolStripMenuItem1_Click(object sender, EventArgs e)
{
    edit pEdit = new edit();
    pEdit.CurFeatureLayer = m_curFeatureLayer;
    pEdit.ShowDialog();
    this.axMapControl1.ActiveView.Refresh();
}
```

创建数据库，根据"工作空间目录"和"数据库名称"，创建个人数据库，如图 12-3 所示。

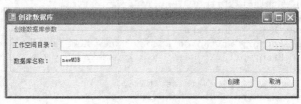

▲图 12-3 创建数据库

创建数据库实现程序如下所示：

```
using System;
using System.Collections.Generic;
using System.ComponentModel;
using System.Data;
using System.Drawing;
using System.Text;
using System.Windows.Forms;
using ESRI.ArcGIS.Geodatabase;
using ESRI.ArcGIS.DataSourcesGDB;
using ESRI.ArcGIS.esriSystem;
namespace GeoManage
{
    public partial class CreateDatabase : Form
    {
        public CreateDatabase()
        {
            InitializeComponent();
        }
        string filePath;

        public string FilePath
        {
            get { return filePath; }
```

```csharp
        set { filePath = value; }
    }
    string fileName;

    public string FileName
    {
        get { return fileName; }
        set { fileName = value; }
    }
    IWorkspace newWorkspace;
    //新建工作空间
    public IWorkspace NewWorkspace
    {
        get { return newWorkspace; }
        set { newWorkspace = value; }
    }
    //浏览工作空间目录
    private void btDirect_Click(object sender, EventArgs e)
    {
        if (this.folderBrowserDialog1.ShowDialog() == DialogResult.OK)
        {
            this.txtDirect.Text = this.folderBrowserDialog1.SelectedPath;
        }
    }
    //创建数据库
    private void btCreate_Click(object sender, EventArgs e)
    {

        filePath = this.txtDirect.Text.ToString().Trim();
        fileName = this.txtDbName.Text.ToString().Trim();
        if (filePath == "") return;
        IWorkspaceFactory workspaceFactory = new AccessWorkspaceFactoryClass();
        IWorkspaceName workspaceName = workspaceFactory.Create(filePath, fileName, null, 0);
        IName name = workspaceName as IName;
        IWorkspace workspace = (IWorkspace)name.Open();
        newWorkspace = workspace;
        this.Close();
        this.Dispose();
    }
    private void btCancel_Click(object sender, EventArgs e)
    {
        this.Close();
        this.Dispose();
    }
}
```

导入数据，根据选择的 Shape 文件数据导入当前工作空间中，系统默认导入空间数据库中的"数据集名"为 Shape 文件名，如图 12-4 所示。

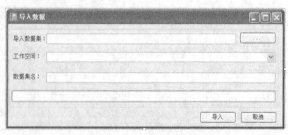

▲图 12-4 导入数据

导入数据实现程序如下：

```csharp
using System;
using System.IO;
using System.Collections.Generic;
using System.Collections;
using System.ComponentModel;
using System.Data;
using System.Drawing;
using System.Text;
using System.Windows.Forms;
using ESRI.ArcGIS.Geodatabase;
using ESRI.ArcGIS.Carto;
using ESRI.ArcGIS.DataSourcesFile;
using ESRI.ArcGIS.esriSystem;
namespace GeoManage
{
    public partial class importdata : Form
    {
        ArrayList curWorkspaceList;
        //当前工作空间列表
        public ArrayList CurWorkspaceList
        {
            get { return curWorkspaceList; }
            set { curWorkspaceList = value; }
        }

        public importdata()
        {
            InitializeComponent();
        }
        //选择 SHAPE 数据文件
        private void btFName_Click(object sender, EventArgs e)
        {
            this.openFileDialog1.Title = "shp 文件";
            this.openFileDialog1.Filter = "shp(*.shp)|*.shp";
            if (this.openFileDialog1.ShowDialog() == DialogResult.OK)
            {
                string fileName = this.openFileDialog1.FileName;
                this.txtImportData.Text = fileName;
                FileInfo fileInfo = new FileInfo(fileName);
                this.txtNewName.Text = fileInfo.Name.Substring(0, fileInfo.Name.Length - 4);
            }
        }
        //初始化 Combox 控件，将工作空间列表中的工作空间名作为子项
        private void importdata_Load(object sender, EventArgs e)
        {
            IWorkspace workspace;
            for (int i = 0; i < curWorkspaceList.Count; i++)
            {
                workspace = curWorkspaceList[i] as IWorkspace;
                this.cbWorkspace.Items.Add(workspace.PathName);
            }
            this.cbWorkspace.Text = this.cbWorkspace.Items[0].ToString();
        }
        //退出窗体
        private void btCancel_Click(object sender, EventArgs e)
        {
            this.Close();
```

```csharp
        this.Dispose();
    }
    //导入空间数据
    private void btSave_Click(object sender, EventArgs e)
    {
        if (txtNewName.Text.Trim() == "")
            return;
        IWorkspace workspace;
        int index=-1;
        for (int i = 0; i < curWorkspaceList.Count; i++)
        {
            workspace = curWorkspaceList[i] as IWorkspace;
            if (workspace.PathName == this.cbWorkspace.Text.Trim())
            {
                index = i;
                break;
            }
        }
        if (index >= 0)
        {
            workspace = curWorkspaceList[index] as IWorkspace ;
            //shp
            IWorkspaceFactory workspaceFactory = new ShapefileWorkspaceFactory Class();
            IWorkspace shpWorkspace;
            FileInfo fileInfo = new FileInfo(this.txtImportData.Text);
            shpWorkspace = workspaceFactory.OpenFromFile(fileInfo.DirectoryName , 0);
            IFeatureClass shpFeatureClass ;
            IFeatureWorkspace shpFeatureWorkspace = shpWorkspace as Ifeature Workspace;
            shpFeatureClass= shpFeatureWorkspace.OpenFeatureClass(fileInfo.Name);

            //workspace
            UID pUID=new UIDClass ();
            IFeatureWorkspace featureworkspace = workspace as IFeatureWorkspace;

            switch (shpFeatureClass.FeatureType )
            {
                case  esriFeatureType.esriFTSimple :
                    pUID.Value ="esriGeoDatabase.Feature";
                    break ;
                case  esriFeatureType.esriFTSimpleJunction  :
                    pUID.Value ="esriGeoDatabase.SimpleJunctionFeature";
                    break ;
                case  esriFeatureType.esriFTComplexJunction  :
                    pUID.Value ="esriGeoDatabase.ComplexJunctionFeature";
                    break ;
                case  esriFeatureType.esriFTSimpleEdge   :
                    pUID.Value ="esriGeoDatabase.SimpleEdgeFeature";
                    break ;
                case  esriFeatureType.esriFTComplexEdge  :
                    pUID.Value ="esriGeoDatabase.ComplexEdgeFeature";
                    break ;
            }
            IFeatureClass newFeatureClass;
            //创建新数据集
            newFeatureClass= featureworkspace.CreateFeatureClass(this.txtNewName.
            Text, shpFeatureClass.Fields, pUID, null, shpFeatureClass.FeatureType,
            "shape", "");

            int featureCount = shpFeatureClass.FeatureCount(null);

            this.pbImport.Minimum  = 1;
            this.pbImport.Maximum = featureCount;
```

```csharp
            IFeature feature;
            IWorkspaceEdit workspaceEdit;
            IFeatureCursor featureCursor,insertFeatureCursor;
            workspaceEdit = workspace as IWorkspaceEdit;
            //开始编辑
            workspaceEdit.StartEditing(true);
            workspaceEdit.StartEditOperation();
            //导入数据记录
            featureCursor = shpFeatureClass.Search(null, true);
            feature = featureCursor.NextFeature();
            insertFeatureCursor = newFeatureClass.Insert(false );
            int count = 0;
            while (feature != null)
            {
                count++;
                this.pbImport.Value = count;

                insertFeatureCursor.InsertFeature(feature as IFeatureBuffer);
                feature = featureCursor.NextFeature();
            }

            //结束编辑
            workspaceEdit.StopEditOperation();
            workspaceEdit.StopEditing(true);
            this.Close();
            this.Dispose();
        }
    }
}
```

元数据管理，提供了元数据增加、删除以及更改的功能，如图 12-5 所示。

▲图 12-5　元数据管理

元数据管理实现程序如下：

```csharp
using System;
using System.Collections.Generic;
using System.ComponentModel;
using System.Data;
using System.Drawing;
```

```csharp
using System.Text;
using System.Windows.Forms;
using ESRI.ArcGIS.Geodatabase;
using ESRI.ArcGIS.Carto;

namespace GeoManage
{
    public partial class schema : Form
    {
        TreeNode curNode;

        public TreeNode CurNode
        {
            get { return curNode; }
            set { curNode = value; }
        }
        IWorkspace curWorkspace;

        public IWorkspace CurWorkspace
        {
            get { return curWorkspace; }
            set { curWorkspace = value; }
        }

        IFeatureLayer curFeatureLayer;

        public IFeatureLayer CurFeatureLayer
        {
            get { return curFeatureLayer; }
            set { curFeatureLayer = value; }
        }
        public schema()
        {
            InitializeComponent();
        }

        private void schema_Load(object sender, EventArgs e)
        {
            refreshListBox();
        }
        //刷新字段列表
        private void refreshListBox()
        {
            IFields fields;
            IField field;
            IFieldEdit fieldEdit;
            fields = curFeatureLayer.FeatureClass.Fields;
            this.lbField.Items.Clear();
            for (int i = 0; i < fields.FieldCount; i++)
            {
                field = fields.get_Field(i);
                if (field.Type == esriFieldType.esriFieldTypeGUID || field.Type ==
                esriFieldType.esriFieldTypeOID || field.Type == esriFieldType.
                esriFieldTypeGeometry || field.Type == esriFieldType.esriFieldTypeBlob)
                {

                }
                else if (field.Name.ToLower() == "shape_length" || field.Name.ToLower() ==
                "shape_area" || field.Name.ToLower() == "id")
                {

                }
```

```csharp
            else
            {
                this.lbField.Items.Add(field.Name);
            }
        }
    }
    //保存新增、更新字段
    private void btSaveField_Click(object sender, EventArgs e)
    {
        IFields fields = curFeatureLayer.FeatureClass.Fields;

        esriFieldType  strType;
        IField  field =new FieldClass ();
        IFieldEdit2 fieldEdit;
        int findIndex =-1;
        findIndex =curFeatureLayer.FeatureClass.Fields.FindField(this.txtName.Text);

        if (findIndex >=0)
        {
            field = curFeatureLayer.FeatureClass.Fields.get_Field(findIndex);

        }
        fieldEdit = field as IFieldEdit2;
        fieldEdit.Name_2 = this.txtName.Text;
        switch (this.cbType.Text)
        {
            case "字符串":
                fieldEdit.Type_2 = esriFieldType.esriFieldTypeString;
                fieldEdit.Length_2 =int.Parse ( this.txtPrecision.Text);
                break;
            case "整型":
                fieldEdit.Type_2 = esriFieldType.esriFieldTypeInteger;
                fieldEdit.Length_2 =int.Parse ( this.txtPrecision.Text);
                break;
            case "单精度":
                fieldEdit.Type_2 = esriFieldType.esriFieldTypeSingle;
                fieldEdit.Precision_2 =int .Parse ( this.txtPrecision.Text);
                break;
            case "双精度":
                fieldEdit.Type_2 = esriFieldType.esriFieldTypeDouble;
                fieldEdit.Precision_2 =int .Parse ( this.txtPrecision.Text);
                break;
            case "日期":
                fieldEdit.Type_2 = esriFieldType.esriFieldTypeDate;
                break;
        }
        if (findIndex >= 0)
        {
            //curFeatureLayer.FeatureClass.Fields.get_Field(findIndex )= field;
        }
        else
        {
            curFeatureLayer.FeatureClass.AddField(field);
        }
        refreshListBox();
    }
    //选择要更改的字段
    private void lbField_DoubleClick(object sender, EventArgs e)
    {
        IFields fields = curFeatureLayer.FeatureClass.Fields;
        IField field;
```

```csharp
            for (int i = 0; i < fields.FieldCount; i++)
            {
                field = fields.get_Field(i);
                if (field.Name == lbField.Items[lbField.SelectedIndex].ToString())
                {
                    txtName.Text = field.Name;
                    switch (field.Type)
                    {
                        case esriFieldType.esriFieldTypeString:
                            this.cbType.Text = "字符串";
                            this.txtPrecision.Text = field.Length.ToString();
                            break;
                        case esriFieldType.esriFieldTypeInteger:
                            this.cbType.Text = "整型";
                            this.txtPrecision.Text = field.Length.ToString();
                            break;
                        case esriFieldType.esriFieldTypeSingle:
                            this.cbType.Text = "单精度";
                            this.txtPrecision.Text = field.Precision.ToString();
                            break;
                        case esriFieldType.esriFieldTypeDouble:
                            this.cbType.Text = "双精度";
                            this.txtPrecision.Text = field.Precision.ToString();
                            break;
                        case esriFieldType.esriFieldTypeDate:
                            this.cbType.Text = "日期";
                            break;
                    }
                }
            }
        }

        private void cbType_SelectedIndexChanged(object sender, EventArgs e)
        {
            txtPrecision.Enabled = true;
            switch (this.cbType.Text)
            {
                case "字符串":
                    txtPrecision.Text = "255";
                    break;
                case "整型":
                    txtPrecision.Text = "8";
                    break;
                case "单精度":
                    txtPrecision.Text = "12";
                    break;
                case "双精度":
                    txtPrecision.Text = "38";
                    break;
                case "日期":
                    txtPrecision.Enabled = false;
                    break;
            }
        }
        //删除字段
        private void btDelField_Click(object sender, EventArgs e)
        {
            IFields fields = curFeatureLayer.FeatureClass.Fields;
            IField field = new FieldClass();
```

```
            int findIndex = -1;
            findIndex = curFeatureLayer.FeatureClass.Fields.FindField(this.txtName.Text);

            if (findIndex >= 0)
            {
                field = curFeatureLayer.FeatureClass.Fields.get_Field(findIndex);

            }

            if (findIndex >= 0)
            {
                curFeatureLayer.FeatureClass.DeleteField (field);
            }
            refreshListBox();
        }

        private void btCancel_Click(object sender, EventArgs e)
        {
            this.Close();
            this.Dispose();
        }


    }
}
```

数据编辑，提供了点、线、面的添加、删除、移动和复制等功能，如图 12-6 所示。

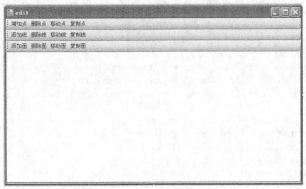

▲图 12-6　数据编辑

数据编辑实现程序如下：

```
using System;
using System.Collections.Generic;
using System.ComponentModel;
using System.Data;
using System.Drawing;
using System.Text;
using System.Windows.Forms;
using ESRI.ArcGIS.Geodatabase;
using ESRI.ArcGIS.Carto;
using ESRI.ArcGIS.Geometry;
using ESRI.ArcGIS.Display;
using ESRI.ArcGIS.Controls;

namespace GeoManage
{
    public partial class edit : Form
```

```csharp
{
    public edit()
    {
        InitializeComponent();
    }

    IFeatureLayer curFeatureLayer;

    public IFeatureLayer CurFeatureLayer
    {
        get { return curFeatureLayer; }
        set { curFeatureLayer = value; }
    }
    string geoType;
    string opType;
    IActiveView m_ActiveView;
    IFeature m_Feature;
    IPointCollection curPointCollection;
    IGeometry m_Geometry=null ;
    //当前点移动反馈对象
    IMovePointFeedback m_MovePointFeedback = new MovePointFeedbackClass();
    //当前线移动反馈对象
    IMoveLineFeedback m_MoveLineFeedback = new MoveLineFeedbackClass();
    //当前面移动反馈对象
    IMovePolygonFeedback m_MovePolygonFeedback = new MovePolygonFeedback Class();

    INewLineFeedback m_NewLineFeedback = null;
    INewPolygonFeedback m_NewPolygonFeedback = null;

    private void edit_Load(object sender, EventArgs e)
    {
        IFeatureClass featureClass;
        featureClass = curFeatureLayer.FeatureClass;
        System.Drawing.Point point = new System.Drawing.Point(0, 0);
        if (featureClass.ShapeType == esriGeometryType.esriGeometryPoint)
        {
            this.tsPoint.Visible = true;
            this.tsPoint.Location = point;
            this.tsLine.Visible = false;
            this.tsPolygon.Visible = false;
            geoType = "point";
        }
        else if (featureClass.ShapeType == esriGeometryType.esriGeometryPolyline )
        {
            this.tsPoint.Visible = false;
            this.tsLine.Visible = true;
            this.tsLine.Location = point;
            this.tsPolygon.Visible = false;
            geoType = "polyline";

        }
        else if (featureClass.ShapeType == esriGeometryType.esriGeometryPolygon )
        {
            this.tsPoint.Visible = false;
            this.tsLine.Visible = false;
            this.tsPolygon.Visible = true;
            this.tsPolygon.Location = point;
            geoType = "polygon";
        }
        this.axMapControl1.Map.AddLayer(curFeatureLayer as ILayer);
        m_ActiveView = this.axMapControl1.ActiveView;
    }
```

```csharp
//添加新实体
private void addFeature(IGeometry geometry)
{
    IWorkspaceEdit workspaceEdit;
    IFeatureCursor insertFeatureCursor;
    IDataset dataset =curFeatureLayer.FeatureClass as IDataset ;
    workspaceEdit = dataset.Workspace as IWorkspaceEdit;
    try
    {
        //开始编辑
        workspaceEdit.StartEditing(true);
        workspaceEdit.StartEditOperation();

        IFeature feature = curFeatureLayer.FeatureClass.CreateFeature();
        feature.Shape = geometry;
        feature.Store();

        //结束编辑
        workspaceEdit.StopEditOperation();
        workspaceEdit.StopEditing(true);
    }
    catch (Exception ex)
    {
    }
    insertFeatureCursor = null;
    workspaceEdit = null;
}
//添加新实体
private void addFeature(IGeometry geometry, IFeature copyFeature)
{
    IWorkspaceEdit workspaceEdit;
    IFeatureCursor insertFeatureCursor;
    IDataset dataset = curFeatureLayer.FeatureClass as IDataset;
    workspaceEdit = dataset.Workspace as IWorkspaceEdit;
    try
    {
        //开始编辑
        workspaceEdit.StartEditing(true);
        workspaceEdit.StartEditOperation();

        IFeature feature = curFeatureLayer.FeatureClass.CreateFeature();
        if (curFeatureLayer.FeatureClass.ShapeType == esriGeometryType.esriGeometryPoint)
        {
            feature.Shape = geometry;
        }
        else if (curFeatureLayer.FeatureClass.ShapeType == esriGeometryType.esriGeometryPolyline)
        {
            feature.Shape = geometry;
        }
        else if (curFeatureLayer.FeatureClass.ShapeType == esriGeometryType.esriGeometryPolygon)
        {
            feature.Shape = geometry;
        }
        //添加属性字段值
        int index1=-1, index2 = -1;
        IFields fields = copyFeature.Fields;
        IField field ;

        for (int i = 0; i < fields.FieldCount ; i++)
```

```csharp
            {
                field = fields.get_Field(i);
                if (field.Type != esriFieldType.esriFieldTypeGUID && field. Type !=
                esriFieldType.esriFieldTypeOID && field.Type != esriFieldType.
                esriFieldTypeGeometry && field.Type != esriFieldType.esriFieldTypeBlob
                && field.Name.ToLower() != "shape_ length" && field.Name.ToLower() !=
                "shape_area")
                {
                    index1 = -1;
                    index1 = feature.Fields.FindField(field.Name);
                    feature.set_Value (index1 ,copyFeature.get_Value (i));
                }
            }
            feature.Store();

            //结束编辑
            workspaceEdit.StopEditOperation();
            workspaceEdit.StopEditing(true);
        }
        catch (Exception ex)
        {
        }
        insertFeatureCursor = null;
        workspaceEdit = null;
    }
    //转换像素到地图单位
    private double ConvertPixelsToMapUnits(IActiveView pActiveView, double pixelUnits)
    {
        tagRECT pRect = pActiveView.ScreenDisplay.DisplayTransformation.get_ Device
        Frame();
        int pixelExtent = pRect.right - pRect.left;

        double realWorldDisplayExtent = pActiveView.ScreenDisplay.Display
        Transformation.VisibleBounds.Width;
        double sizeOfOnePixel = realWorldDisplayExtent / pixelExtent;
        return pixelUnits * sizeOfOnePixel;
    }
    //删除实体
    private void delFeature(IPoint point )
    {
        IFeatureCursor featureCursor = selectFeature(point);
        IFeature feature;
        feature = featureCursor.NextFeature();
        IWorkspaceEdit workspaceEdit;
        IDataset dataset = curFeatureLayer.FeatureClass as IDataset;
        workspaceEdit = dataset.Workspace as IWorkspaceEdit;
        try
        {
            //开始编辑
            workspaceEdit.StartEditing(true);
            workspaceEdit.StartEditOperation();
            while (feature != null)
            {
                //删除实体
                feature.Delete();
                feature = null;
                feature = featureCursor.NextFeature();

            }
            featureCursor.Flush();
            //结束编辑
            workspaceEdit.StopEditOperation();
```

```csharp
            workspaceEdit.StopEditing(true);
        }
        catch (Exception ex)
        {
        }
    }
    //保存实体
    private void saveFeature(IFeature feature)
    {
        IWorkspaceEdit workspaceEdit;
        IWorkspace workspace;
        IDataset dataset = curFeatureLayer.FeatureClass as IDataset;
        workspace = dataset.Workspace;
        workspaceEdit = workspace as IWorkspaceEdit;
        //开始编辑
        workspaceEdit.StartEditing(true);
        workspaceEdit.StartEditOperation();
        //保存实体
        m_Feature.Store();
        //结束编辑
        workspaceEdit.StopEditOperation();
        workspaceEdit.StopEditing(true);
        m_MovePointFeedback = null;
        m_MoveLineFeedback = null;
        m_MovePolygonFeedback = null;
    }
    //选择实体
    private IFeatureCursor  selectFeature(IPoint point)
    {
        ITopologicalOperator topologicalOperator;
        topologicalOperator = point as ITopologicalOperator;
        IGeometry geometry;
        double bufferLength =ConvertPixelsToMapUnits (m_ActiveView ,8);

        geometry = topologicalOperator.Buffer(bufferLength);

        ISpatialFilter spatialFilter = new SpatialFilterClass();
        IFeatureSelection feaureSelection;
        spatialFilter.SpatialRel = esriSpatialRelEnum. esriSpatialRelEnvelopeIntersects;
        spatialFilter.Geometry = geometry;
        spatialFilter.GeometryField = curFeatureLayer.FeatureClass. ShapeFieldName;
        spatialFilter.WhereClause = "";
        IFeatureCursor featureCursor ;
        featureCursor =curFeatureLayer.FeatureClass.Search (spatialFilter , false );
        return featureCursor;
    }
    private void axMapControl1_OnMouseDown(object sender, ESRI.ArcGIS.Controls.IMapControlEvents2_OnMouseDownEvent e)
    {
        IPoint point = m_ActiveView.ScreenDisplay.DisplayTransformation.ToMapPoint(e.x, e.y);
        IGeometry geometry;
        object missing1 = Type.Missing;
        object missing2 = Type.Missing;
        IFeatureCursor featureCursor;
        IFeature feature;
        IRgbColor rgbColor;
        ISimpleMarkerSymbol simpleMarkerSymbol;
        IMarkerElement markerElement;
        IEnvelope envelope;
```

```csharp
            IElement element;
            IGraphicsContainer graphicsContainer;

            switch (opType)
            {
                case "addpoint":
                    geometry = point as IGeometry;
                    addFeature(geometry);
                    break;
                case "addline":
                    if (m_NewLineFeedback == null)
                    {
                        m_NewLineFeedback = new NewLineFeedbackClass();
                        m_NewLineFeedback.Display = m_ActiveView.ScreenDisplay;
                        m_NewLineFeedback.Start(point);
                    }
                    else
                    {
                        m_NewLineFeedback.AddPoint(point);
                        m_NewLineFeedback.Refresh(0); ;
                    }
                    //设置结点符号
                    rgbColor = new RgbColorClass();
                    rgbColor.Red = 255;
                    rgbColor.Green = 0;
                    rgbColor.Blue = 0;

                    simpleMarkerSymbol = new SimpleMarkerSymbolClass();
                    simpleMarkerSymbol.Color = rgbColor as IColor;
                    simpleMarkerSymbol.Size = 4;
                    simpleMarkerSymbol.Style = esriSimpleMarkerStyle.esriSMSSquare;

                    markerElement = new MarkerElementClass();
                    markerElement.Symbol = simpleMarkerSymbol as IMarkerSymbol;
                    element = markerElement as IElement;
                    element.Geometry = point as IGeometry;

                    //获取结点 element 的范围，以便确定刷新范围
                    envelope = new EnvelopeClass();
                    element.QueryBounds(m_ActiveView.ScreenDisplay as IDisplay, envelope);

                    graphicsContainer = m_ActiveView as IGraphicsContainer;
                    graphicsContainer.AddElement(element, 0);
                    m_ActiveView.PartialRefresh(esriViewDrawPhase.esriViewGraphics,
                    element, envelope);
                    break ;
                case "addpolygon":
                    if (m_NewPolygonFeedback  == null)
                    {
                        m_NewPolygonFeedback = new NewPolygonFeedbackClass ();
                        m_NewPolygonFeedback.Display = m_ActiveView.ScreenDisplay;
                        m_NewPolygonFeedback.Start(point);
                    }
                    else
                    {
                        m_NewPolygonFeedback.AddPoint(point);
                        m_NewPolygonFeedback.Refresh(0); ;
                    }
                    //设置结点符号
                    rgbColor = new RgbColorClass();
                    rgbColor.Red = 255;
                    rgbColor.Green = 0;
```

```csharp
                rgbColor.Blue = 0;

                simpleMarkerSymbol = new SimpleMarkerSymbolClass();
                simpleMarkerSymbol.Color = rgbColor as IColor;
                simpleMarkerSymbol.Size = 4;
                simpleMarkerSymbol.Style = esriSimpleMarkerStyle.esriSMSSquare;

                markerElement = new MarkerElementClass();
                markerElement.Symbol = simpleMarkerSymbol as IMarkerSymbol;
                element = markerElement as IElement;
                element.Geometry = point as IGeometry;

                //获取结点 element 的范围,以便确定刷新范围
                envelope = new EnvelopeClass();
                element.QueryBounds(m_ActiveView.ScreenDisplay as IDisplay, envelope);

                graphicsContainer = m_ActiveView as IGraphicsContainer;
                graphicsContainer.AddElement(element, 0);
                m_ActiveView.PartialRefresh(esriViewDrawPhase.esriViewGraphics,
                    element, envelope);
                break;
            case "delpoint":

                delFeature(point);
                break;
            case "delline":
                delFeature(point);
                break;
            case "delpolygon":
                delFeature(point);
                break;
            case "movepoint":
                featureCursor = selectFeature(point);
                if (featureCursor != null)
                {
                    m_Feature = featureCursor.NextFeature();
                    if (m_Feature != null)
                    {
                        //设置显示对象,并启动移动
                        m_MovePointFeedback.Display = m_ActiveView.ScreenDisplay;
                        m_MovePointFeedback.Start(m_Feature.Shape as IPoint, point);
                    }
                }
                break;
            case "moveline":
                 featureCursor = selectFeature(point);
                if (featureCursor != null)
                {
                    m_Feature = featureCursor.NextFeature();
                    if (m_Feature != null)
                    {
                        //设置显示对象,并启动移动
                        m_MoveLineFeedback.Display = m_ActiveView.ScreenDisplay;
                        m_MoveLineFeedback.Start(m_Feature.Shape as IPolyline, point);
                    }
                }
                break;
            case "movopolygon":
                featureCursor = selectFeature(point);
                if (featureCursor != null)
                {
                    m_Feature = featureCursor.NextFeature();
```

```csharp
            if (m_Feature != null)
            {
                //设置显示对象,并启动移动
                m_MovePolygonFeedback.Display = m_ActiveView.ScreenDisplay;
                m_MovePolygonFeedback.Start(m_Feature.Shape as IPolygon, point);
            }
        }
        break;
    case "copypoint":
        featureCursor = selectFeature(point);
        if (featureCursor != null)
        {
            feature = featureCursor.NextFeature();
            if (feature != null)
            {
                m_Feature = feature;
                this.axMapControl1.Map.SelectFeature(curFeatureLayer, feature);
            }
        }
        break;
    case "copyline":
        featureCursor = selectFeature(point);
        if (featureCursor != null)
        {
            m_Feature = featureCursor.NextFeature();
            if (m_Feature != null)
            {
                //设置显示对象,并启动移动
                m_MoveLineFeedback.Display = m_ActiveView.ScreenDisplay;
                m_MoveLineFeedback.Start(m_Feature.Shape as IPolyline, point);
            }
        }
        break;
    case "copypolygon":
        featureCursor = selectFeature(point);
        if (featureCursor != null)
        {
            m_Feature = featureCursor.NextFeature();
            if (m_Feature != null)
            {
                //设置显示对象,并启动移动
                m_MovePolygonFeedback.Display = m_ActiveView.ScreenDisplay;
                m_MovePolygonFeedback.Start(m_Feature.Shape as IPolygon, point);
            }
        }
        break;
    }
}

private void axMapControl1_OnMouseMove(object sender, ESRI.ArcGIS.Controls.IMapControlEvents2_OnMouseMoveEvent e)
{
    IGeometry geometry;
    //将当前鼠标位置的点转换为地图上的坐标
    IPoint point = m_ActiveView.ScreenDisplay.DisplayTransformation.ToMapPoint(e.x, e.y);
    switch (opType)
    {
        case "addline":
            if (m_NewLineFeedback != null)
            {
```

```csharp
                m_NewLineFeedback.MoveTo(point);
            }
            break;
        case "addpolygon":
            if (m_NewPolygonFeedback != null)
            {
                m_NewPolygonFeedback.MoveTo(point );
            }
            break;

        case "movepoint":
            if (m_MovePointFeedback != null)
            {
                //移动对象到当前鼠标光标位置
                m_MovePointFeedback.MoveTo(point);
            }
            break;
        case "moveline":
            if (m_MoveLineFeedback != null)
            {
                //移动对象到当前鼠标光标位置
                m_MoveLineFeedback.MoveTo(point);
            }
            break;
        case "movepolygon":
            if (m_MovePolygonFeedback != null)
            {

                //移动对象到当前鼠标光标位置
                m_MovePolygonFeedback.MoveTo(point);
            }
            break;
        case "copyline":
            if (m_MoveLineFeedback != null)
            {
                //移动对象到当前鼠标光标位置
                m_MoveLineFeedback.MoveTo(point);
            }
            break;
        case "copypolygon":
            if (m_MovePolygonFeedback != null)
            {

                //移动对象到当前鼠标光标位置
                m_MovePolygonFeedback.MoveTo(point);
            }
            break;


    }
}

private void axMapControl1_OnMouseUp(object sender, ESRI.ArcGIS.Controls.
IMapControlEvents2_OnMouseUpEvent e)
{
    IGeometry geometry;
    IGeometry resultGeometry;
    IPoint point=new PointClass ();
    switch (opType)
    {
```

```csharp
            case "movepoint":
                if (m_MovePointFeedback != null)
                {
                    //停止移动
                    resultGeometry = m_MovePointFeedback.Stop() as IGeometry;
                    m_Feature.Shape = resultGeometry;
                    saveFeature(m_Feature);
                }
                break;
            case "moveline":
                if (m_MoveLineFeedback != null)
                {
                    //停止移动
                    resultGeometry = m_MoveLineFeedback.Stop() as IGeometry;
                    m_Feature.Shape = resultGeometry;
                    saveFeature(m_Feature);
                }
                break;
            case "movepolygon":
                if (m_MovePolygonFeedback != null)
                {
                    //停止移动
                    resultGeometry = m_MovePolygonFeedback.Stop() as IGeometry;
                    m_Feature.Shape = resultGeometry;
                    saveFeature(m_Feature);
                }
                break;
            case "copyline":
                if (m_MoveLineFeedback != null)
                {
                    //停止移动
                    resultGeometry = m_MoveLineFeedback.Stop() as IGeometry;

                    addFeature(resultGeometry, m_Feature);
                }
                break;
            case "copypolygon":
                if (m_MovePolygonFeedback != null)
                {
                    //停止移动
                    resultGeometry = m_MovePolygonFeedback.Stop() as IGeometry;

                    addFeature(resultGeometry, m_Feature);
                }
                break;
        }
        m_ActiveView.Refresh();
    }

    private void tsbAddPoint_Click(object sender, EventArgs e)
    {
        opType = "addpoint";
    }

    private void tsbDelPoint_Click(object sender, EventArgs e)
    {
        opType = "delpoint";
    }

    private void tsbMovePoint_Click(object sender, EventArgs e)
```

```csharp
    opType = "movepoint";
    m_MovePointFeedback = new MovePointFeedbackClass();
}

private void tsbCopyPoint_Click(object sender, EventArgs e)
{
    opType = "copypoint";
}

private void tsbAddLine_Click(object sender, EventArgs e)
{
    opType = "addline";
}

private void tsbDelLine_Click(object sender, EventArgs e)
{
    opType = "delline";
}

private void tsbMoveLine_Click(object sender, EventArgs e)
{
    opType = "moveline";
    m_MoveLineFeedback = new MoveLineFeedbackClass();
}

private void tsbCopyLine_Click(object sender, EventArgs e)
{
    opType = "copyline";
}

private void tsbAddPolygon_Click(object sender, EventArgs e)
{
    opType = "addpolygon";
}

private void tsbDelPolygon_Click(object sender, EventArgs e)
{
    opType = "delpolygon";
}

private void tsbMovePolygon_Click(object sender, EventArgs e)
{
    opType = "movepolygon";
    m_MovePolygonFeedback = new MovePolygonFeedbackClass();
}

private void tsbCopyPolygon_Click(object sender, EventArgs e)
{
    opType = "copypolygon";
}

private void tsPoint_ItemClicked(object sender, ToolStripItemClickedEventArgs e)
{
    ToolStripButton tsButton ;
    foreach (ToolStripItem tsItem in tsPoint.Items)
    {
        if (tsItem.Equals(e.ClickedItem))
        {
            tsButton = tsItem as ToolStripButton;
```

```csharp
                    tsButton.CheckState = CheckState.Checked;
                }
                else
                {
                    tsButton = tsItem as ToolStripButton;
                    tsButton.CheckState = CheckState.Unchecked;
                }
            }
        }

        private void tsLine_ItemClicked(object sender, ToolStripItemClickedEventArgs e)
        {
            ToolStripButton tsButton;
            foreach (ToolStripItem tsItem in tsLine.Items)
            {
                if (tsItem.Equals(e.ClickedItem))
                {
                    tsButton = tsItem as ToolStripButton;
                    tsButton.CheckState = CheckState.Checked;
                }
                else
                {
                    tsButton = tsItem as ToolStripButton;
                    tsButton.CheckState = CheckState.Unchecked;
                }
            }
        }

        private void tsPolygon_ItemClicked(object sender, ToolStripItemClickedEventArgs e)
        {
            ToolStripButton tsButton;
            foreach (ToolStripItem tsItem in tsPolygon.Items)
            {
                if (tsItem.Equals(e.ClickedItem))
                {
                    tsButton = tsItem as ToolStripButton;
                    tsButton.CheckState = CheckState.Checked;
                }
                else
                {
                    tsButton = tsItem as ToolStripButton;
                    tsButton.CheckState = CheckState.Unchecked;
                }
            }
        }

        private void axMapControl1_OnDoubleClick(object sender, ESRI.ArcGIS. Controls. ImapControlEvents2_OnDoubleClickEvent e)
        {
            IPoint point = m_ActiveView.ScreenDisplay.DisplayTransformation.ToMapPoint(e.x, e.y);
            IGeometry resultGeometry;
            IGroupElement groupElement;
            ISimpleLineSymbol simpleLineSymbol;
            IRgbColor rgbColor;
            IElement element;
            switch (opType)
            {
                case "copypoint":
```

```csharp
            if (m_Feature != null)
            {
                addFeature(point as IGeometry ,m_Feature );
            }
            break;
    case "addline":
        groupElement = new GroupElementClass();

        resultGeometry = m_NewLineFeedback.Stop();
        addFeature(resultGeometry);
        m_NewLineFeedback = null;
        IPolyline polyline = resultGeometry as IPolyline;
        //设置线型
        rgbColor = new RgbColorClass();
        rgbColor.Red = 0;
        rgbColor.Green = 120;
        rgbColor.Blue = 0;

        simpleLineSymbol = new SimpleLineSymbolClass();
        simpleLineSymbol.Color = rgbColor as IColor;
        simpleLineSymbol.Width = 1.5;
        simpleLineSymbol.Style = esriSimpleLineStyle.esriSLSSolid;

        ILineElement lineElement = new LineElementClass();
        lineElement.Symbol = simpleLineSymbol as ILineSymbol;
        element = lineElement as IElement;
        element.Geometry = polyline as IGeometry;
        groupElement.AddElement(element);

        break;
    case "addpolygon":
        groupElement = new GroupElementClass();

        resultGeometry = m_NewPolygonFeedback.Stop();
        addFeature(resultGeometry);
        m_NewPolygonFeedback = null;
        IPolygon  polygon = resultGeometry as IPolygon ;
        //设置线型
        rgbColor = new RgbColorClass();
        rgbColor.Red = 0;
        rgbColor.Green = 120;
        rgbColor.Blue = 0;

        simpleLineSymbol = new SimpleLineSymbolClass();
        simpleLineSymbol.Color = rgbColor as IColor;
        simpleLineSymbol.Width = 1.5;
        simpleLineSymbol.Style = esriSimpleLineStyle.esriSLSSolid;

         IRgbColor rgbColor2 = new RgbColorClass();
        rgbColor2.Red = 0;
        rgbColor2.Green = 120;
        rgbColor2.Blue = 0;

        ISimpleFillSymbol simpleFillSymbol = new SimpleFillSymbolClass();
        simpleFillSymbol.Outline = simpleLineSymbol;
        simpleFillSymbol.Style = esriSimpleFillStyle.esriSFSCross;
        simpleFillSymbol.Color = rgbColor2;

        IPolygonElement polygonElement = new PolygonElementClass();
        element = polygonElement as IElement;
```

```
                    element.Geometry = polygon as IGeometry;
                    groupElement.AddElement(element);
                    break;
            }
            this.axMapControl1.Map.ClearSelection();
            m_ActiveView.Refresh();
        }
    }
}
```

12.4 本章小结

本章用一个综合示例，综合使用了数据库、数据编辑相关的知识点，读者可以通过本示例，加强对这些接口、类、方法、属性掌握。

第四篇

常见疑难解答与经验技巧集萃

第 13 章　空间数据库连接与释放
第 14 章　空间数据库加载
第 15 章　程序出错和异常
第 16 章　其他经验技巧

　　本篇介绍一些开发过程中常见的异常、数据库连接与释放、数据加载,以及一些经验技巧。本篇的例子主要是笔者开发过程中经常碰到的问题,以及网上一些常见的提问。

第 13 章　空间数据库连接与释放

ArcEngine 空间数据库连接 arcEngine 可以接受多种数据源，如企业数据库、个人数据库、Shapefile 文件、AutoCAD dwg 文件以及影像图文件等。

13.1　Shapefile 文件

Shapefile 是文件型的空间数据格式，以文件的形式在磁盘上存储空间数据和属性数据。下面的示例代码是打开位于 D:\Data 文件夹下的文件名为 Cities 的 Shapefile 要素类。对于 Shapefile 来说，工作空间就是它所在的文件夹，打开工作空间需要使用对应的工作空间工厂，即 ShapefileWorkspaceFactoryClass，然后调用 IWorkspaceFactory 的 OpenFromFile 方法，就可以得到一个工作空间，这也是设计模式中工厂方法的体现。工作空间工厂的打开方法返回的是一般意义的工作空间，根据具体数据还需要进行接口转换。因为 Shapefile 是矢量数据，所以把工作空间接口跳转到 IFeatureWorkspace，从而读取其中的要素类。这一点对于接下来的几个数据格式来说，也是同样的打开方式。

```
IWorkspaceFactory pWorkspaceFactory;
pWorkspaceFactory = new ShapefileWorkspaceFactoryClass();
IFeatureWorkspace pFeatWS;
pFeatWS = pWorkspaceFactory.OpenFromFile(@"D:\Data\", 0) as IFeatureWorkspace;
//打开一个要素类
IFeatureClass pFeatureClass = pFeatWS.OpenFeatureClass("Cities");
```

13.2　Coverage 数据格式

Coverage 是 ArcInfo workstation 的原生数据格式。该格式是基于文件夹存储的，这是因为在 windows 资源管理器下，它的空间信息和属性信息是分别存放在两个文件夹里。coverage 是一个非常成功的早期地理数据模型，20 多年来深受用户的欢迎，很多早期的数据都是 coverage 格式的。ESRI 不公开 coverage 的数据格式，但是提供了 coverage 格式转换的一个交换文件（interchange file，即 E00），并公开了数据格式。但是 ESRI 为推广其第三代数据模型 geodatabase，从 ArcGIS 8.3 版本开始，屏蔽了对 coverage 的编辑功能。如果需要使用 coverage 格式的数据，可以安装 ArcInfo workstation，或者将 coverage 数据转换为其他可编辑的数据格式。Coverage 是一个集合，它可以包含一个或多个要素类。Coverage 数据的工作空间也是它所在的文件夹。由于 Coverage 可以包含多个要素类，因此得到工作空间后，在打开具体的要素类时可以使用 "Coverage 名称:要素类名称"，例如下面代码中的 "basin:polygon"。

```
IWorkspaceFactory pFactory = new ArcInfoWorkspaceFactoryClass();
IWorkspace pWorkspace = pFactory.OpenFromFile(@"D:\ArcTutor\TopologyData", 0);
```

```
IFeatureWorkspace pFeatWorkspace = pWorkspace as IFeatureWorkspace;
IFeatureClass pFeatureClass = pFeatWorkspace.OpenFeatureClass("basin:polygon");
```

13.3 Geodatabase 数据格式

Geodatabase 作为 ArcGIS 的原生数据格式，体现了很多第三代地理数据模型的优势。Personal Geodatabase 基于 Microsoft Access 一体化存储空间数据和属性数据。Enterprise Geodatabase 通过大型关系数据库+ArcSDE 实现，ArcSDE 作为中间件，把关系数据库中的普通表转换为空间对象。Personal Geodatabase 数据的工作空间指的是扩展名为 mdb 的文件。以下是打开位于 Monto.mdb 中的 Water 要素类的代码。

```
IWorkspaceFactory pFactory = new AccessWorkspaceFactoryClass();
IWorkspace pWorkspace = pFactory.OpenFromFile(@"D:\ArcTutor\Monto.mdb", 0);
IFeatureWorkspace pFeatWorkspace = pWorkspace as IFeatureWorkspace;
IFeatureClass pFeatureClass = pFeatWorkspace.OpenFeatureClass("Water")
```

13.4 ArcSDE（Enterprise Geodatabase）数据库连接

对应的工作空间为数据库连接，关系数据库是 Oracle 时，连接参数需要 5 个，分别是 SERVER、INSTANCE、USER、PASSWORD 和 VERSION。SERVER 是指服务器的主机名，INSTANCE 是指服务名或端口号，USER 是数据库的用户名，PASSWORD 数据库对应用户的密码，VERSION 是指 Enterprise Geodatabase 多版本机制中的某个版本，默认的一个版本是"SDE.DEFAULT"。如果关系数据库是 SQLServer，那么连接参数还需要 Database 参数。

下面是打开 Enterprise Geodatabase 中 ControlPoint 点要素类的代码，关系数据库为 Oracle9i。

```
IWorkspaceFactory pWorkspaceFactory = new SdeWorkspaceFactoryClass();
IPropertySet propSet = new PropertySetClass();
propSet.SetProperty("SERVER", "actc");
propSet.SetProperty("INSTANCE", "5131");
propSet.SetProperty("USER", "apdm");
propSet.SetProperty("PASSWORD", "apdm");
propSet.SetProperty("VERSION", "SDE.DEFAULT");
IWorkspace pWorkspace = pWorkspaceFactory.Open(propSet, 0);
IFeatureWorkspace pFeatWS = pWorkspace as IFeatureWorkspace;
IFeatureClass pFeatureClass= pFeatWS.OpenFeatureClass("ControlPoint");
```

13.5 TIN 不规则三角网

TIN 全称不规则三角网,也叫不规则三角表面，是采用一系列不规则的三角点来建立表面。例如，每一个采样点有一对（x,y）坐标和一个表面值(z 值)，这些点被一组互不重叠的三角形的边所连接，从而构成一个表面。TIN 数据是空间分析和三维分析重要的数据格式，以文件的形式在磁盘上存储。TIN 的工作空间是所在的文件夹，下面是打开 D:\ArcTutor\3DAnalyst 文件夹下名称为 mal 的 TIN 的代码。

```
IWorkspaceFactory pWSFact = new TinWorkspaceFactoryClass();
IWorkspace pWS = pWSFact.OpenFromFile(@"D:\ArcTutor\3DAnalyst\", 0);
ITinWorkspace pTinWS = pWS as ITinWorkspace;
ITin pTin = pTinWS.OpenTin("mal");
```

13.6 栅格数据

栅格数据也是 GIS 数据中很重要的一部分，ArcGIS 中最常用的文件类型有 GRID、TIFF、ERDAS IMAGE 等，这几种栅格数据的工作空间也是所在的文件夹。要打开栅格数据，需要使用栅格工作空间工厂（RasterWorkspaceFactory），然后使用 IRasterWorkspace 接口的打开栅格数据集的方法，即可打开一个栅格数据集。在打开栅格数据集时，如果数据格式为 ESRI GRID，那么 OpenRasterDataset() 方法的参数为栅格要素集的名称；如果数据格式为 TIFF，那么该方法的参数为完整的文件名，即要加上.tif 扩展名，例如 OpenRasterDataset("hillshade.tif")。下面为打开 GRID 格式的栅格数据的代码。

```
IWorkspaceFactory rasterWorkspaceFactory = new RasterWorkspaceFactoryClass();
IRasterWorkspace rasterWorkspace =
rasterWorkspaceFactory.OpenFromFile(@"D:\data\grid", 0) as IRasterWorkspace;
IRasterDataset rasterDataset= rasterWorkspace.OpenRasterDataset("ca_hillshade");
```

13.7 CAD 数据

对于 CAD 数据，也可以通过 AO 直接访问。访问 CAD 数据的方式与 Coverage 类似，但须注意要使用 CAD 的工作空间工厂。打开一个 dxf 的 CAD 数据，在打开要素类时须使用"cad 文件名：要素类名称"，注意 cad 文件名要包含扩展名，否则会报错。以下是打开位于 D:\ArcTutor\Editor\ExerciseData\EditingFeatures 文件夹下的 buildings.dxf 中的多边形要素类的代码。

```
IWorkspaceFactory pCadwf = new CadWorkspaceFactoryClass();
IWorkspace pWS =
pCadwf.OpenFromFile(@"D:\ArcTutor\Editor\ExerciseData\EditingFeatures", 0);
IFeatureWorkspace pCadFWS = pWS as IFeatureWorkspace;
IFeatureClass pFeatClass = pCadFWS.OpenFeatureClass("buildings.dxf:polygon");
```

13.8 一般关系表

对于关系表中的数据，也可以通过 ArcGIS 直接读取，这为数据的共享提供了极大的便利。对于一些业务上的非空间数据，通过使用 OLE 方式可以很方便地实现数据的访问，业务数据可以位于各种关系数据库中。以下代码片段演示了如何访问位于 Microsoft Access 中的 Custom 表。当然也可以访问 Oralce 或 SQL Server 中的数据，只要变化一下连接字符串（CONNECTSTRING）即可。

```
//创建一个连接
IPropertySet pPropset;
pPropset = new PropertySetClass();
pPropset.SetProperty("CONNECTSTRING",
@"Provider=Microsoft.Jet.OLEDB.4.0;Data Source=E:\Company.mdb;Persist Security
Info=False");
//创建一个新的OleDB工作空间并打开
IWorkspaceFactory pWorkspaceFact;
IFeatureWorkspace pFeatWorkspace;
pWorkspaceFact = new OLEDBWorkspaceFactoryClass();
pFeatWorkspace = pWorkspaceFact.Open(pPropset, 0) as IFeatureWorkspace;
ITable pTTable = pFeatWorkspace.OpenTable("Custom");
```

以上为 ArcGIS 最常用的几种访问数据的方法，对访问的数据可以进行 GIS 分析、数据处理和空间可视化等操作。在获取到数据以后可以把数据加到图层里，也可以对数据进行检索或维护。

13.9 ArcSDE 客户端负载连接方式

服务器端负载连接（最常用的连接方式）如下。

服务器名称（Server）：SDE 服务器的主机名称。

服务端口（Service）：安装 SDE 时选择的端口，默认是 5131 或 esri_sde。

数据库（Database）：可根据不同的 DBMS 决定是否要填。若是 Oracle 系列不用填，若是 SQL Server 则需要填写。

用户名（UserName）：×××。

密码（password）：×××。

客户端负载连接如下。

服务器名称（Server）：不用填写。

服务端口（Service）：SDE：s 数据库类型。例如是 Oracle9i，则 SDE:Oracle9i。

数据库（Database）：可根据不同的 DBMS 决定是否要填。若是 Oracle 系列不用填，若是 SQLServer 则需要填写。

用户名（UserName）：需要填写。

密码（password）：密码@服务器名称。例如：pwd@server。

两种连接方式的异同：

客户端负载连接是通过 SDE 访问数据表，并在本地完成对数据的各种操作（像空间分析、编辑等）；而服务器端负载连接则是通过 SDE 访问数据表后，在服务器端完成对数据的各种操作，再把操作结果返回客户端。

因此即便服务器上的 SDE 服务没有启动，采用客户端负载连接的方式也可以访问和操作 SDE 数据库，而服务器端负载连接只有在 SDE 服务启动后才能访问和操作 SDE 数据库。ArcCatalog 里面的连接设置和 ArcEngine 开发里面的设置道理一样。

13.10 ArcSDE 连接 Oracle 数据库

ArcSDE 连接 Oracle 数据库有两种连接方式：直接连接和应用服务器连接。无论使用哪一种连接方式，都需要对数据库进行配置。常采用的方法是使用 Oracle 10g 的客户端软件进行配置。

首先，在 Oracle Net Configuration Assistant 中配置服务名，然后在 ArcCatal0g 中进行连接。使用直接连接方式输入用户名的密码时，需要使用以下格式：用户名@网络服务名。通过直接连接客户端，可以和 Oracle 10g 实现连接，而不需要 ArcSDE（专用服务器进程的功能已经在 ArcGIS Desktop 中实现了）。采用这种方式进行连接和访问数据库的速度比较快。而使用应用服务连接，直接通过端口进行操作，在服务器端需要单独开启一个专用服务器管理器进程，这种连接访问数据库的速度相对较慢。直接输入用户名的密码即可，不需要再添加网络服务名。

ArcSDE 是由以下 3 部分组成：ArcSDE 服务器管理进程、专用服务器进程、ArcSDE 客户端。ArcSDE 服务器管理进程，负责维护 ArcSDE 和监听来自客户端的连接请求。ArcSDE 启动就是启动 ArcSDE 服务器管理进程，利用管理员账户管理 ArcSDE 与 RDBMS 连接，处理客户端的连接请求。专用服务器进程，是由 ArcSDE 服务器管理进程创建，用于每一个特定的客户端应用程序与数据库的连接。ArcSDE 客户端，是通过 ArcSDE 服务器管理进程和专用服务器进程建立和 RDBMS 的连接，从而实现对数据库的操作。

13.11 ArcSDE 连接释放

对 ArcSDE 的连接数是有限制的，如果一次进行大量数据的导入，往往会出现各种各样的 SDE ERROR，这类错误很多，如 FDO_E_SE_DB_IO_ERROR 和 FDO_E_SE_OUT_OF_LOCKS 等。

首先分析一下 SDE 的连接机制。当 SDE 安装后，SDE 与 DBMS 之间有一个 giomgr 进程负责管理双方的通信，即根据请求来建立 gsrvr.exe 进程，gsrvr 进程处理与 SDE 有关的查询、存储和删除等操作，并且能处理多个连接请求。但是传输的数据量多了、连接多了，gsrvr 的各种问题就出来了，此时需要在一个要素类处理完后释放这个连接，即关闭 gsrvr 进程。在 NET 中是使用 System.Runtime.InteropServices.Marshal.ReleaseComObject 方法来释放，但有的时候使用了该方法，可能还有连接存在，此时需要将代码中 new 出来的所有 ArcEngine 对象都用该方法释放掉。

13.12 自动关闭空闲 SDE 连接

在 SDE 的应用中，如果连接的客户比较多，可能会出现 SDE 连接占用 CPU 资源很严重，甚至出现死机或部分客户端连接断开的情况。

SDE 提供有 TCPKEEPALIVE 参数用来监控客户端连接情况。当 TCPKEEPALIVE 为 TRUE 时，在两个小时内（默认），如果客户端没有向 SDE 发送请求，SDE 连接则自动关闭；如果该参数为 false，SDE 连接则仍然占用。在默认安装的情况下，该参数值为 false，因此可以根据系统的实际情况调整该参数，以达到更好的性能。

第 14 章 空间数据库加载

14.1 通过设置属性加载个人数据库

首先通过 IPropertySet 接口定义要连接数据库的一些相关属性，在个人数据库中为数据库的路径，例如。

```
IPropertySet Propset = new PropertySetClass();
Propset.SetProperty("DATABASE",@"D:\test\Ao\data\sh\MapData.mdb" );
```

定义完属性并设置属性后，就可以进行打开数据库的操作了。在 ArcEngine 开发中，存在 IWorkspaceFactory、IFeatureWorkspace、IFeatureClass、IFeatureLayer 等几个常用的用于打开和操作数据空间地物的接口。IWorkspaceFactory 是一个用于创建和打开工作空间的接口，它是一个抽象的接口，我们在具体应用时，要用对应的工作空间实例化它，方式如下。

```
IWorkspaceFactory Fact = new AccessWorkspaceFactoryClass ();
```

完成了工作空间的实例化后，就可以根据上面设置的属性打开对应的 Access 数据库。打开方式如下。

```
IFeatureWorkspace Workspace = Fact.Open(Propset,0) as IFeatureWorkspace;
IFeatureClass Fcls = Workspace.OpenFeatureClass("District");
IFeatureLayer Fly = new FeatureLayerClass();
Fly.FeatureClass = Fcls;
MapCtr.Map.AddLayer (Fly);
MapCtr.ActiveView.Refresh();
```

其中 District 为地物类的名字，MapCtr 为 AE 中 MapControl 的对象。上面的通过属性设置加载数据空间的方式还可用于 SDE 数据库，在 SDE 数据库加载时会介绍。

以下为通过设置属性加载 Access 数据库的完整 C#代码。

```
public void AddAccessDBByPro()
{
    IPropertySet Propset = new PropertySetClass();
    Propset.SetProperty("DATABASE",@"D:\test\Ao\data\sh\MapData.mdb" );
    IWorkspaceFactory Fact = new AccessWorkspaceFactoryClass ();
    IFeatureWorkspace Workspace = Fact.Open(Propset,0) as IFeatureWorkspace;
    IFeatureClass Fcls = Workspace.OpenFeatureClass ("District");
    IFeatureLayer Fly = now FeatureLayerClass();
    Fly.FeatureClass = Fcls;
    MapCtr.Map.AddLayer(Fly);
    MapCtr.ActiveView.Refresh();
}
```

14.2 通过名称加载个人数据库

通过名称加载个人数据，通常使用 IWorkspaceName 接口提供的方法和属性，这里先给出完整的代码。

```
public void AddAccessDBByName()
{
    IWorkspaceName  pWorkspaceName = new WorkspaceNameClass() ;
                    pWorkspaceName.WorkspaceFactoryProgID= "esriDataSourcesGDB.
                    AccessWorkspace Factory";
    pWorkspaceName.PathName = @"D:\test\Ao\data\sh\MapData.mdb";
    IName n = pWorkspaceName as IName ;
    IFeatureWorkspace Workspace = n.Open() as IFeatureWorkspace;
    IFeatureClass Fcls = Workspace.OpenFeatureClass ("District");
    IFeatureLayer Fly = new FeatureLayerClass();
    Fly.FeatureClass = Fcls;
    MapCtr.Map.AddLayer (Fly);
    MapCtr.ActiveView.Refresh();
}
```

打开 Access 工作空间后，接下来的代码是一样的，都是找到对应的地物类，赋给相应的层，通过 MapControl 控件添加对应的层，然后刷新地图。现在讲解一下上面的代码，首先是创建一个个人数据库工作空间名，再指定工作空间名的 ProgID，以确定打开的是什么类型的工作空间。例如在打开 Access 个人数据库时，使用的是下面的代码。

```
  IworkspaceName pWorkspaceName = new WorkspaceNameClass() ;
  pWorkspaceName.WorkspaceFactoryProgID= "esriDataSourcesGDB.AccessWorkspaceFactory";
pWorkspaceName.PathName = @"D:\test\Ao\data\sh\MapData.mdb";
```

属性 WorkspaceFactoryProgID 可以确保工作空间是 AccessWorkspaceFactory，即个人数据库，同时指定要打开数据库的路径。为了打开数据库，通过 AE 的类图可以看到，打开工作空间必须使用 IName 接口，所以接着定义 IName 对象，把工作空间名转换成 IName 类型，并赋值给 IName 对象，然后通过 IName 对象的 Open()方法打开相应的工作空间，代码如下所示。

```
IName n = pWorkspaceName as IName ;
IFeatureWorkspace Workspace = n.Open () as IFeatureWorkspace;
```

14.3 SDE 数据库

ESRI 公司为了使空间数据能保存在关系数据库中，并且能很好地查询相关的空间属性，而开发了一个中间件，使用 SDE 能很好地将空间数据保存在关系数据库中，如 Orcale、SQL Server 等。要了解 SDE 具体的细节，可查找相关的资料，这里只介绍一下 SDE 数据库连接加载。SDE 数据库的联机分为直接连接和通过 SDE 连接。当服务器的性能比较好的时候，可以采用 SDE 连接，否则采用直接连接，这样可以减轻服务器的工作。建议采用直接连接。其实，SDE 连接方式和直接连接的方式只是一个属性参数设置的问题。与个人数据库采用属性连接的方式一样，可先定义一个属性对象，设置属性参数，接着定义一个工作空间，并用 SdeWorkspaceFactoryClass()实例化它，然后加在加载图层即可。至于加载图层的代码，与加载个人数据库中图层的方法一样。其实不止加载这两种数据类型，加载其他类型的数据时也是采用相同的方法加载图层，只是工作空间采用不同的实例而已。

```csharp
public void AddSDELayer(bool ChkSdeLinkModle)
{
    //定义一个属性
    IPropertySet  Propset = new PropertySetClass();
    if (ChkSdeLinkModle==true)          // 采用 SDE 连接
    {
        //设置数据库服务器名，服务器所在的 IP 地址
        Propset.SetProperty ("SERVER", "192.148.2.41");
        //设置 SDE 的端口，这是安装时指定的，默认安装时"port:5151"
        Propset.SetProperty ("INSTANCE", "port:5151");
        //SDE 的用户名
        Propset.SetProperty ("USER", "sa");
        //密码
        Propset.SetProperty ("PASSWORD", "sa");
        //设置数据库的名字，只有 SQL Server  Informix 数据库才需要设置
        Propset.SetProperty ("DATABASE", "sde");
        //SDE 的版本，在这里为默认版本
        Propset.SetProperty ("VERSION", "SDE.DEFAULT");
    }
    else        // 直接连接
    {
        //设置数据库服务器名，如果是本机，可以用"sde:sqlserver:."
        Propset.SetProperty ("INSTANCE", "sde:sqlserver:zhpzh");
        //SDE 的用户名
        Propset.SetProperty ("USER", "sa");
        //密码
        Propset.SetProperty ("PASSWORD", "sa");
        //设置数据库的名字，只有 SQL Server  Informix 数据库才需要设置
        Propset.SetProperty ("DATABASE", "sde");
        //SDE 的版本，在这里为默认版本
        Propset.SetProperty ("VERSION", "SDE.DEFAULT");
    }
    //定义一个工作空间，并实例化为 SDE 的工作空间
    IWorkspaceFactory Fact = new SdeWorkspaceFactoryClass();
    //打开 SDE 工作空间，并转化为地物工作空间
    IFeatureWorkspace Workspace = (IFeatureWorkspace )Fact.Open(Propset,0);
    /*定义一个地物类，并打开 SDE 中的管点地物类，写的时候一定要写全，如 SDE 中有一个管点层，就不能写成
IFeatureClass Fcls = Workspace.OpenFeatureClass ("管点"); 这样，而一定要写成下面的样子*/
    IFeatureClass Fcls = Workspace.OpenFeatureClass ("sde.dbo.管点");
    IFeatureLayer Fly = new FeatureLayerClass ();
    Fly.FeatureClass = Fcls;
    MapCtr.Map.AddLayer (Fly);
    MapCtr.ActiveView.Refresh ();
}
```

14.4 分图层加载 CAD 图层

我们可以把 CAD 图分为点、线、面、标注等加载到 MapControl 中。与加载其他数据一样，首先要定义一个工作空间，并用 CadWorkspaceFactoryClass () 实例化它，当得到了工作空间后，就

可以打开相应的工作空间，然后再打开指定的层类型。完整的代码如下。

```
public void AddCADByLayer()
{
    //定义工作空间，并用 CadWorkspaceFactoryClass()实例化它
    IWorkspaceFactory Fact = new CadWorkspaceFactoryClass();
    //打开相应的工作空间，并赋值给要素空间，OpenFromFile()
    //中的参数为 CAD 文件夹的路径
    IFeatureWorkspace Workspace = Fact.OpenFromFile(@"I:\test\",0) as IFeatureWorkspace;
    /*打开线要素类，如果要打开点类型的要素，需要把下面的代码改成：
    IFeatureClass Fcls = Workspace.OpenFeatureClass ("modle.dwg:point");
```

由此可见，modle.dwg 为 CAD 图的名字，后面加上要打开的要素类的类型，中间用冒号隔开。对于多义线，可以用以下代码打开。

```
    IFeatureClass Fcls = Workspace.OpenFeatureClass ("modle.dwg:polyline");
    IFeatureLayer Fly = new FeatureLayerClass ();
    Fly.FeatureClass = Fcls;
    MapCtr.Map.AddLayer (Fly);
    MapCtr.ActiveView.Refresh ();
}
```

14.5 整幅 CAD 图的加载

当要加载整幅 CAD 图时，需要使用下面的代码，这和加载地物类有一定的区别，详细的介绍请看代码中的注释。

```
public void AddWholeCAD()
{
    //下面的两行代码是定义一个 CAD 工作空间，然后打开它，但这次不是赋值给
    //IFeatureWorkspace 对象，而是赋值给 IWorkspace 定义的对象
    IWorkspaceFactory Fact = new CadWorkspaceFactoryClass();
    IWorkspace Workspace = Fact.OpenFromFile(@"I:\test\",0);
    //定义一个 CAD 画图空间，并把上面打开的工作空间赋给它
    ICadDrawingWorkspace dw = Workspace as ICadDrawingWorkspace;
    //定义一个 CAD 的画图数据集，并且打开上面指定的工作空间中的一幅 CAD 图，
    //然后赋值给 CAD 数据集
    ICadDrawingDataset ds = dw.OpenCadDrawingDataset ("modle.DWG");
    //通过 ICadLayer 类，把上面得到的 CAD 数据赋值给 ICadLayer 类对象的
    //CadDrawingDataset 属性
    ICadLayer CadLayer = new  CadLayerClass();
    CadLayer.CadDrawingDataset = ds;
    //利用 MapControl 加载 CAD 层
    MapCtr.Map.AddLayer (CadLayer);
    MapCtr.ActiveView.Refresh ();
}
```

上面的代码通过将 CAD 数据 DWG 整个添加到地图控件中，加载时 ArcEngine 就会将 DWG 数据文件中同一类型的数据作为一个图层添加到地图控件中。

为了在 MapControl 中加载 CAD 层，必须使用 ICadLayer 控件的对象，因为在 MapCtr.Map.AddLayer ()方法中只能是 ICadLayer 的对象。

第 15 章　程序出错和异常

15.1 释放资源异常问题

ArcEngine 中存在着一些错误，其中最有名的就是对象不能释放资源的问题。比如打开了一个 Shapefile，除非程序关闭，否则就没有办法释放对 Shapefile 的控制。还有读取 ServerStyle 文件的时候，那个 StyleItemEnum 只能使用一次，第 2 次就会出错。如果在短时间内搜索多个 FeatureClass，这样就会产生多个 FeatureCursor，那么就会出现打开的游标数目过多，或者不能打开更多的表这样的错误。这些错误都是由于对象不能释放资源所造成的。

为此，可以使用以下方法来释放(C#)。

```
//其中 relObj 就是要释放的对象
System.Runtime.InteropServices.Marshal.ReleaseComObject(relObj);
```

如果想要确保被释放了，则可循环调用该方法，直到这个方法返回 0 为止。

15.2 表结构操作错误

-2147220649-FDO_E_TABLE_DUPLICATE_COLUMN 错误：表示使用 AddField 方法添加的字段，在表中已经存在。

-2147220961-FDO_E_NO_SCHEMA_LICENSE 错误：表示无权修改表结构，应更换有编辑权限的 License。

-2147219878-FDO_E_FIELD_CANNOT_DELETE_WEIGHT_FIELD 错误：表示不能删除 Geometric Network 中与权重相关的字段。

-2147219877-FDO_E_FIEID_CANNOT_DELETE_REQUIRED_FIELD 错误：表示不能删除待删除的字段。如果将字段设为 Required，则可使用 IFieldEdit 接口的 Required 属性移除 Required 状态，然后才能删除该字段。

-2147215862-FDO_E_SE_DBMS_DOES_NOT_SUPPORT 错误：表示不能从数据库中删除字段。对 GeoDatabase 中的有些字段是不能删除的，如 ObjectID、Shape、Shape_length、Shape_area 等字段。

15.3 要素编辑的错误

在 ArcEngine 应用程序中，如果使用的 License 权限不足，就会出现 "Objects in this class cannot be updated outside an edit session" 错误。如果在应用程序中通过 ArcSDE 进行编辑，则可使用下面的代码为程序增加 Geodatabase 的编辑权限。

```
licenseStatus = CheckOutLicensesesriLicenseProductCode.esriLicenseProductCodeEngine
GeoDB;
```

编辑的 License 有下面几类。

Registered As Visioned without the option to move edits to base。该 License 能够实现的操作包括 Undo 和 Redo 操作、长事务编辑、为设计和工程使用命名版本、使用 Geodatabase 归档和进行数据库复制等。注意该 License 不能做的事有：创建拓扑、从拓扑中添加或删除要素、添加和删除拓扑规则、创建几何网络、从几何网络中添加或删除要素类。

registered as visioned with the option to move edits to base。该 License 不能够做的事有：编辑参与拓扑和几何网络的要素类、数据库归档、数据库复制。

not registered as versioned 是最原始的状态，能够实现复杂数据类型，包括拓扑和几何网络的编辑与更新。因为 Default 版本是数据库中最关键的，需要经常更新，因此需要对 Default 版本定期备份。

一般只要选择够用的 License 就可以了。如果多选，则有可能造成 License 冲突，而出现各种异常。

15.4 Network I/O Error 异常

I/O 异常是 ArcEngine 开发中连接 SDE 数据库最常碰到的问题，原因在于 SDE 进程 gsrvr.exe。一个 SDE 进程一次只能在 SDE 库中产生有限个要素游标，并且每个游标上能够传输的数据也是有限的，否则 gsrvr.exe 会撑爆。但一个 gsrvr.exe 遍历多个要素类则不会产生问题，所以此时只产生一个要素游标，即使传输上百万条数据也不会产生问题。但如果是遍历多个要素类，就会产生多个游标，问题就会出现。因此建议遍历 10 多个要素类后强制关闭 SDE 连接，然后再产生连接继续操作。

15.5 数据插入错误

在往 MDB 和 SDE 两种要素类插入数据时，不一定需要使用 IWorkspaceEdit 接口来开启和关闭一个 Session。但是如果 SDE 的要素类被注册了版本，则必须使用该接口，否则 CPU 会高达 100%，并会爆出 "the operation in invalid on a closed state" 的错误。但是在将多个 MDB 导入一个 MDB 的时候，如果使用了 IWorkspaceEdit 接口，就会出现某几个图层无法用 ArcMap 或 ArcCatolog 打开的情况。

15.6 索引被占用异常

当往一个 MDB 中导入多个 MDB 数据时，往往在第 2 个 MDB 导入时会发生 "***_shape_index 被占用" 的情况。这是因为在遍历和插入数据的时候游标没有被释放的缘故，此时应使用 System.Runtime.InteropServices.Marshal.ReleaseComObject 方法将对象释放。在本书介绍符号库的代码中有演示例子，读者可参阅。

15.7 SDE 导入空间数据错误

"Underlying DBMS error(ORAA-0001:Unique constraint(SDE.GDB_OC_PKC)violated" 错误，是由于 oracle 序列 R3 产生的值在表 gdb_objectclasses 中已经有记录造成的。为此可在 SDE 中选中 SDE

连接点后右击，然后选择"刷新"，即可再导入数据。

15.8 HRESULT:0x80040228 异常

连接 SDE，有的时候会碰到 HRESULT:0x80040228 异常，此异常项目中使用了 Engine 的组件，而实际上并未安装 Engine，或者 Lisence 有问题。一般直接添加一个 Lisence 控件就可以消除该异常。

15.9 HRESULT:0x80040213 异常

此异常往往是在加载数据库 MDB 的时候找不到 IPropertySet 属性指定的 MDB 引发的。此时需要核对属性以及 MDB 路径是否正确。

15.10 HRESULT:0x80040205

由于调试机器与部署机器文件系统格式不一样，比如调试机器是 FAT32，而部署机器是 NTFS，因此当往 NTFS 文件系统中写 MDB 数据库文件的时候，由于权限不足就会引发此异常。此时给 NTFS 文件系统的目录添加足够的权限即可。

15.11 HRESULT:0x80010105 (RPC_E_SERVERFAULT)

在做空间查询的时候可能引发此异常，这是因为未指定查询的空间关系所致。添加如下的代码即可：

```
spatialfilter.SpatialRel = esriSpatialRelEnum.esriSpatialRelIntersects;
```

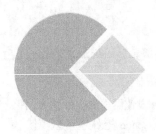

第 16 章 其他经验技巧

16.1 ArcEngine 中的先闪烁后刷新现象

在应用程序开发的过程中,有的时候需要对实体对象进行闪烁和刷新,如果采用下面的代码。

```
m_activeView.Refresh();
m_mapControl.FlashShape(feature);
```

在执行时图形会先闪烁,然后才刷新背景。显然这不是我们想要的结果,原因在于图形显示和闪烁不在同一个线程中执行,刷新背景比较复杂,需要的时间较长,而闪烁图形的动作很快就能完成。而使用 m_activeView.UpdateWindow 后,就可以让代码在此等待,等到刷新确实搞定了,才开始执行 FlashShape。代码如下所示。

```
m_activeView.Refresh();
m_activeView.UpdateWindow();    //这条语句是关键
m_mapControl.FlashShape(feature);
```

16.2 ArcEngine 中几种数据的删除方法和性能比较

在查询结果中删除数据。

```
private void Delete1(IFeatureClass PFeatureclass)
    {
       IQueryFilter pQueryFilter = new QueryFilterClass();
       pQueryFilter.WhereClause = "objectID<=" + DeleteNum;
       IFeatureCursor pFeatureCursor = PFeatureclass.Search(pQueryFilter, false);
       IFeature pFeature = pFeatureCursor.NextFeature();
       while (pFeature != null)
       {
         pFeature.Delete();
         pFeature = pFeatureCursor.NextFeature();
       }
       System.Runtime.InteropServices.Marshal.ReleaseComObject(pQueryFilter);
    }
```

更新游标删除。

```
private void Delete2(IFeatureClass PFeatureclass)
    {
       IQueryFilter pQueryFilter = new QueryFilterClass();
       pQueryFilter.WhereClause = "objectID<=" + DeleteNum;
```

```
        IFeatureCursor pFeatureCursor = PFeatureclass.Update(pQueryFilter, false);
        IFeature pFeature = pFeatureCursor.NextFeature();
        while (pFeature != null)
        {
          pFeatureCursor.DeleteFeature();
          pFeature = pFeatureCursor.NextFeature();
        }
        System.Runtime.InteropServices.Marshal.ReleaseComObject(pQueryFilter);
    }
```

使用 DeleteSearchedRows 删除。

```
private void Delete4(IFeatureClass PFeatureclass)
    {
        IQueryFilter pQueryFilter = new QueryFilterClass();
        pQueryFilter.WhereClause = "objectID<=" + DeleteNum;
        ITable pTable = PFeatureclass as ITable;
        pTable.DeleteSearchedRows(pQueryFilter);
        System.Runtime.InteropServices.Marshal.ReleaseComObject(pQueryFilter);
    }
```

使用 ExecuteSQL 删除。

```
private void Delete4(IFeatureClass PFeatureclass)
    {
        IDataset pDataset = PFeatureclass as IDataset;
        pDataset.Workspace.ExecuteSQL("delete from " + PFeatureclass.AliasName + " where objectid<=" + DeleteNum);
    }
```

测试代码如下。

```
IFeatureLayer pFeatureLayer = axMapControl1.Map.get_Layer(0) as IFeatureLayer;
IFeatureClass PFeatureClass = pFeatureLayer.FeatureClass;
System.Diagnostics.Stopwatch MyWatch = new System.Diagnostics.Stopwatch();
MyWatch.Start();
Delete1(PFeatureClass)
//Delete2(PFeatureClass);
//Delete3(PFeatureClass);
//Delete4(PFeatureClass);
//Delete5(PFeatureClass);
MyWatch.Stop();
MessageBox.Show("删除时间:" + MyWatch.ElapsedMilliseconds.ToString() + "毫秒");
```

测试结果如表 16-1 所示。

表 16-1　　　　　　　　　　　测试结果

测试方法	第 1 次时间（单位 ms）	第 2 次时间（单位 ms）
1	5214	5735
2	299	290
3	59	28
4	26	26

结论：使用 ExecuteSQL，可以直接进行数据的删除操作，所以删除最快，数据库的使用效率最高。DeleteSearchedRows 和 ExecuteSQL 属于批量删除，性能较优。而在查询结果中删除，速度最慢。

16.3 数据游标

ArcEngine 使用何种游标（Cursor）来管理记录子集，取决于数据源。ICursor 和 IFeatureCursor 一样，ICursor 用于操作表，而 IFeatureCursor 则用于操作要素类。即 Cursor 是一种为了特定的目的——操作存储在传统数据库表中的记录子集而建立的类结构；FeatureCursor 的记录子集则是存储在 Shapefile 文件、个人 Geodatabase 和企业 Geodatabase 中。

Cursor 和 FeatureCursor 有 3 种类型的游标，常用的是 SearchCursor，用于查询操作以返回一个满足查询条件的记录子集，该游标是一种只读的游标，用于遍历获取的信息。InsertCursor 专门用于往一个表中插入一条记录，UpdateCursor 则用于更新或删除记录。

Cursor 类用于产生一个与数据库表进行交互的对象，是一个非实例化对象，因此必须使用另一个对象来获得一个 Cursor 类的实例。在 ArcEngine 中，表类用于产生一个 Cursor 类的实例。表类包含有 3 种方法，能够产生一个 Cursor 类的实例。如 ITable 接口拥有 3 种方法，能够返回特定类型的 Cursor。ITable 接口的 Search、Insert 和 Update 等方法能够用于返回 Cursor 实例，这些方法的名字与返回的 Cursor 类型相对应。

FeatureCursor 与 Cursor 类相似，其区别在于前者是操作地理数据集，而后者是操作传统数据库表。FeatureCursor 类也是通过一个 FeatureClass 对象的方法产生的非实例化对象，IFeatureClass 接口也包含 Search、Insert 和 Update 等方法，用于返回一个 IFeatureCursor 实例。

16.4 投影变换

投影变换大多采用 IGeometry 接口的 Project 方法，这个方法简单，但是是使用 GeoStar 组件做投影变换的最慢的方法。在 GeoStar 的投影变换组件中，有一个 GeoProjectedCoordinateSystem 组件类，该类实现 IGeoProjectedCoordinateSystem2 接口，通过该接口可以进行高效的投影变换。具体步骤如下：

创建 GeoProjectedCoordinateSystem 组件；
获取源和目标空间参考对象；
获取要变换的 Geometry 对象；
通过 Geometry 得到坐标串和解释串；
用 GeoProjectedCoordinateSystem 组件进行投影变换。

一般情况下，线和面对象的 Geometry 使用这种方法投影变换，加上读写的时间，比使用 Geometry 的 Project 方法要快 15 到 20 倍。

16.5 ITopologicalOperator

在空间分析中通常要使用 ITopologicalOperator 接口，但是这个接口出现问题的几率也是非常高。有的时候既没有报错，又没有结果，比如 merge 方法。下面的代码演示了如何处理这类问题。

```
ICursor cursor ;
featureSelect.SelectionSet.Search(null,false,out cursor );
IFeatureCursor featureCursor =cursor as IFeatureCursor ;
IFeature featureFirst =featureCursor.NextFeature();
m_EditWorkspace.StartEditOperation();
IGeometry geometryFirst =featureFirst.Shape;
```

```
ITopologicalOperator2  topo =(ITopologicalOperator2) geometryFirst;
//这里是重点，先强制检查下面 3 个步骤，然后进行操作，成功的可能性高一些
topo.IsKnownSimple_2=false;
topo.Simplify();
geometryFirst.SnapToSpatialReference();
//这是准备合并的图斑使用的
ITopologicalOperator2  topo2;
IGeometry  geometryNext;
IFeature featureNext=featureCursor.NextFeature();
geometryNext=featureNext.Shape;
While(featureNext!=null)
{
    topo2=geometryNext as ITopologiccalOperator2;
    top2.IsKnownSimple_2=false;
    geometryNext.SnapTOSpatialReference();
    geometryFirst=topo.Union(geometryNext);
    featureNext=featureCursor.NextFeature();
    geometryNext=featureNext.Shape;
}
```

16.6 缓冲区查询

缓冲区查询是 GIS 应用程序中最常使用的功能，但是经常会出现缓冲区查询的图形参数是正确的，但却出现了异常的情况，为此（系统性能许可的情况下）可以采用一些多次执行缓冲区查询的方式来避免出现这类问题。

比如写个缓冲区查询方法，在内部做 10 次循环，如果缓冲区出现异常，则扩大缓冲区半径增加 0.1 倍，直到得到正确的结果。

16.7 插入记录效率

往要素类中插入记录的方法有两种，一种是 IFeature.Store，另一种是 IFeatureCursor.Insert（IfeatureBuffer）和 IFeatureCursor.Flush 方法。由于后者使用了缓存，因此速度比前者快。